FISHING PLACES,
TRADITIONS AND

MW01517892

ЕS

An interdisciplinary approach is the hallmark of *Fishing Places, Fishing People*. The book proposes a radically different way of thinking about current problems in Canadian small-scale fisheries and lays the groundwork for an alternative management approach to the fisheries. Comprising entirely new material, the collection brings together the work of many highly regarded scholars – historians, biologists, sociologists, anthropologists, consultants, geographers, and ecologists – to discuss this topical issue. Drawing on case studies from across Canada, they demonstrate that there are many shared issues in the various small-scale fisheries of this country, and they place Canadian small-scale fisheries in their historical context as well as in that of global ecological and policy concerns.

DIANNE NEWELL is a professor of history at the University of British Columbia and author of *Tangled Webs of History: Indians and the Law in Canada's Pactific Coast Fisheries*.

ROSEMARY E. OMMER is a professor of history at Memorial University of Newfoundland and author of *From Outpost to Outport: A Structural Analysis of the Jersey-Gaspé Fishery, 1767–1886*.

EDITED BY DIANNE NEWELL AND
ROSEMARY E. OMMER

Fishing Places, Fishing People

Traditions and Issues in Canadian Small-Scale Fisheries

UNIVERSITY OF TORONTO PRESS
Toronto Buffalo London

Printed in Canada

ISBN 0-8020-4116-7 (cloth)
ISBN 0-8020-7959-8 (paper)

Printed on acid-free paper

Canadian Cataloguing in Publication Data

Main entry under title:

Fishing places, fishing people : traditions and issues in Canadian small-scale
   fisheries

Includes bibliographical references.
ISBN 0-8020-4116-7 (bound)   ISBN 0-8020-7959-8 (pbk.)

1. Fisheries – Canada.   I. Newell, Dianne, 1943–.   II. Ommer, Rosemary E.

SH223.F57 1999   338.3'727'0971   C98-931414-6

Stuart Daniel of Starshell Maps, Victoria, BC, drew the maps and illustrations in
Figures 4.1, 5.1, 6.1, 7.2, 7.3a, 7.3b, 7.4, 9.1, 10.1, 11.1, 11.2, and 14.1.

This book has been published with the help of a grant from the Humanities and Social
Sciences Federation of Canada, using funds provided by the Social Sciences and
Humanities Research Council of Canada.

University of Toronto Press acknowledges the financial assistance to its publishing
program of the Canada Council for the Arts and the Ontario Arts Council.

Contents

Acknowledgments

The editors wish to acknowledge the help and support of all the contributing authors in the preparation of this book. A volume like this is always difficult to prepare, particularly when the participants have contributed not only by writing their own papers but by critiquing those of others, thus helping to ensure that the book is a coordinated and coherent whole. Their enthusiasm and dedication are deeply appreciated.

We are also grateful to the University of Toronto Press, in particular to Gerry Hallowell, and to Emily Andrew, who, along with Gerry, provided guidance throughout the preparation of the manuscript. Our thanks to John Parry, who edited this complex manuscript with exceptional skill and precision. We are also grateful to the two anonymous readers who assessed an earlier version of this book for the Press and the Aid to Scholarly Publications Program and whose comments helped us refine the chapters and the final focus and organization of the volume. Sylvie Guenette of the Fisheries Centre, University of British Columbia, provided useful comments on one chapter. To June Bull and Lugard de Decker of the Centre for Studies in Religion and Society at the University of Victoria, where Rosemary Ommer spent her sabbatical year, 1997, and to Linda Beck of the Institute of Social and Economic Research at Memorial University, go our thanks for secretarial assistance in the preparation of the text. The various forms of assistance provided by the International Social Sciences Institute, University of Edinburgh, where Dianne Newell visited in 1996 for part of her sabbatical, are also gratefully acknowledged.

Thanks are owed to Stuart Daniel (Starshell Maps), Victoria, who created several of the maps and redrew other maps, charts, and graphs, and to Patricia A. Thornton of Concordia University and *Labour/Le Travail*, for permission to use the poem 'The Strait of Belle Isle.'

DIANNE NEWELL
ROSEMARY OMMER

FISHING PLACES, FISHING PEOPLE:
TRADITIONS AND ISSUES IN CANADIAN SMALL-SCALE FISHERIES

1

Introduction: Traditions and Issues

DIANNE NEWELL AND ROSEMARY E. OMMER

This book tells a story of a range of fisheries and fishing communities across Canada. Using mainly a series of case studies from a wide range of disciplines, it examines the historical roots of different types of fishing communities: freshwater (riverine and lacustrine), marine (east and west coasts), including those in which fishing was a primary focus of settlement (as in Newfoundland), or a supplementary part of the economy (as in the 'fur trade economy'), or an integral part of an indigenous culture and economy (as with the First Nations of coastal British Columbia and central Canada).

This is not a comprehensive regional coverage of the fisheries, since we do not see 'region' as a crucial distinguishing feature of fishing communities' resource strategies. Indeed, we consider that regional analysis would blur the interesting similarities between small fishing communities wherever in Canada they are located. We seek to emphasize the fisheries strategies of *all* small-scale fishing communities with historical roots in this way of life.

We have also sought to demonstrate the link that community fisheries had to the wider commercial world and to point to the ways in which they were affected by commercial pressures and external markets over time. The widening of the resource 'management world' from the community to the state (provincial and national) brought growing difficulties to small-scale community-ordered fisheries, which is perhaps not surprising, given the difficulties involved in matching the 'macro,' state view with detailed ecological knowledge, which must be built from the grass roots up – perhaps in this context 'bottom up' would be more appropriate. We demonstrate this challenge through historical and present-day studies of state management, state 'science,' and efforts at community organization and co-management; we also link this perspective to various kinds of 'knowledge' about the resource: local ecological knowledge, whether indigenous or not, and formal fisheries science. Finally, we have revisited the current fisheries crises on both coasts, with an eye to the future and with a glimpse of the wider global context within which Canadian fisheries now must operate.

Fronting on the Atlantic, Arctic, and Pacific, Canada has the longest ocean coastline of any continental country in the world today, as well as the greatest percentage of its surface covered in fresh water. Such unrivalled marine and freshwater aquatic resources have meant that Canada's fisheries have always been of major significance to its human history. Until the twentieth century, for example, occupation of the vast subarctic and arctic territories would have been impossible without an abundant annual supply of fish. Indeed, among the first of Canada's Aboriginal groups to be named in the historical records constructed by Europeans were regional bands located in temporary, although all-important summer fishing camps in the subarctic. Even Europe's Canadian fur trade was born in the North Atlantic cod, seal, and whale fisheries. The international nature of Canadian fisheries is attested to in Harold Innis's classic 1940 study, *The Cod Fisheries* (1954), which he subtitled 'the history of an international economy.' Thus early codfish posts on the coasts of Newfoundland, Labrador, and the Gaspé were as much outposts of empire as were the central subarctic fur-trade posts of the Hudson's Bay Company (Ommer 1991; Ray 1998).

On the Pacific slope, mainly because of the large-scale exploitation of salmon (*Orcorhynchus*), a proliferation of Aboriginal societies could live at a level and population density well above the average of the world's non-agricultural societies – and many of the agricultural ones. Salmon canneries with links to world markets were the first factories when British Columbia became a province in 1871 (Newell, 1993). In a different mode, coastal and inland sports fishing grounds across Canada have been internationally publicized in popular books and magazines since at least the 1860s. Without doubt, marine and freshwater fisheries continued to shape lives and livelihoods in hundreds of Native and non-Native communities from sea to sea.

It is therefore surprising how little public awareness exists about the long-standing importance of fisheries in Canada's social and economic development. No broad-based study that covers all the various modalities of Canadian fisheries has ever been produced. Instead, fish production has been taken for granted, assumed to be either a large, centralized modern industry, a lighthearted hobby, or a local 'quaint' relic of a 'traditional' past. The social, cultural, and economic significance of Canadian small-scale, sustainable fisheries consequently has been grossly underestimated.

Recently, however, a series of catastrophic events in Canada have transformed the nation's fisheries into national and international news. There has
been litigation from coast to coast over Aboriginal and treaty rights to aquatic
resources and fishing sites. Public concern has been expressed over a variety of
environmental problems, including industrial contamination of international
boundary waters; oil spills in coastal waters; massive flooding and damming in
connection with hydro-electric projects, which obliterate spawning habitats and
release the soil's soluble mercury into the rivers; and atmospheric ozone depletion, which appears to be killing the phytoplankton – the primary link in the
marine food chain. Throughout much of Canada there is an ongoing battle
between Natives and non-Natives, and in Pacific waters between Americans
and Canadians, and between them and the Japanese, for the threatened salmon
stocks that originate in Canadian waters. Finally, the spectacular collapse of the
northern cod (*Gadus morhua*) of Atlantic Canada, with the failure of experts
either to agree on the causes or to predict the rebound of a major stock being
driven into collapse by a fishery that followed mainstream scientific advice,
has raised awareness across the nation and the world about the uncertainty of
science and the fragility of what was usually thought of as an inexhaustible
resource. West coast fisheries communities fear that they may be the next
Newfoundland.

What has befallen Canadian fish and fish-based communities is not inevitable, nor is it historically unique. There are tragedies here, but they are not
necessarily to be understood in the manner in which they are usually presented
to us: the so-called tragedy of the commons. The idea that fisheries are inherently common property resources that necessitate government-formulated regulations to prevent harvesters from destroying them – 'the freedom in a
commons brings ruin to all' – is at best a questionable proposition. As we show
below, the actions of the Canadian state have sometimes played a key role in
producing the decline of the country's various fisheries, while local fisheries
communities and fisheries scientists have often been struggling to hold the
problem in check.

The experts who have joined us to produce this pioneering, multidisciplinary
stocktaking enterprise for Canadian fisheries examine fisheries and communities in innovative ways. They find striking similarities over time and across
space in the way human consciousness enters the ecology, and vice versa. They
also expose remarkable, rich differences. Both the similarities and differences
are revealing of the interface between humans and fish within the ecosystem of
which they are both part. We invite readers to probe these essays for what they
say about community, landscape/waterscape, and memory, and about the ability

of local people 'living the fishing,' in Paul Thompson's famous words (1983), to see, understand, and value the resource environment. It is possible, with this collection of new research by North American experts, to discover the capacity of fisheries communities and cultures to incorporate deceptively simple uses of places, where culture, ecology, economy, subsistence, and society are intertwined – are effectively indistinguishable from one another. This was true in the past, and it continues to inform the present, despite the contemporary crisis for all fishing communities. We ask you to understand the essential need of these authors to explore the complexities and contradictions of how landscape/waterscape is shaped and interpreted by humans biologically, culturally, economically, geographically, legally, politically, socially, and even intellectually and spiritually.

The term 'ecosystem,' coined in the 1930s, refers to a unit that includes all the elements in a given area interacting with the physical environment. It implies a functional, balanced, and dynamic relation between system components, and it is essentially holistic. Though the notion of a balanced ecological state is no longer automatically accepted as a scientific proposition, ecologically based patterns do exist, are coherent, and can be studied and understood as rational. For human beings, top predators in the fishing environment, ocean resources represent a constellation of options. The specific timing and locale of different activities of fish and other animal species occur during the year and over longer cycles too, and all human fishing communities have used, and in places continue to use, these biological rhythms and spatial patterns to fashion their own round of activities over the season, the year, and the longer cycles of the natural world.

The changing intellectual history of theories about the relationship of humans to nature, 'man's conquest of the natural world,' and the common-property 'myth' of fisheries – the 'fisherman's problem' (Berkes 1985; McEvoy 1986) – informs many of the studies presented here. The chapters by Cadigan, Gallaugher and Vodden, and many others demonstrate that, one way or another, the conquest-of-nature strategy, which is a habit of thought about controlling nature with applications of science and technology, has been the dominant industrial mode of attaining flexibility, of coping, and is failing us now as the fish stocks and fish habitats of the world face biological and commercial collapse.

For all Canadian First Nations today, competing claims to territory and access to resources such as fish arise from very different assumptions about the fundamental notion of ownership and the relationship of groups to government. This crucial issue is addressed in this volume by Newell, Ray, Thoms, Tough, and Usher and Tough. Chapters by Cadigan, Manore and Van West, and Ommer show that something similar seems to be the case among local non-Aboriginal communities historically.

If today we are truly in danger of *systemic* shift in several fish-based ecosystems, it is more important than ever to reacquaint ourselves with the ecology of fish and the efficient nature of the way in which some human groups and communities were, and still are, able to conduct fisheries on a sustainable basis. However, interpreting 'landscape' and unearthing the memory of the fishing community are hardly straightforward tasks, especially, as Usher and Tough discovered in their chapter, when it comes to reconstructing historical data about subsistence fisheries. Nor it easy to comprehend the subtleties of community life, today or in the past. Both the separations and the interconnections of individuals in fishing communities emerge in the chapters by Barbara Neis, from her interviews with women about their roles in the economy of outport Newfoundland, and by Dianne Newell, from her fieldwork with Native commercial herring spawn-on-kelp harvesters on the north coast of British Columbia, for example.

Fishing communities were once, and some manage to continue to be, dynamic, but the nature of the dynamic, is changing. Patricia Thornton (1979) has spoken of the dynamic equilibrium of historic non-Native Labrador fishing communities, and Dianne Newell, in this volume, of the dynamic tradition of Pacific coast Indian fisheries. The authors here invite us to discover the nature of representative small-scale fisheries and the ecological awareness and functional sophistication with which people have managed to interact with the environment, creating their own styles of coping with change. Coping has involved a complex array of strategies: the sort of physical or intergenerational migration of part of the household discussed by Neis, Ommer, Manore and Van West, and Sinclair, Squires, and Downton; participation in new commercial fisheries for export to foreign markets, addressed by Cadigan, Newell, and Tough; the integration of a range of commercial and subsistence activities (including in the twentieth century government unemployment and retraining programs) examined by Cadigan, Neis, Ray, Sinclair, Squires, and Downton, Tough, and Thoms; the fine-tuning of techniques, explored in the essays by Cadigan and Newell; and the pooling of labour and equipment, as Manore and Van West discovered among Lake Huron poundnet fishers and local farmers. These findings point to the need to destultify our categorization of coping mechanisms in these difficult and sometimes dangerous resource environments.

Commercial life was once wedded to fishing people's seasonal round, regardless of end markets. Company store operations, whether at a remote Pacific coast salmon cannery, a Hudson's Bay post in the Albany River or the

Nipigon River drainages of northern Ontario, or a fishing outport in Bonavista, were as seasonal as the lives of people who conducted the fisheries and developed symbiotic relations with merchants. For all parties, as Cadigan, Ommer, and Ray demonstrate, constant adjustments and compromises were called for, and always with the possibility of failure. But the point is this: risks were spread over the year and over longer-term cycles within a known and claimed geographical area. As the Pauly chapter points out, our historic fisheries lasted for centuries because catches were limited relative to stock size and location. All that changed with the introduction of modern industrial capitalism and its attendant technological ability to locate almost any fish, anywhere, the systemic shift to urban-industrial wage labour structures, and the development of specific state policies for managing people and resources, which Barbara Neis identifies in her chapter as 'social patriarchy.'

These shifts to an advanced industrial fishery in Atlantic Canada in the 1950s brought with them, among other things, the kind of state management that promoted high-technology offshore fleets and marginalized the traditional seasonality and spatially dispersed location of production, which left part of the stock not susceptible to fishing, that had been the *modus vivendi* of the outport. Part-time and seasonal work and inshore fisheries were seen as undesirable, inefficient, not very productive, and hence something to be eradicated. A similar pattern may be seen for the fisheries of the Pacific coast. Canadian fisheries policy and regulation since the late 1960s has increasingly promoted full-time, highly mobile, capital-intensive, deep-sea industrial fisheries, along with aquaculture (domestication and importation of species), over inshore and part-time or small-scale commercial fisheries of wild stocks. At the same time, even well-meaning efforts to cushion the blow for the traditional fishing communities have often fostered dependence and loss of culture (Newell 1993, Sinclair 1985).

Of course, despite the sharp divisions, narrow definitions, and tight controls imposed by laws and regulations under the various fisheries acts and regulations in Canada, the policies are often ambiguous and contradictory when seen as a whole. Nor is it the case that local people necessarily conformed to the new rules at all times or in all respects. Nor, in fact, have fisheries officers, by necessity or by design, consistently or thoroughly enforced the laws and regulations. In truth, the history of modern fisheries management, especially for marine fisheries, is as much about the economics and technology of enforcement as it is about the economics and technology of harvesting, processing, or marketing. Local traditions, social practice, folklore and narratives, and customs for handing down knowledge, gear, and harvesting spots have proved to be incredibly tenacious – 'embedded' is the term used by McCay and Sinclair, Squires, and

Downton. This should not have surprised us, since embeddedness has helped to sustain our fisheries and fishing communities for centuries.

The creation and imposition of resource conservation and management expertise under modern industrial capitalism, beginning in the mid-nineteenth century, make up a history of reorganization and restructuring grafted onto the old merchant capital strategies of long-distance management and long-distance markets, which had always been a feature of colonial fisheries (Copes 1979–80). A new, expert, 'objective' reality, stressing individualism and professionalism, has reconfigured the old canvas of 'subjective, traditional' ecological knowledge and social practice, which had emphasized group dynamics and interdependence. The essays by Cadigan, Neis et al., Newell, and Thoms, among others, address this critical issue.

In line with the new mindset that talks about fish as 'product,' today's 'common property' fisheries are conducted and managed as categories and conceptually bounded spaces: British Columbia, Newfoundland, and northern Ontario are dealt with as separate boxes, inside of which are their various aquatic resource boxes. Ocean waters are subdivided and labelled, 'postal coded' to assist in the international, federal, and provincial management of fish. Within each, boats are coded, fishers are coded, fish are coded. Hutchings, in his chapter, argues that stock assessment reviews typically rely too heavily on information enclosed in three separate 'boxes': the fishing box, the biology box, and the environment box. Ideally, from a management perspective, fish would swim obediently inside bureaucratically created limits (such as 2J3KL) rather than ecological ones. Pacific coast roe-herring fisheries managers are scratching their heads trying to find ways to entice mature female herring to produce premier-grade spawn-on-kelp 'product' on cue and on target. This is business, not biological, language. It refers to fish as a disposable item, rather than a living part of an ecosystem.

What is true for fish is also true for people: 'Indian fish' formally denotes Native harvests for customary use. Customarily, of course, Native people both ate and traded their fish, especially on the Pacific Slope, where they also needed it for distinctive social and ceremonial purposes. Over the decades, 'customary use' became redefined by law as use for food, then more narrowly as food for immediate members of families, when, where, by what means, and if it suited fisheries managers. Fish taken by Native people for any purpose other than 'household need,' or without an Indian food fish licence and compliance with all the restrictions on harvesting and use that that entailed, could land them in jail (Newell 1993). This idea of harvesting based on need requires careful reflection; once Native economies became coupled with open-ended international markets, household 'need' could involve harvesting more than the local ecosystem could support.

Professional fisheries managers set marine fisheries apart from riverine fisheries and lake fisheries – so much so that some fisheries scientists model the anadromous salmon fishery of the Skeena River as a 'boxcar' fishery in which the First Nations and sports fisheries of the river occupied the 'last car' on the spawning run. Fish are dealt with separately from their physical environment and from the human cultural environment. Such compartmentalization in the stock review process, Hutchings argues in his chapter, has been harmful to the sustainability of resources and the human communities that depend on them. Producers are alienated from managers, and fisher folk are assumed to have nothing in common with the social scientists who study them or with the biologists who study fish. In British Columbia, Aboriginal fishers are regarded as a third-party interest in a dubious triangle completed by commercial and sports interests.

Compartmentalization has been endemic in commercial fisheries: Harold Innis made the point many years ago when he spoke of the cod fisheries as 'inherently divisive,' and our history is one of continuing exacerbation, not resolution, of that problem (Ommer 1991). There are alternatives. In this volume, Evelyn Pinkerton reviews the history of cooperative management; Daniel Pauly discusses the efficacy of marine protected areas (MPAs), which work as artificial refuges for fish; Neis et al. examine traditional ecological knowledge (TEK) and provide a Newfoundland fisheries case study; and Gallaugher and Vodden demonstrate how BC coastal fishing communities recently came together in a series of forums to reconstruct a sense of common cause.

Fish, like people, have their own food webs and food cascades, life cycles and ecological niches. Some species, like Alaska cod and Pacific salmon, have complex migratory patterns, which, it is believed, are still not fully understood. For Daniel Pauly, however, this mystique about fish is a cliché, a smokescreen for disciplinary irresponsibility. Debate over the sustainability of fisheries often centres around the binary question of whether overharvesting or environmental change is responsible for large-scale alterations in abundance and productivity in 'modern' fisheries (another cliché, Pauly says). Villagarcia, Haedrich, and Fischer make their own scientific contribution to this particular debate for the groundfish of the northeast Newfoundland Shelf using a multi-species perspective to detect variation at the fish-community level. A firm basis for the anger and frustration that many fisheries scientists experience exists in the impossible demands in the past of government fisheries managers and their political bosses for a scientific consensus on fisheries matters. These unrealistic demands, in fact, form the basis of Jeff Hutchings's reading on the shocking collapse of the northern cod.

This book takes us across the usual boundaries in fisheries management and fisheries scholarship. By concentrating on scale as a diagnostic feature, it draws together a multiplicity of fisheries and fishing communities and begins to incorporate the work of fisheries biologists as well as that of anthropologists, historians, geographers, sociologists, and others. Not surprisingly, the 'fit' is not perfect. The chapters written by natural scientists reflect the difference in starting points, methods, and styles of presentation: their geographical scale is sometimes broader and their data are more quantitative than those that used by most social scientists. But there are many signs in the book that current policy dilemmas require that disciplines move closer together; and there are also clear indications of what joint questions need to be asked and how interdisciplinary and historical study might contribute to the answers: the multi-author chapters, in particular, represent interdisciplinary collaboration.

It is also clear that the issues surrounding Canadian fisheries are often much older and much more complex than the usual characterization of them would suggest. This volume, focusing on scale and concerned with traditions, sets the path for future fruitful endeavour: it cannot hope to be definitive, but we think it a productive and challenging first step. Above all, we hope that it provides the outlines of a new configuration of the debate, which will force us to revise current thinking, with its inadequate representation of the hazards through which we have been steering our policies and planning for fishing places and fishing people.

WORKS CITED

Berkes, Fikret. 1985. 'Fishermen and the "Tragedy of the Commons."' *Environmental Conservation* 12: 199–206.

Copes, Parzival. 1979–80. 'The Evolution of Marine Fisheries Policy in Canada.' *Journal of Business Administration* 11, nos. 1–2: 125–48.

Innis, Harold. 1954. *The Cod Fisheries: The History of an International Economy*, 2nd ed. First pub. 1940. Toronto: University of Toronto Press.

McEvoy, Arthur F. 1986. *The Fisherman's Problem: Ecology and Law in the California Fisheries, 1850–1980*. Cambridge: Cambridge University Press.

Newell, Dianne. 1993. *Tangled Webs of History: Indians and the Law in Canada's Pacific Coast Fisheries*. Toronto: University of Toronto Press.

Ommer, Rosemary. 1991. *Outpost to Outport: A Structural Analysis of the Jersey–Gaspé Fishery, 1767–1886*. Montreal: McGill-Queen's University Press.

Ray, Arthur J. 1998. *Indians in the Fur Trade: Their Role as Hunters, Trappers, and Middlemen in the Lands Southwest of Hudson Bay, 1660–1870*. 2nd ed. First pub. 1974. Toronto: University of Toronto Press.

Sinclair, Peter S. 1985. *From Traps to Draggers: Domestic Commodity Production in Northwestern Newfoundland, 1850–1982*. St John's: ISER, Memorial University.

Thompson, Paul, with Tony Wailey and Trevor Lummis. 1983. *Living the Fishing*. London: Routledge and Kegan Paul.

Thornton, Patricia A. 1979. 'Dynamic Equilibrium: Settlement, Population and Ecology in the Strait of Belle Isle, Newfoundland, 1840–1940.' PhD dissertation, University of Aberdeen.

PART ONE:
COMMUNITY ROOTS AND COMMERCE

THE STRAIT OF BELLE ISLE
Rosemary E. Ommer

Sure, we can study kings and princes,
Watch the sweep of Empire catch
Even this lonely Shore within its grasp:
For dry men in old chambers
Have copied it on parchment for our future eye.

But if we seek to disembark,
If we leave behind the merchant's brig
(Turn our backs upon her),
Ask instead
Of local men, of living and of dying on this Shore,
Of cooks and servant girls
And youngsters;
If we seek to find
The generations of the Strait;
If we try to comprehend
Their interwoven life and land and sea –
Where is the charting of their days?

The answers are not written in a well-formed hand,
But found amid the gravestones and the wooden homes,
Amid the seamed and weather-beaten faces of the old,
Learned from a life where cliffs and strand
(The ocean and the land)
In subtle balance with each other's wealth
Jointly supply the riches of the poor.

Here's an integrity of earth and folk
Complex and finely-turned to fit the balance of their days:
This is their ledger and their life's accounts.

2

Rosie's Cove: Settlement Morphology, History, Economy, and Culture in a Newfoundland Outport

ROSEMARY E. OMMER

I have lived in Newfoundland, give or take a few years 'away,' for a quarter of a century, and over that time I have come to appreciate the mostly unwritten yet still tangible history of the ordinary outport and its people. The outport does record its past, using the land, rather than the printed page, as its parchment. The story is there to see, if you know, or learn, what to look for – and the story is important. It is the basis of how things have come to be as they currently are: a concern that underpins much of this book and allows us to understand the human context for many of the issues dealt with here.

An outport is much more than a collection of wooden buildings clustered round a beach, or a string of houses along an undulating, rocky shoreline (Figure 2.1). It is not just a windy seaside village, picturesque in some cases, untidy and apparently run-down in others. It is the imprint on the landscape of a culture, a history, a way of 'living the fishing,' as Paul Thompson's evocative title has it (Thompson et al. 1983). As a consequence of its embeddedness in local culture, attitudes, and European history in the New World, the fishery here has always been more than an industry to rural Newfoundlanders (see McCay, Sinclair, Squires, and Downton, this volume), though some people argue that the old way of life is gone, irretrievable and, moreover, not worth mourning because it was a harsh and impoverished existence.

What was outport life really like? To examine it I have created a generic outport – an 'ideal type' – and set it out on the printed page, as I might explain it verbally to you were we actually visiting such a place.[1] Stand beside me, then, in your imagination, here on the shore road where it crests the brow of a hill, and look down on Rosie's Cove, my surrogate for hundreds of settlements – communities – on the coast of island Newfoundland.

What are we looking at? Two rather impressive churches, a store, a fish plant

Figure 2.1   Rose Blanche, Newfoundland, March 1983. *Photo*: Peter R. Sinclair

that is closed, a wharf with the boats tied up beside it, and a collection of houses, painted white or bottle green or rust brown mostly, though here and there we see Wedgwood blue or ocean sand or primrose yellow. The clapboard siding on some buildings is narrow wooden strips, others (more modern) have broader ones, and some have vinyl siding – easy to keep up, but disapproved of by the heritage people. Buildings like the one by the stream with the trees beside it – with a single storey, narrow board, steep-pitched roof, and 'summer kitchen' or linhay out back – are very old. It appears that one of them is being used as a barn now.

Actually, there is a historical progression of house types that is worth remembering, because it helps us to read the chronology of the place (Mills 1977: 83–97). The single-storey to one-and-a-half-storey house dates back to the first century of settlement; then comes the two-storey, still with steep-pitched roof; after that the mansard roof at the beginning of the twentieth century; then the flat-roofed saltbox (a style that became possible only when tarred roof coverings, which are easier to maintain, came here in the 1920s); then the modern standard bungalow with its mother-in-law doorstep (a steep drop, no steps, because no one ever dreams of using the front door), which dates from around the 1970s; and, finally, the 1990s copy of an up-market suburban St John's designer home.

All these homes, modern or 'traditional,' have probably been built and remodelled by the owners, not by a building company, because Newfoundlanders have not lost the skills needed for home construction, nor the ability to help kin and neighbours on this job – which is why the highest percentage of homeownership in Canada is found in this otherwise not affluent province (see Sinclair, Squires, and Downton, this volume). I am reminded, as we stand here, of how twenty years ago I sat on the Field on the Bank in Campbells' Creek, Port au Port, and watched, listened, and took notes while Joe MacDonald showed me how to build a house, using his box of Swan Vestas matches to illustrate his lecture. His grandson is now probably helping *his* son build his own place, with all the extended family taking part.

Not all the houses in Rosie's Cove, you will notice, are right on the shoreline – but do not be fooled. The purpose of an outport was, and still is, to have access to the sea. Shore frontage used to be how people did it, until the government wharf let the settlement expand inland. Mind you, even in the old days, not all houses faced the ocean. Many a front door and living-room did not: the sea is, after all, workplace ... and dangerous. Access to it, however, defined the early morphology of the Cove.

Moreover, that vast expanse out there of (to our uneducated eyes) featureless ocean beyond the headland was never undifferentiated in the eyes of the local people, here or in any marine environment. To this day, these people know the good spots for setting nets or traps, for catching lobster or scallops. There are nasty currents, tidal rips, hidden rocks; there are bearings to be taken on, for example, a line from Lizzie's Point out there to the single fir tree on the northern headland at Joe's Field if you want to position yourself accurately for that prime cod trap berth. This is highly respected and intimately known territory (see Neis et al., this volume).

Actually, it is essential to think in terms of 'sea tenure' for an outport in the same way as we usually think of land tenure, because even though marine access these days may come most often via the government wharf rather than the stage on some one family's part of a beach, there have always been particular spots at sea that have been commonly recognized as belonging to individual families. 'Belonging,' that is, in the sense of right of access, not really privately owned (and see Manore and Van West, this volume). For example, in Rosie's Cove in the early days before the population expanded, the cod trap berths – whose orange marker buoys you could have seen out there at the headland until the 1992 government moratorium – were inherited. They were passed on from father to son along with the fishing 'room': storage building, stage, flakes, gear, boat, and all the other accoutrements of the fish production unit. On this shore the custom was that the oldest son inherited and his brothers then fished with him as crew (Nemec 1972; Thornton 1979: 141–65). When the population got too large for that, the people of the Cove established a system of drawing lots annually for the prime berths, so that all families would have access to the wealth of the sea and would share in the benefits of the fishery.

This kind of arrangement speaks to something intrinsic to the culture of the place, one of the things that makes the term 'community' the correct one to use when we talk about this Cove and the actual places it represents. There was, and to some degree still is, a kind of communalistic, egalitarian ethic here that none the less operated in combination with the individual, patriarchal family-based fishing enterprise (see Neis, this volume). It is worth remembering that the old Fisherman's Protective Union, formed in the early years of the twentieth century, had as its motto 'To Each His Own,' reflecting nicely the curious blend of individualism and community that is part of what an outport is. We see that same ethic, what E.P. Thompson would have called a 'moral economy,' in practices such as the kind of job-sharing that lets someone work long enough to get his or her 'EI' (employment insurance) stamps, and then (unofficially, of course) hand on the job to someone else (Thompson 1991; Thornton 1986: 19–20). This custom is, to local people, common sense. It is, however, misunderstood in the urban world as 'cheating,' rather than as a strategy, evolved over the years in a rapidly modernizing world, for survival in hard times.

Taken as strategy, it speaks potently to the combination of rationality and social bonding that is inherent in outport culture. It does not, however, relate readily to the formal, capitalist, urban industrial world, with its ethic of individual achievement, which runs counter to, and does not usually comprehend, the purpose of such local customs. There is an urgent need for our society to understand how these strategies have come to be developed, so that they can be evaluated in terms of what they really are and then either accepted or altered in such a way as to meet (or at least not confound) the goal for which they were created: survival (Ommer 1994: 15–16).

The sea that we are looking at in Rosie's Cove is not a faceless 'commons,' in the simplistic sense of merely being shared in common, as far as this community is concerned. Neither are the so-called vacant lots and bogs between the houses, or the wilderness areas back inland, featureless waste lands. What to you and me is scrub and barrens on a bald landscape to the people of the Cove is pasture, a place to harvest blueberries, partridgeberries, and bakeapples, or to set trap lines, though now paths for all-terrain vehicles (ATVs) and favourite spots for hunting moose would be more likely (see Figure 2.2). Of course, trout-

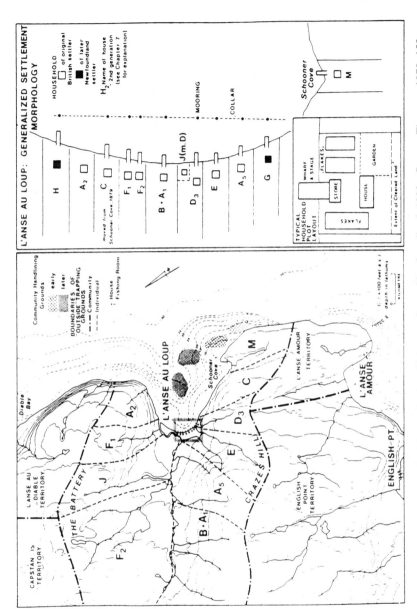

Figure 2.2 Community territorial organization, L'Anse au Loup, Newfoundland, 1890s. *Source:* Thornton 1979: 159. Permission to reproduce this figure gratefully acknowledged.

and salmon-fishing spots are well known too, and berrying places are moved from time to time, the old patches being abandoned and left to recover while new spots are burned over for the next year's harvest (Sean Cadigan, pers. comm., n.d.). In other words, this land is husbanded by the community, both singly and jointly, in a patchwork of local ownership and cooperation that speaks to a whole way of life for people who live from the sea.

As we go down the hill, let me tell you how this type of cove was first settled. This place grew out of the early migratory English fishery, whose fleets came across annually from the West Country. The captain of the first vessel into a cove by custom became the legal representative of law and order for that season – the 'fishing admiral.' West Country or Channel Island merchants hired indentured male apprentices to catch and process the fish over the summer. Caretakers were left over winter to protect stages, flakes, and anything else left behind.

In this particular make-believe cove of ours, let us suppose that a winter caretaker, Henry Tracy, met a cook – Rosie Ash – off one of the boats from another cove, and married her. That is how Rosie's Cove got its name. They settled here (so my story goes) and had four sons and a daughter; shortly after the birth of the last child, Henry was drowned at sea, and Rosie and the boys kept the family fishery going. In due course, the children married people from up and down the shore and settled in their turn in various parts of the cove. That is how this place has come to have two churches – one Methodist, one Anglican, in this case. Their kids grew up and married, and the little settlement continued to expand.

By the nineteenth century, the migratory fishery had given way to small-boat family operations. At first, everyone used to winter at the 'bottom of the bay' where we are today, because it was sheltered and they could cut wood, lay their trap lines, and survive through the winter months. Every summer, they all moved out there to Lizzie's Point (probably named after Henry's daughter) on the headland, where their other home was built close to where they fished. Joe Tracy, Henry's oldest son, was agent for an old West Country fishing firm – let's call it Strawbuck's – and he set up the company store out on the Point. He did well, and eventually set up a merchant business 'on his own account,' working first in partnership with Strawbuck's and later independently (see Thornton 1990: 150–1). His son Tom moved the store into the winter settlement, though he also kept the old place out at Lizzie's Point for warehousing. Soon Tracy's was doing business up and down the whole shore with stores at other places such as Tracy's Cove, Shallop Cove, and Narrow Tickle. The family business survived for nearly one hundred years until, in the 1890s, when poor years in the fishery and market crises hurt them badly, they finally went bankrupt (see Ommer and Sweeny 1992). After that, folks around here sold to another firm up the shore who dealt with the big firms in St John's.

People moved around a lot in the old days – to the headland, into the bay, summer and winter, like I said, in what geographers might call 'maritime transhumance,' or the seasonal migration of people from one marine resource niche to another. This often meant that they had to travel far beyond the bay. Until the 1970s, when the international animal rights movement forced its closure, the seal fishery, which took men away at the end of the winter 'to the front,' was always there, to a greater or lesser extent (and see Cadigan, this volume). At first, sealing was run from local communities, but by the second half of the nineteenth century an organized sealing industry was being managed from St John's, using large, iron-hulled vessels, which could penetrate deeper into the ice. For some places on the Island there was the banks cod fishery, which took men away for weeks at a time – sometimes for ever, if they went to crew on a Nova Scotian saltbanker out of Lunenburg and then stayed there and made a new life. There was also the Labrador fishery, while it lasted; sometimes whole families went 'down the Labrador' in the summer and perhaps settled there for good. Rosie's Cove still feels that it has a special relationship with L'Anse aux Islets on the Labrador coast, because a couple of Tracy families moved there in the 1870s to stay, so the place is kin.

In hard times, young men went far away – they 'rode the rails' in the 1930s depression, looking for work in Canada and the United States, working on the Canadian Great Lakes boats, or building skyscrapers in New York, or picking apples in the Annapolis Valley of Nova Scotia. Young women, if they didn't stay and marry locally, went as servants to Montreal and Toronto and came home with their heads filled with fancy ideas, or brought their 'come from away' husbands home with them, to start married life in the linhay until the boys could help them build their own house in time for the first baby (see Neis, this volume).

In recent times, young men have gone on the offshore oil rigs, or to the Alberta tar sands, and right now there are some young Newfoundland boys from the Cove sailing on the big draggers off the coast of Namibia, now that there are no fish left here; but they do not feel right about that, their mother says. Many such migrants have sent money home – they may have been 'away,' but they were still part of the family, and part of the family income (see Sinclair, Squires, and Downton, this volume). And, of course, when there was no work to be had on the mainland, home they would come, to fish and work the 'gardens' (those little patches of vegetables there), to keep body and soul together. 'You'll never starve here' has been a proud boast of Newfoundlanders for generations.

However, you had to be skilled in reading the land and sea to wrest a living from a place like this, and families were creative about where, when, and what

Figure 2.3   Gardens, Trepassey, Newfoundland, date unknown. *Photo*: Peter R. Sinclair

they fished, seasonally exploiting a whole range of ecological niches and a variety of species of animal life. 'Making the fish' (salting and then sun-drying it for market) was a skilled process too, carried out on the home-made stage at the beach where the fish were gutted and then taken to the flakes for salting and drying. Badly done, the fish would rot and not sell; and particular fish had to be cured in particular ways for particular markets. Most of the inshore salted cod was produced this way until the mid-twentieth century, and in some parts of the coast it was done by the women. In places such as Petty Harbour near St John's, you can still see remnants of long lines of flakes looking for all the world like an open-air, pre-industrial shop-floor assembly line, but in Rosie's Cove people had their flakes on their land, or arching across the road. In the late 1960s, when I first visited rural Newfoundland, I sometimes had to drive under giant flakes that crossed the road. Today, if there are any flakes at all in a place like Rosie's Cove, they are small, maybe covered with wire mesh to let the air get through, not the fine wood slats and fir brush of the old days.

People used traditional, pre-industrial methods for getting produce from their gardens (Figure 2.3). Flexibility was the key here. The women grew their root crops in built-up 'lazy beds' to get more soil depth and drainage, and they kept animals family by family, or, in some cases, one family would have a cow and others some sheep and they would exchange the surplus milk for the extra wool and arrange pasturing and fodder as necessary throughout the community (Cadigan 1994: 21–2). As well, of course, people were berrying and hunting, preserving the family food, building homes, milling cloth, and making clothes – the hundred and one tasks that made up community life. 'Occupational pluralism' was practised in an annual, seasonal round, as expressed in Figure 2.4, in which the community crafted for itself a complex and sophisticated set of adaptive strategies that ideally wove ecology and economy into a seamless way of life and maintained a balance – a 'dynamic equilibrium' – between people and their environment (Thornton 1979; Tizzard 1979: 203–311). And when they died they were laid to rest in a small cemetery, in the plot of land closest to the village that had enough soil depth for decent burials.

Merchants had to be as flexible as the fishing families. Strawbuck's would have had agents all round the Island so that it could get the maximum amount of fish every season to send to market. When this shore was blocked with ice, another was still open; Labrador fish were good for the Caribbean market, South Coast fish for the Mediterranean.[2] Of course Strawbuck's sometimes dealt in more than fish, if that helped it or the agents it needed, especially in the early days, so Tracy's accounts talk about seals, salmon, and even furs and berries on occasion.

Just stand and look at the old store, boarded up and disused as it now is, and think that once it was the centre of commerce for the Cove, selling everything, and buying everything too. It was the spot where this tiny, isolated settlement was connected to the international economy. That sounds grandiose, I know, but it is true. There would have been a direct link between this old building we're standing beside now, Tracy's wharf behind it (long since gone), and Zante (in Greece), Rio de Janeiro, the Turks and Caicos Islands (for salt), Alicante, Naples, Southampton, and Liverpool. It used to be an outpost of empire, this place – now it's a little outport. And the store used to be an international articulation point – now it's a Newfoundland craft shop and restaurant. The house next door, by the way, that smart bed and breakfast with the gingerbread trim round the front porch, was the merchant's home.

You can see this sort of heritage and this kind of transformation all round Atlantic Canada. In Paspébiac in the Gaspé there is a very good gourmet restaurant in the one-time home of Charles Robin, the founder of the huge Jersey merchant firm of Charles Robin and Company that dominated the fish trade in the Gulf of St Lawrence from the late eighteenth century to the end of the nineteenth. The community at Paspébiac was dominated by the Robin company – there were barns, a forge, a carpenter's shop, warehouses, and even a 'bache-

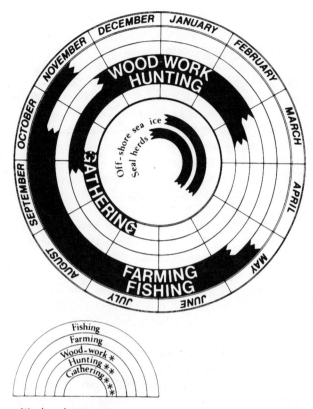

*Wood-work; Fire-wood, fencing materials, saw-logs and other building materials.
**Hunting; Birds, terrestrial mammals
***Gathering; Wild fruits and berries.

Figure 2.4   Rural northeastern Newfoundland: activity cycle of a subsistence household. All boundaries are approximations and represent decreased activity. *Source*: Sanger 1977: 137. Permission to reproduce this figure gratefully acknowledged.

lor's hall' where the apprentices to the company stayed when they were brought over from Jersey on a three-year indenture to train as crew, or clerks, or the like. Indeed, the outport was the industrial shop floor (Ommer 1991: 112–14). That was really big business, not small stuff like Tracy's.

But big or small, fish merchants ran their business in much the same way, building flexibility into their operations. Old account books with the entries for fish and for supplies, written in elegant copperplate handwriting, show us how the credit system ran – the infamous 'truck system,' by which merchants are said to have enslaved fishers all round the coasts here. There was certainly plenty of opportunity for abuse in the system, and many families were 'tied' to the merchant store, but merchant credit was a way by which both merchants and fishing families got round the difficulty of making a living in an isolated and cash-scarce part of the world. For a firm such as our hypothetical Tracy's, it meant the ability to keep 'their' fishers here year-round without having to pay them twelve months' wages for four months' fishing, which the firm could not have afforded. For local families (the firm's own kinfolk, let us not forget), it meant a way of getting basic necessities that they could not produce here, such as salt, molasses, tea, material for clothing, flour, nails, and saws. People traded their fish and whatever else they could to Tracy's in return for those things, as you can see on the pages of any merchant's ledger.

So the credit system was a trade-off of interests in that way, but also in another important way. Tracy had to 'carry' fishers through bad years in the fishery, which happened either because the fish did not 'run' in some years, possibly because of resource depletion (see Cadigan, this volume), or because the market price of fish fell and so expected returns failed to come in. Fishers were, of course, vulnerable to merchants' abuse of a pricing mechanism in which the merchant set the price of the fish sold to him and, through that, the price of the supplies that he sold to fishing families, but merchants were also vulnerable to poor fishing seasons, to people (or other firms) who failed to pay up, and to the vagaries of the market-place. Some merchants did abuse the system badly. Others, such as Tracy's, actually overextended credit to keep folk going in the bad years of the 1880s and 1890s, and went bankrupt in the process (Ommer and Sweeny 1992).

The twentieth century has been one of ongoing crises in the fishery in Newfoundland (Ommer 1994). With the end of the Great War, the old alliance between merchant and settler was badly shaken. The best sign of that was that by the 1920s the price of a barrel of flour (what people *imported* as a basic necessity), which had tracked the price of a quintal of fish (what they *exported* for their livelihood) for over one hundred years, ceased to be correlated with it (Ommer and Hong 1990). In the 1930s, things got worse, and only the Second World War brought any respite to the Newfoundland economy at large. American bases were set up here, and then, after the war, Newfoundland joined Canada in 1949. Together, these events promoted the integration of the new provincial economy with North American advanced industrial capitalism and, as a consequence, began to alter people's expectations and way of life.

In postwar outports, the old inshore small-boat family fishery began to give

Figure 2.5   A small fish plant, Brigus South, Newfoundland, date unknown. *Photo*:
Peter R. Sinclair

way to a government-promoted and -regulated industry (Wright 1997). There
developed a large-vessel offshore fishery run by a few big firms, capital intensive and using increasingly more effective search-and-kill technology (Hutchings, this volume). At the same time, household saltfish production died away
and fish was frozen and processed in small fish plants in many outports such as
Brigus South, shown in Figure 2.5. The pluralist household economy changed,
as transfer payment cheques took the place of merchant credit, sealing came
under international scrutiny and died out, and, one by one, the underpinnings of
the traditional way of life were chopped away.

The fishery became an 'employer of last resort' in a province plagued by
chronic un- or under-employment. Outports started to take on some of the worries and troubles of single-industry towns, losing the ability to sustain themselves or to function as expressions of a cherished culture. In Rosie's Cove, as
elsewhere, families started to worry. The fish that they caught were getting

smaller, then there were fewer of them, then there were no 'mother fish' – the big old fish that are essential to the healthy survival of the species (Cadigan, Villagarcía et al., this volume). Finally, in 1992, came the collapse of the groundfish stocks and the moratorium on northern cod.

What we see on the wharf today in 1996 are not nets being hung up to dry, or people leaving the fish plant at the end of a shift. We see boats beached and upside down on the slipway, unattended; we see crab and lobster pots, not nets, on the wharf. The single boat out there at the headland is catching crab (*Chionecetes opilio*, see McCay, this volume), and the longliners in the summer are taking tourists out to watch the whales. Plants are idle most of the time. It is true that the houses are all freshly painted – that is a sign of money from The Atlantic Groundfish Strategy (TAGS) and too much free time – and women are making partridgeberry, blueberry, and bakeapple preserves, or hooking rugs with scenes of dories and cod flakes, to sell to passing tourists like you and me. Newfoundland crafts, the sign tells you, are now exempt from provincial sales tax (see Neis et al., Sinclair, Squires, and Downton, this volume).

Meanwhile, a whole culture, one in which ecology and economy worked hand in hand, is dying before our eyes. Rosie's Cove and all the other small outports are suffering the consequences of losing the old economy that allowed people to work their land and sea together in intimate understanding of the local ecology, and it is not clear what can or will replace it. Rosie's Cove looks lovely in the late autumn sunshine, but it is a tragic beauty.

The story told here reflects the reality of many outports in Newfoundland. These communities are the physical expression of a kind of cultural rootedness that I have learned to understand and to respect. My grandsons – the sons of my son who grew up here and his wife, who is a Newfoundlander with roots that go back many generations – should have a sense of belonging to this place, but sadly that is not certain. For this young family, like so many others, has had to move to the mainland to make a living. There are Newfoundlanders to be found, across Canada and beyond, for whom Rosie's Cove or its equivalent is a cherished memory of something that they have had to leave, and may have lost forever.

NOTES

1 Of course, the danger in this kind of overview is that it cannot reflect the wealth of variety that Newfoundland outports possess. None the less, this is my encapsulation of that which (I have been told by outport people during my more than twenty-five years of working there) is to be considered essential to the culture, the 'soul,' of rural Newfoundland.

2 For details of fish markets see Ryan 1986 and Ommer 1990b; for merchant strategies, see Ommer 1990b, 1991.

WORKS CITED

Cadigan, Sean T. 1994. 'The Historical Role of Marginal Agriculture in Sustaining Coastal Communities on the Bonavista Peninsula.' Occasional paper. St John's: Eco-Research Project, Memorial University.

Mills, David B. 1977. 'Development of Folk Architecture in Trinity Bay, Newfoundland.' In J.J. Mannion, ed., *The Peopling of Newfoundland*, 77–101. Toronto: University of Toronto Press.

Nemec, Thomas. 1972. 'I Fish with My Brother: The Structure and Behaviour of Agnatic-Based Fishing Crews in a Newfoundland Irish Outport.' In R.R. Andersen and Cato Wadel, eds., *North Atlantic Fishermen: Anthropological Essays on Modern Fishing*, 9–34. St John's: ISER, Memorial University.

Newfoundland and Labrador, 1986. *Building on our Strengths: the Final Report of the Royal Commission on Employment and Unemployment*. St John's. Queen's Printer.

Ommer, Rosemary E. 1990a. 'Introduction.' In Rosemary E. Ommer, ed., *Merchant Credit and Labour Strategies*, 9–15. Fredericton: Acadiensis Press.

– 1990b. 'Merchant Credit and the Informal Economy: Newfoundland, 1918–1929.' *Historical Papers*, Journal of the Canadian Historical Association: 167–89.

– 1991. *From Outpost to Outport: A Structural Analysis of the Jersey–Gaspé Cod Fishery, 1767–1886*. Montreal: McGill-Queen's University Press.

– 1994. 'One Hundred Years of Fishery Crises in Newfoundland.' *Acadiensis* 23, no. 2: 5–20.

Ommer, Rosemary E., and Robert Hong. 1990. 'The Newfoundland Fisheries: The Crisis Years, 1914–1937.' Paper presented to the Atlantic Canada Studies Conference, Orono, Maine.

Ommer, Rosemary E., and Robert C.H. Sweeny. 1992. 'Which Ties Bound Who to Whom? Lessons Learned from Computerizing Merchant Ledgers for Bonavista Bay, Newfoundland, 1889–1891.' Paper presented to the Australia and New Zealand Conference on Canadian Studies, Wellington, New Zealand.

Ryan, Shannon. 1986. *Fish Out of Water: The Newfoundland Saltfish Trade, 1814–1914*. St John's: Breakwater Books.

Sanger, Chesley. 1977. 'The Evolution of Sealing and the Spread of Settlement in Northeastern Newfoundland.' In J.J. Mannion ed., *The Peopling of Newfoundland*, 136–51. Toronto: University of Toronto Press.

Thompson, E.P. 1991. *Customs in Common*. London: Merlin Press.

Thompson, Paul, with Tony Wailey and Trevor Lummis. 1983. *Living The Fishing*. London: Routledge and Kegan Paul.

Thornton, Patricia A. 1979. 'Dynamic Equilibrium: Settlement, Population and Ecology in the Strait of Belle Isle, 1840–1940.' PhD dissertation, University of Aberdeen.

– 1986. *Jack of All Trades*. Technical Report for the Newfoundland Royal Commission on Employment and Unemployment. St John's: The Commission.

– 1990. 'The Transition from the Migratory to the Resident Fishery in the Strait of Belle Isle.' In Rosemary E. Ommer, ed., *Merchant Credit and Labour Strategies*, 138–66. Fredericton: Acadiensis.

Tizzard, Aubrey M. 1979. *On Sloping Ground: Reminiscences of Outport Life in Notre Dame Bay, Newfoundland*. St John's: Memorial University.

Wright, Miriam. 1997. 'Newfoundland and Canada: The Evolution of Fisheries Development Policies, 1940–1966.' PhD dissertation, Memorial University of Newfoundland.

3

Familial and Social Patriarchy in the Newfoundland Fishing Industry

BARBARA NEIS

Feminist researchers have contributed to our contemporary understanding of fishery economies. They have critiqued the androcentrism of pre-1980 analyses and in doing so have introduced new concepts for analysing women's active participation throughout fishery economies and the diverse gender relations and differing sexual divisions of labour that can be found within them. Feminist accounts have also challenged male anthropologists' common assumption of universal patterns of male dominance and female passivity within fishing households and communities (Allison, Jacobs, and Porter 1989; Cole 1991; Nadel-Klein and Davis 1988; Neis 1988a; Porter 1983, 1985, 1993).

For the Newfoundland fishing industry, Marilyn Porter (1985) provides significant evidence to support the view that a rigid sexual division of labour did not imply, in any simple sense, male dominance. Wives of boat owners in inshore fishing communities apparently exercised considerable control within the spheres of their kitchens and, in the past, shore-based work. These conclusions are also supported by other accounts (Davis 1983, 1988; Murray 1979). This multifaceted feminist critique has tended, however, to rely on data collected from groups of women with the most power – the wives of boat owners as opposed to crewmen, women who live in endogamous communities near their female kin, or women employed in unionized as opposed to non-unionized fish plants – in contexts where women's power is strongest and most visible – moments of protest as opposed to quiescence (Cole 1991; Davis 1983, 1988; Murray 1979; Neis 1988a, 1994; Porter 1988).

In fishing economies, the state, households, workplaces, and communities are terrains of struggle for women. But the women bring differing class, marital, and other resources to these struggles. Some exercise significant power within some fishing households, and others have successfully challenged discriminatory measures and aspects of male control through the state and within their communities. A full understanding of gender relations in fishing economies needs to include, however, those women who are economically and socially marginal: women on welfare, widows, daughters who were forced out of the fishery, and non-unionized fish-plant workers. Documented processes of class differentiation within fishing economies make it particularly important to avoid generalizing from the experiences of the wives of boatowners to cover other groups of women (Sinclair 1985). In all classes, however, there may be some who are silenced by the threat of violence, of sexual abuse, or of lost livelihood in the form of either a job or support from the welfare state. The feminist litera-ture on the Newfoundland fishery says little about the effects of state policies on women's options and experiences; the few sources that do address this dimension often overlook patriarchal ideologies and dominance and the poten-tial for violence within these households, as well as their relationship to policy initiatives and outcomes (Connelly and MacDonald 1991–2; Wright 1995).

This gap in the existing literature and the current context of crisis and restruc-turing in fishery economies throughout the world have prompted this chapter. I hope to highlight some of the risks that currently confront women in fishery economies and draw attention to shortcomings in some proposed 'solutions' to these crises. The combined influence of resource depletion and attempts to fur-ther enclose the fishery commons are whittling away at those with the right and the ability to derive their livings from the sea. If current trends continue, there will be few fishers' wives in the future because there will be far fewer fishers (McGoodwin 1990).

My analysis uses recent research on the fishery and fragmentary accounts from individual women in Newfoundland and Labrador fishing communities, past and present, whose experiences do not fit the image of strong women, embedded in caring fishing families and confidently in control of their own, separate spheres. I examine some of the ideologies and practices that have lim-ited women's access to control over the fishery resource, and to the wealth that it produces, while making these women responsible for production- and repro-duction-related work within the fixed-gear, coastal household–based fishing community (often called the inshore fishery). I have woven some of the women's voices and experiences into the text to remind us, as Sally Cole (1991) does, that these women are not merely 'passive victims.'

Two concepts guide my analysis – 'familial patriarchy,' which refers to 'the experience of the past in which power and authority over women and chil-dren were largely exercised in the home,' by men (Ursel 1992: 2), and 'social patriarchy,' or the state-sponsored social welfare institutions such as family allowance, minimum wage, unemployment insurance, and old age pensions associated with twentieth-century industrialized nations. Within social patriar-chy, the state supports and controls women and children through theories, ideologies, and practices such as moral regulation that shape these social welfare institutions (Valverde 1991; Wright 1995). Ursel's 1992 analysis of the transition from familial to social patriarchy in three Canadian provinces identifies the state as mediator between the spheres of reproduction (the family) and production. She argues that this process of mediation is contradictory, with the result that state initiatives tend to sustain, but in the longer term erode, familial patriarchy. I document a similar contradictory relationship between familial and social patriarchy within the Newfoundland inshore fishery.

Familial Patriarchy

Thomas Hutchings, the merchant's agent, commented 'as you can see I'm crediting each household with the work of the women and children it will feed this winter.' ... In that instant Mary Bundle resolved she would marry Thomas Hutchings. If Mary could add nothing to her possessions without a man – then a man she would have. (Morgan 1992: 66)

The Newfoundland fishery began as a migratory, primarily male enterprise based on master–servant relationships and a fixed wage for the season.[1] During the nineteenth century, a British settler fishery displaced the migratory one and spread up the northeast coast of the island. Conditions of costly credit, ecological uncertainties, poorly managed resources, better gender balance, greater labour abundance, and seasonality transformed the fishery into an enterprise reliant on familial labour, supplemented when necessary by hired labour (Cadigan 1995; Ommer 1990). During this period, the hired labour often included young men and women of other families, many of whom would later establish fishing households of their own.

Sean Cadigan (1995) has explored how these fishing households developed, with some encouragement from the state, into a form of familial patriarchy. Fish harvesting and processing became decentralized and production and reproduction activities fused. Production of fish for exchange was combined with subsistence agriculture, gathering activities, and daily and intergenerational caring. A fairly rigid division of labour characterized work. Generally speaking, men broke the land for subsistence agriculture in spring, fished in summer and fall, and worked in the woods, at their gear, and in other fisheries in winter. Women cared for children, husbands, the elderly, and the infirm, prepared meals, gardened, and made clothes and other household necessities. In some communities, women also took primary responsibility for drying the fish for sale to the merchants, as Marilyn Porter's study shows (1985), though their involvement in fish processing varied by region.

Economic surplus from the fishery was largely controlled by the merchants. They controlled marketing, as well as the prices of the minimal requirements for household production and of fish. They paid largely in goods ('truck') rather than cash, and fishers were often indebted to them. Merchants' accounts contain no separate ledger entries for the male fishers' wives and children: their labour was recorded under the names of their husbands and fathers (Sweeny, pers. comm., 1993). Some women may have controlled occasional cash surpluses (Porter 1985). However, it is not insignificant that while exchanges between fishers and merchants were based partly on the results of wives and children's work, women had no formal right to determine the type or scale of purchases made by male fishers. Men could make debts they were subsequently unable to pay and, as many did, walk away from them. Without direct access to credit, women had no such 'privilege' but paid part of the price for this debt avoidance in higher costs for goods (Sweeny, pers. comm., 1993). However, women do not seem to have been held liable for the men's debts (Cadigan 1991: 205).

Husbands and fathers dominated women and children's access to the wealth from the fishery through their control over houses, land, fishing technology, and access to the fishery resource itself (Faris 1972; Firestone 1967). As well, there were patrilineal inheritance laws, state-sanctioned regulations governing access to the fishery resource (the Newfoundland Fishery Regulations), and laws that made husbands and fathers responsible for supporting women and children without providing guarantees for the provision of this support (Cadigan 1991; Cullum, Baird, and Penney 1993; McCay 1976; Martin 1979). The centrality of women's work to the fishery in some communities reinforced women's economic dependence on men by leaving them with little time to produce agricultural and other products for exchange (Cadigan 1991). It also reinforced men's economic dependence on women.

A review of Conception Bay court cases from the late eighteenth and early nineteenth centuries reveals that a widow was generally left without property because her deceased husband's house(s), land, and fishing gear would pass to sons or other male relatives. Sean Cadigan's research suggests that property was normally left to sons, 'usually with some provision that they care for their mother or mother-in-law.' The women thus depended on the goodwill of these male relatives for support, which could be 'disastrous for a widow' (Cadigan 1995: 67). There were, however, cases where daughters sued to protect some share of their fathers' estates and others where women appeared in court 'to defend their household's interests' (Cadigan 1991: 206). The court records also show that some women who worked as servants in the fishery encountered violence and that women who were made pregnant by their employers had to sue for support. Women's wages were also lower than those of their male counterparts (Cadigan 1995: 71). In the context of patriarchal control and economic vulnerability within the household and patriarchal law outside, 'the family's basic struggle for survival ... ensured an essential solidarity between men and women in households despite the presence of male violence' (Cadigan 1991: 210). As the opening quotation for this section suggests, women learned early in their lives that their comfort and economic survival depended on the strength of their ties to men and on the willingness of those men to act responsibly. However, as immigration to Newfoundland declined, and the ratio of men to women shrank, forming these necessary ties would become increasingly difficult.

During early settlement, men outnumbered women, and women, along with their children, were valuable assets, providing the basis for the establishment of a household fishery (Porter 1985). However, after male immigration declined in the nineteenth century and population growth became a product of local reproduction and hence greater gender balance, census data show that female populations in fishing districts tended to stabilize at a level below male populations. This was particularly true for women in the age group twenty to twenty-four. According to the 1935 population census for Newfoundland, for example, the ratio of women to men in this age group for eleven fishing districts was 770.8 women per 1,000 men (*Tenth Census*, 1937: Vol. 1, Table 32, 168–71).[2] A pattern of female removal from fishing communities appears to have been particularly strong during periods of economic stress, as was the case in Bonavista during the early twentieth century. There, the proportion of the population classed as females curing fish fell from 22.1 per cent to 8.3 per cent between 1891 and 1911 (Ommer 1990: 172–3). Fishing communities differed significantly in technologies used, the combination of fishing with other activities, periods of male absenteeism, and the sexual division of labour. These factors, as well as parental, church, and ethnic concerns regarding suitable marriage partners for children, might influence migration patterns for young people. Benoit (1982: 63) argues that in the Stephenville area during the early twentieth century women were rarely allowed to leave their community, and, in contrast with inshore communities elsewhere, young unmarried women appear to have remained in their fathers' or uncles' households until marriage.

Generally speaking, familial patriarchy demonstrated greater attachment to sons than daughters. The traditional sexual division of labour, preference for sons, and male inheritance practices pushed many women out of fishing households, forcing them to find work elsewhere. Their mothers were left to share the burden of household and fishery work with the remaining daughters, if any, and with their husbands and sons. Young men lived at home and worked in their fathers' boats until they could fish on their own or had sons old enough to form a new crew (Faris 1972). Until recently, knowledge of the fishing grounds learned from their fathers and local rights to fishing berths acquired by virtue of being male and coming from a fishing family seem to have been critical to their survival as fishers (Butler 1983; Neis 1992). Where technology was costly and could not be easily divided among sons, as in the Labrador schooner fishery, the access of some men to the fishery seems to have depended on their willingness to forgo marriage altogether (Britan 1979). A shortage of young, marriageable women in the area would have contributed to this pattern of bachelorhood.

Though they lived alongside the fishery resource, young women did not have opportunities to gain the knowledge required to fish successfully (Silk 1995). Nor did they have the means to share in the wealth from that resource unless they were married to a fisher or worked for the local merchant or a boatowning family. As argued by Antler (1977), women followed men into the fishery. If there was no man whom they could (or would) follow, many had to leave fishing communities to find paid work. Young daughters, particularly those in poorer families, were expected to 'ship out,' work for local fish merchants for low wages (Antler 1981), or move into urban areas to work in factories or as domestic servants in private houses. The term 'ship out' derives from the practice of engaging for service as a member of a fishing crew for a specified wage and period of time (Story, Kirwin, and Widdowson 1990: 472). For young girls, it meant agreeing to live with another fishing family and carry out both domestic and fishery-related tasks for a small monthly wage. A 1970s study of women and work in Newfoundland described their work in an unspecified period prior to the Second World War: 'A girl was sent into service for two main reasons: to relieve her parents of her support, and to learn all the skills pertaining to household and fishery in order that she might make a good fishers' wife. Girls were paid about $2.50 a month in winter and $5.00 in summer, because of their help in making fish. A minimum of clothing and board and lodging was also provided. The work was often hard and long; the girls were completely at the mercy of their employers as to the work they were required to do and the treatment they received' (Batten et al. 1974: 13).

Young girls shipped out well into the twentieth century. Some men also did so, but a male was more likely to work on a stronger economic and social footing as a 'shareman,' taking a share of the catch, and working within his father's family enterprise. The nineteenth-century practice of offering women lower wages continued into the twentieth century. In 1935, annual male incomes from fishing, forestry, and trapping averaged $143.20. In contrast, the above quote suggests that the annual wage of shipped girls around this time was probably about $40.

The overlap between household and processing work, and the importance of

fishery work to some women's identities during the period of familial patriarchy, are well documented in the existing literature. Despite patriarchal and
patrilineal practice, women who married boatowning fishers often enjoyed their
work and identified themselves with the fishery. A fairly extreme sexual division of labour and the central role of women's labour allowed women, in Marilyn Porter's words, to secure for themselves 'places of security, dignity and
considerable independence' (1993: 175). Some boatowners said that their wives
did more than 50 per cent of the work, acknowledging that the wives were the
'skippers of the shore crew' (Murray 1979; Porter 1985).

   Not all women, however, could describe their homes during the era of familial patriarchy as 'places of security and dignity.' Widows, and daughters from
poorer families, had few employment opportunities. The mother of one woman
interviewed was left widowed at a young age, with small children. She qualified
for a small quarterly pension from the government, which she supplemented
with money earned drying the fish of local fishers whose wives were ill or
unable to do this work. She was paid a small sum for each 'quintal' (that is,
112 lb. of dried fish) that she and her children produced. From these small earnings, she fed, clothed, and housed herself and her children. They lived in a
rented house and were often hungry. Though they lived alongside an abundant
resource, they rarely had any fish to eat. The daughters left home early to find
work as domestic servants in town (Memorial University Folklore and Language Archives [MUNFLA] C14605).

   Women who left their homes to work as domestics, or in the Labrador fishery
at the turn of the century, or in fish plants in the 1950s could end up pregnant
and abandoned. Public concern about the morality of young women who
worked with men does not seem to have been accompanied by willingness to
ensure financial support for them and their children (Cullum, Baird, and Penney
1993; Patey n.d.; Smallwood Papers 1952).

Social Patriarchy

Though Newfoundland had some social welfare institutions prior to joining
Canada in 1949, schemes were minimal in number and scale of support. Widows' allowances are reported to have been $50.00 a year or $12.50 a quarter
(approximately eight cents a day), while the 'dole' (welfare) was $1.80 a month,
or six cents a day, during the 1930s. Newfoundland introduced a state-operated
old age pension scheme in 1911, but payments were low, limited to those seventy-five or over, and only men were ever eligible (Snell 1993). All that changed
after Confederation, during the era of expanding state investment in social welfare institutions that occurred throughout the country between the 1940s and the

1960s (Ursel 1992). As a province, Newfoundland inherited the family allowance, unemployment insurance (UI), and a different old age pension program; it also participated in the development of new Canadian programs, such as medicare. All these programs were shaped by the implicit assumptions about the ideal family and women's place that dominated public thinking of the day. The result for the Newfoundland fishery was a set of programs that at first transformed and sustained familial patriarchy and then ultimately undermined it.

Social welfare initially combined with the development of corporate-owned fish processing in the postwar period to bolster familial patriarchy. Such programs gave households greater financial independence from merchants and reduced the costs of caring for and raising children. Cash payments reduced a person's debt and reliance on credit. Family allowance cheques were used to buy shoes and clothing for children, reducing the annual financial burden of caring in the household fishery. Family allowances were paid to mothers, not fathers. In a partial subsistence economy, where households had little access to cash, $15 a month (the amount of a cheque covering three children in 1949) seemed like a lot of money to a young mother. Pensions gave adequate incomes to the elderly, with surplus left over to help finance new engines for their sons' boats (MUNFLA C14593). Pension incomes in Newfoundland in 1952–3 averaged approximately $40 a month ($480 a year), whereas, as one woman estimated, at the end of a good fishing season a fishing household would be lucky to clear $200 (*Historical Statistics* 1985: Table B-5; MUNFLA C14605).

The setting up of corporate-owned fish plants in some communities also strengthened familial patriarchy in the early 1950s. These plants permitted households to sell some of their fish fresh, for cash. As suggested by the woman cited above: 'It was the convenience and the readiness of it, it was a great thing to get a cheque from the fish plant every week, because I remember they used to come in from fishing or if it was a queer day or anything go up to ... and get your cheque, $600 and $700, go up to the Royal Bank and change it. Sure that was a godsend compared to waiting till the last of October like you used to have to before and you might not get something then. If you didn't pay your bill, you probably wouldn't have very much coming to you' (MUNFLA C14605).

Though they bolstered familial patriarchy, social welfare institutions and corporate-owned fish plants also worked to transform and undermine it. In order to qualify for family allowance payments, for example, children had to remain in school instead of going in the boat. Because family allowance cheques were paid to women and not to men, women received new financial independence. Cash incomes from government construction and UI meant that more young people could marry. Young men and older fishers who had formerly worked as sharemen for boatowners took better-paying construction jobs in the province and elsewhere in Canada, causing a labour shortage in the household fishery (Antler and Faris 1979; McCay 1976).

Women gradually disappeared from the cod-drying flakes in those communities where they had laboured for generations. Some young women stayed in school and then took jobs in the expanding education and health sectors; others left their communities to work in fish plants. It is difficult to track the number of women employed in the fishery from 1935 onward. Census data do show, however, a decline of almost 50 per cent in men employed in the industry between 1945 and 1951 (Anger, McGrath, and Pottle 1986: 9–10). A gendered analysis of population and net migration of young people from the Northern Peninsula between 1951 and 1961 found high net migration of youth as a whole, and higher net migration for young women than for young men (Sinclair 1985: 54–6).

As the number of corporate-owned fish plants grew, single women, widows, and women who could not count on their husbands to support them fully took the small number of poorly paid fish-plant jobs open to women at the time. This work paid better than domestic service. It provided badly needed second incomes for some of the women and a break from brutal home situations for others. For some women, it also provided a break in routine: 'I was fed up with the hum drum of doing the same things in the house, washing clothes, cooking meals, babying drunken husband when he was home. I had six children, which meant I didn't get out much if at all in 30 years. We didn't have water and sewage in the house at that time, therefore I had to carry water in two galvanized buckets ... Fish plant work I found to be hard work, but compared to housework it was a breeze' (Batten et al. 1974: 25).

With fewer young 'shipped girls' around to help out, and with fewer 'sharemen' whose wives might be available for the 'shore crew,' boatowners' wives had fewer workers to supervise. Their participation in the fishery was also more constrained by their responsibilities for home work and child care. These many changes produced a crisis in the reproduction of the fishery based on familial patriarchy, particularly in regions lacking a fish plant to purchase fresh catch from fishers. The result was growing pressure from fishers and the provincial government for increased social support to maintain the fishery.

Gender ideology and gendered practices pervaded the provincial and federal policies that affected fishing communities during the postwar period. In 1950, Newfoundland introduced its first Minimum Wage Act. Unlike the rest of Canada, where minimum wages were introduced earlier and began as protective legislation for women, the Newfoundland legislation applied until 1955 only to men (Creese 1991–2). After 1955, the legislation took the form common in other provinces – separate minimum wages for men and women, with the rate for women set lower than that of men. This situation did not change until 1 January 1974. Domestic employees working in private homes, among whom would be included girls 'shipped' for the fishery in the 1950s, were not covered until much later (Creese 1991–2; Gillespie 1986; *Historical Statistics* 1985: Table D-4). Thus minimum wage legislation perpetuated an inequity rooted in familial patriarchy (Creese 1991–2: 121). A similar pattern prevailed in fishers' UI and in policies introduced to 'modernize' the fishery.

   Female fishery workers were often married, engaged in seasonal work, and working within family enterprises. All three categories of workers had difficulty qualifying for UI during the early years. Improvements came in the 1950s, however, when the government eliminated formal discriminatory requirements imposed on married women and seasonal workers and qualified male family-based fishing crews for UI.[3] Discrimination against fish-processing workers engaged in the household fishery and women who fished with their husbands persisted much longer.

   In 1957, the federal government created UI for seasonal inshore fishers. Several scholars have persuasively argued that this program perpetuated the historical pattern of crediting women's work in fish processing to male household members and denied women working in household enterprises access to UI (McCay 1988; Wright 1995). Fishers' UI was based on the fiction that fishers were the employees of the merchants and plantowners who bought their fish. During its early years, the new program provided fishers with qualifying 'stamps' based on the amount of fish produced in their households. The system for allocating stamps directly reflected the extra work required to dry fish by paying more for dried fish than the fresh product. Like familial patriarchy, however, the UI generated by the extra work of drying the fish – women's work – went to men. By the same token, the new system made no distinction between fish sold in the form of salt bulk and higher-quality, light-salted dried fish (Canada 1965). The processing of light-salted dried fish was more labour intensive, and the labour involved was often women's. In short, the UI system credited women's work to men by allowing only the fishers to qualify for UI and failed to reward women's extra work. As a result, the system discouraged women from making dried fish and from investing the extra labour required to produce higher-quality dried fish. This situation contributed to the longer-term decline of this branch of the industry, its displacement by corporate-owned fish processing, and the associated weakening of familial patriarchy caused by growing dependence on the corporate sector (Alexander 1977; Neis 1988b).

   Whether women were fishers' wives or daughters, or 'shipped' girls from other households who combined domestic work with drying fish, they were not eligible for UI. Fishers' UI schemes thus perpetuated the allocation of the wealth and benefits resulting from women's labour to men and the defininition of women as dependents of men in the inshore fishery. UI regulations also combined with practices retained from familial patriarchy to exclude women from direct access to the fishery resource and to fishing incomes. They reinforced male control over fishing by disqualifying women who fished with their husbands.

In 1962, Premier Joseph Smallwood convened a provincial conference to address a deepening crisis in inshore fishing communities. The conference was designed to establish provincial fisheries policy for the next decade, and its transcripts reveal the patriarchal assumptions that pervaded policy formation.[4] Smallwood described the meeting as dealing with problems confronting 'fishermen' and the growing numbers of 'unemployed young men' in rural communities:

You remember here, yesterday morning, I said that the things we were to talk about yesterday morning were the problems of a fisherman, not a man as anything else but a man as a fisherman, and then you remember that I pointed out that the fisherman in addition to being a fisherman, are [sic] very often husbands and very often fathers, and certainly they are always citizens and they are people, they are not just fishermen, and as citizens and as people they are interested in all kinds of things besides fish. They just don't spend all their lives thinking of nothing else but fish, they're interested in homes and houses, they're interested in schools, churches, in lodges, in co-op societies, in unions, in sport, in athletics, and they are interested in their families, and they are interested in their communities. Now this morning, what we are trying to do is this, we are trying to take a look at the problems of the fishermen, not as fishermen but their problems as men, their problems as citizens. (Newfoundland Fisheries Conference 1962: Vol. 4, 3)

With the exception of one plant owner and one federal civil servant, there appear to have been no women at the conference. The fishery discussed consisted only of male fishers, fishing companies, and unemployed men. No concerns about the problems confronting unemployed young women were raised. The rare mention of women was confined to condescending references to them as the new owners of 'nylons,' meaning nylon stockings, who had opted out of working on the flakes rather than run the risk of getting runs; as helpers, assisting their husbands; or as young girls in search of the few dollars they could make 'spreading fish.' The premier actually identified the absence of women and children from the flakes as a sign of progress. He was surprised to hear that some of them were still doing this work, which he and some other participants described as 'slavery.'

Smallwood's speech enfolded the women who lived in fishing communities and depended on the fishery resource inside the concepts of 'family' and 'community,' thereby implying that addressing the needs of male fishers and unemployed young men would meet women's needs as well. His comments showed little of the respect and recognition for women's economic contribution to and reliance on the fishery that contemporary accounts would have led us to expect (Murray 1979; Neis and Williams 1996; Porter 1985). Women, their issues, and their work in the fishery were largely invisible.

Fishery support in the 1960s was tied to an overall program of 'modernization.' Wright (1995) has shown that the modernization policies of federal bureaucrats and federal Department of Fisheries publications from this period were permeated with gender ideology. There were ideas about 'man's place' and 'women's place' and the relationship of each to the fishery: men would be trained in new ways of catching fish; women would have little presence in the modern fishery but would devote their time to their families.

Central to modernization for Newfoundland was the resettlement of fishing communities into 'growth centres' equipped with corporate-owned fish plants. Resettlement separated households from the traditional land and forest resources that had always provided most of the necessary food and heating fuel. Store-bought substitutes for subsistence commodities had to be paid for with income from the fishery and, increasingly, from social welfare programs. Many families found themselves even worse off than before (Iverson and Matthews 1968). Modernization proposals also included construction of community stages for fish processing. This new centralized-processing arrangement for communities meant that wives and children could no longer easily combine processing work with child care and other household responsibilities, making them dependents of fishers in a new sense (Antler and Faris 1979).

The development of fishers' UI programs, community resettlement, and community stages reshaped inshore-fishing households into a form closer to that assumed in 'modernization' ideology (and see Wright 1995 for more on gendered assumptions). These policies also reinforced women and children's economic dependence on men and increased the cash requirements of fishing households. The net result was to force fishing households into increasingly intensive exploitation of fishing resources, at the very time when uncontrolled foreign and local overfishing were driving down landings in the inshore fishery.

Encouraged by state subsidies and by a federal government licensing policy for fishers that limited their employment alternatives, fishing households began to shift their investments of time and money from the land – women's sphere – into boats and gear – men's sphere. The household production of saltfish declined, with more and more fish being sold fresh to corporate-owned plants.

The elimination of alternative markets for fish and legislative barriers to fishers' unionization kept down fish prices and hence the income of fishing households (Neis 1988b: Appendix B; *Historical Statistics* 1985: K–8).

Not surprisingly, when more jobs opened in corporate-owned fish-processing plants during the 1970s, economic need and a strong work ethic pushed the wives of boatowners into the paid labour force in search of wages and UI. Between 1957 and 1970, the participation of female production workers in fish processing (measured in terms of full-time equivalent jobs) hovered between 16 and 18 per cent of the total in that sector. Female rates rose after 1970, almost doubling by 1986 (Rowe 1991: 5). As late as 1962, women and men who worked in seasonal fish plants were unable to qualify for UI, even though their work, like that of the fishers, depended on the fishery (Newfoundland Fisheries Conference 1962).

Familial patriarchy had the ideological advantage of making women's subordination appear natural. It also reduced the cost of reproducing fishing households from day to day and over the generations, and it ensured the availability of women as a cheap source of wage labour for both the expanding government sector and private industry (Ursel 1992). With the gradual separation of production and reproduction in the postwar period, familial patriarchy was overlaid with social patriarchal policies and new ideologies about women's place. Women's access to the fishery resource and the wealth that it produced came to be mediated not only by familial patriarchy but also by corporate ownership, the state's expanding resource regulatory regime, and social welfare institutions. In response, women in the industry, like men, began to use the citizenship rights implied in social welfare institutions, such as rights to a minimum income and the right to strike, to challenge discriminatory corporate practices and state policies (Creese 1991–2; Neis 1988b: Appendix B).

In the early 1970s, plantworkers and fishers unionized. With the help of pressure from the national women's movement, unionized women plantworkers in 1974 were able to bring an end to the wage differentials for women and men doing the same work and to the policy of setting women's minimum wage at a different, lower rate than men's (Batten et al. 1974: 40–1; *Historical Statistics* 1985: Table D-4; Inglis 1985: 95).

Higher wages and better prices, in combination with expanded state regulation of fish stocks and overly optimistic predictions concerning the rate and scale of stock recovery, encouraged a brief revival and expansion of the Atlantic Canadian fishery after 1977. The resultant fishery was more diversified, in that it processed more species and generated a greater variety of products (Neis 1991). Conservation efforts delayed the expansion of the offshore corporate fishery for a few years. But optimism, government support, corporate competition for cheap fish, and diversification of species and products encouraged a revival of the inshore fishery. Many new plants were constructed in outports, resulting in women's increased employment in fish processing (Rowe 1991; Task Force 1983). But the revival of the inshore fishery was short-lived. By 1986, some inshore fishers were challenging scientists' claims that the major cod stocks, on which expansion of the industry depended, were recovering (Neis 1992). Inshore fishing households became caught in the trap of needing to invest constantly in more and better gear in order to land the same amount of, or less, fish (Neis et al. 1996).

The uncertainties about the future of the fisheries in the 1980s and the new environment of societal intolerance for discrimination on the basis of sex and marital status prompted women to challenge UI regulations that made it difficult for them to go fishing. One group of Newfoundland women, for example, fought the old practice of allocating women's processing work to men when they claimed income earned drying squid (Women's Unemployment Study Group 1983). In 1980, Wilhelmina Giovannini successfully challenged the UI regulation that classified the incomes of a husband and wife fishing together as 'joint income' so that only the husband could claim benefits (*Awareness* 1984; Connelly and MacDonald 1991–2). Silk's (1995) successful Supreme Court challenge removed a UI regulation that prevented fishers from combining qualifying fishing weeks with those earned from paid work on shore. This change was an important concession for women with young children, who, as junior partners in fishing enterprises and with responsibility for work on shore, probably found it more difficult than men to get enough insurable weeks from fishing to qualify for UI.

In Newfoundland, and elsewhere in Atlantic Canada, growing numbers of women were pushed and pulled into the boats and shore-based employment with fishing enterprises. With husbands and wives now working together, households could count on two incomes to help cover the increasing costs of the fishery and meet their household needs (Larkin 1990). Many women also enjoyed fishing, and they could earn better incomes from it than they could working for a plantowner or in a service-sector job. Some went into the boats because of the reductions in alternative means of qualifying for UI (McGrath 1990; Rowe 1991).

Work in fishing enterprises was primarily an option for the wives of boatowners, not for other women, and it carried risks. Most women entered the fishery as 'part-timers' after licences for such lucrative species as crab and shrimp had already been allocated. Some women who earned full-time licences subsequently had them taken away by the Department of Fisheries and Oceans, or DFO (Woodrow and Ennis 1994). Because they were 'new entrants' and as wives, their legitimacy as 'workers' has not gone unchallenged in fishing communities or within government (Silk 1995). Though the wives may enjoy fishing more than plant work, their inexperience makes them heavily dependent on their husbands, who are their employers. Recent research from southwest Nova Scotia describes the increasing involvement of boatowners' wives in both fishery-related shore work and fishing in the 1980s. As fishing became more costly, competitive, and heavily regulated, women's ability to manage the household, fishery accounts, and onshore repairs and sales, and to correspond with state regulators, became critical. This work, however, is invisible in the accounts of such enterprises (Kearney 1993).

Conclusion

This chapter has explored the long, complex transition from familial patriarchy to social patriarchy in the Newfoundland inshore fishery. It adds to the growing number of feminist studies of women in fishery economies. Unlike many other recent accounts, however, it focuses on the more vulnerable of these women and the structural sources of women's vulnerabilities. Historically informed dis-cussions of this kind can help us understand the legacy of patriarchy and its relationship to women's economic vulnerabilities in the current state of crisis. Moratoriums in almost all the major groundfisheries in Atlantic Canada have placed the future of nearly forty thousand fishery workers and hundreds of com-munities on hold (Williams 1996). It is currently unclear when, if ever, these fisheries will reopen on a commercial basis.

The federal government based its moratorium compensation packages for fishers and plantworkers on previous income patterns, and for that reason the government has replicated gendered income inequities that are partially a prod-uct of familial patriarchy and social patriarchy.[5] In addition, employment possi-bilities for women in future will be even more constrained than in the recent past. In association with The Atlantic Groundfish Strategy (TAGS), DFO is imple-menting policies to reduce harvesting capacity by buying back licences and designating 'core fishers' who will be part of the future fishery. In 1995, only 1.86 per cent of those meeting the eligibility criteria for core fishers were women.

'Downsizing' in this way could remove most women from the boats and close the door to other women who might want to enter in future, thereby re-establishing gendered boundaries in the fishery that were partially eroded by the court victories of the early 1980s. It will also strengthen class boundaries that divide the richer and poorer male fishers and hence their wives.

Fisheries women are also threatened by a proposal to 'downsize' fish pro-cessing by closing up to one-half of the fish plants along the northeast coast

(Tripartite Committee 1992). This would destroy many jobs and, by eliminating second incomes from inshore fishing households, could threaten the future of many smaller enterprises. The federal government may, in addition, privatize more sectors of the fishery by parcelling out access to the resource in the form of individual, transferable quotas (ITQs) to vessel owners (see McCay, this volume). Elsewhere, ITQs have contributed to the rapid concentration of ownership of fishery resources and eliminated the rights of crew members and their families to independent access to them (McCay and Creed 1990; Sissenwine and Mace 1992). These dubious initiatives could further undermine the economic and ecological basis for familial patriarchy in its current form – households dependent on multiple incomes in a community-based fishery (Connelly and MacDonald 1991–2).

In the current economic context in Canada, most households require multiple incomes to meet their needs. This situation makes it difficult for the men to migrate to jobs outside the province and send home enough money to support their families. Moving a family means finding multiple jobs. Chronic unemployment and low incomes make it harder for fishery households to help the next generation escape poverty through postsecondary education and employment elsewhere. If families have to choose which children to help, will they support their daughters or their sons? Whereas fishery communities based on familial patriarchy used to hold on to their sons and relinquish many of their daughters to work elsewhere, those that have lost their economic base may become poverty traps for women. Not surprisingly, a 1980s study of net out-migration in the economically depressed and fishery-dependent region of the Northern Peninsula found that while female net out-migration among young people exceeded the figure for males in the 1950s, by the 1980s male net out-migration exceeded female (Sinclair and Felt 1993). Recent Newfoundland development projects such as the Hibernia offshore oil platform and Voisey's Bay mining development in Labrador create employment mainly for men.

Recent research described middle-generation women as 'trapped' in their communities by the fishery crisis (Davis 1995). As argued by Marilyn Porter, these women are somewhat caught in a 'circular trap which revolves *through* the household,' forcing them to 'commit untold effort to ensuring the survival of the household, because without it, their position as women outside households or women in women-headed households would be, in most cases, economically pitiful' (1993: 148). Younger women were committed to leaving, but their ability to do so will depend on their access to employment and their local familial commitments (Davis 1995).

In future, the number of lone-parent families headed by women in rural communities may grow. If the current social welfare regime persists, many of these women will be forced onto welfare, as 'brides of the state' (as a friend of mine, a single mother on welfare, describes herself). In Newfoundland, as in other provinces (excluding Ontario), single mothers can be denied welfare if they are thought to be cohabiting with a man – evidently the state is a jealous husband. The state, however, is also a reluctant husband. Current UI and welfare may be replaced for the more marginal workers by a minimum basic income. A program of household income support might increase the basic earnings of some women. Such a plan is potentially dangerous, however, for vulnerable women in abusive situations, for it could deprive them of access to the individual income to which they are entitled, such as UI benefits, because of the often-false assumption in such programs that household income is shared equally (Gregory 1987).

The 'brides of the state' scenario suggests that the current fishery crisis may finalize Newfoundland's transition from familial to social patriarchy. Alternately, familial patriarchy may live on in poorer households, where family members may be forced to rely more on household- than on individual-based income. The basis for this familial patriarchy will, however, be significantly weakened in fishing communities with collapsed economies. The value of land and houses in a depressed rural community with no fish plant, where neither men nor women have retained the right to fish for a living, will be extremely low. Thus for women the role of traditional patterns of male inheritance and the benefits of marriage will decrease. All this will encourage women to establish households without men in order to maximize their incomes (while also maximizing their responsibility for child care), making them again 'brides of the state' (Davis 1993). Familial patriarchy will be strengthened, temporarily at least, in 'professionalized' fishing households, where boatowners' assets are enhanced by the government's gift of exclusive licences and 'individual quotas.' Having few local employment alternatives, the wives of professional fishers will probably devote their time to unpaid shore work in their husbands' fishing enterprises, rather than to paid employment outside the household (Kearney 1993).

NOTES

1 The master–servant law is a complex of British laws exported to British colonies that 'balanced claims by servants and apprentices (for unpaid wages, ill-treatment, etc.) against penal sanctions and other remedies demanded by masters (for leaving work or other forms of breach of contract, insubordination, etc.)' Hay and Craven 1993: 176.
2 I have classified Bonavista North and South, Ferryland, Fogo, Fortune Bay and Hermitage, Green Bay, Placentia and St Mary's, Trinity North and South, Twillingate, and

White Bay as fishing districts. Many of the women who left ended up in urban centres. The ratio of women to men in the districts of St John's East and West in 1935, for example, was 1,322 per 1,000, and the ratio for Newfoundland as a whole for 20- to 24-year-olds was 926.5 per 1,000. Some women returned to their communities if they found husbands.

3 From a preliminary archival review it is unclear to me whether or not independent eligibility for wives was ever discussed in the development of fishers' UI, but the system seems to have reflected the strong antagonism towards eligibility for either married women or seasonal workers; workers employed in family enterprises were equally suspect (National Archives of Canada, RG 27). There was a provision for dependants in UI payments, but its existence did not adequately reflect the value of women's work (Pierson 1990: 89).

4 See also Newfoundland Fisheries Commission (1963). This commission was established following a resolution adopted at this conference. It appears to have engaged only one woman in the proceedings and made no reference to women in its findings or recommendations.

5 Like their wages before the moratorium, women's compensation cheques are generally lower than men's. However, a guaranteed bi-weekly cheque set at a minimum of $225 a week has provided greater income security to some women than they had in the declining fishery immediately before the moratorium. A few of these women used this security to try to escape abusive relationships with their spouses (Walsh 1993).

WORKS CITED

Alexander, David. 1977. *The Decay of Trade*. St John's: ISER, Memorial University.
Allison, Charlene, Jacobs, S., and Porter, M. 1989. *Winds of Change: Women in Northwest Commercial Fishing*. Seattle: University of Washington Press.
Anger, Dorothy, McGrath, D., and Pottle, S. 1986. *Women and Work in Newfoundland*. Background Report to the Royal Commission on Employment and Unemployment. St John's: Queen's Printer.
Antler, Ellen. 1977. 'Women's Work in Newfoundland Fishing Families.' *Atlantis: A Women's Studies Journal* 2: 106–13.
– 1981. 'Fisherman, Fisherwoman, Rural Proletariat: Capitalist Commodity Production in the Newfoundland Fishery.' PhD dissertation, University of Connecticut.
Antler, Ellen, and Faris, J. 1979. 'Adaptation to Changes in Technology and Government Policy: A Newfoundland Example.' In R. Anderson, ed., *North Atlantic Maritime Cultures*, 129–54. The Hague: Mouton.
Awareness for Women in the Fishery Report. 1984. Sydney, NS: Grey Literature.

Batten, Elizabeth, Gray, D., Hallett, C., Lewis, A., and Lewis, J. 1974. *Working Women in Newfoundland.* St John's: Centre for Newfoundland Studies, Memorial University.

Benoit, Cecilia. 1982. 'The Poverty of Mothering: A Case Study of Women in a Newfoundland Community.' MA thesis, Memorial University.

Britan, Gerald. 1979. '"Modernization" on the North Atlantic Coast: The Transformation of a Traditional Newfoundland Fishing Village.' In R. Anderson, ed., *North Atlantic Maritime Cultures*, 65–81. The Hague: Mouton.

Butler, Gary. 1983. 'Culture, Cognition, and Communication: Fishermen's Location-Finding in L'anse-a-Canards, Newfoundland.' *Canadian Folklore canadien* 5: 7–21.

Cadigan, Sean. 1991. 'Economic and Social Relations of Production in the Northeast-Coast of Newfoundland, with Special Reference to Conception Bay 1785–1855.' PhD dissertation, Memorial University.

– 1995. *Hope and Deception in Conception Bay: Merchant–Settler Relations in Newfoundland, 1785–1855.* Toronto: University of Toronto Press.

Canada. 1965. Commission of Enquiry into the Atlantic Salt Fish Industry. *Extracts from Its Verbatim Reports of Public Hearings Held in Newfoundland, Nova Scotia, New Brunswick and Quebec from Feb.1 to Feb. 15, 1965.* St John's: Centre for Newfoundland Studies, Memorial University.

– 1993. Department of Employment and Immigration. 'NCARP Option by Age and Sex.' St John's: Strategic Planning and Analysis.

Cole, Sally. 1991. *Women of the Praia: Work and Lives in a Portuguese Coastal Community.* Princeton, NJ: Princeton University Press.

Connelly, Patricia, and MacDonald, M. 1991–2. 'State Policy, the Household, and Women's Work in the Atlantic Fishery.' *Journal of Canadian Studies* 26, no. 4: 18–32.

Creese, Gillian. 1991–2. 'Sex Equality and the Minimum Wage in British Columbia.' *Journal of Canadian Studies* 4: 120–40.

Cullum, Linda, Baird, M., and Penney, C. 1993. 'A Woman's Lot: Women and Law in Newfoundland, 1890–1949.' In Linda Kealey, ed., *Pursuing Equality, Historical Perspectives on Women in Newfoundland and Labrador*, 67–123. St John's: ISER, Memorial University.

Davis, Donna. 1983. *Blood and Nerves: An Ethnographic Focus on Menopause.* St John's: ISER, Memorial University.

– 1988. '"Shore Skippers and Grass Widows": Active and Passive Women's Roles in a Newfoundland Fishery.' In J.N. Klein and D.L. Davis, eds. *To Work and to Weep: Women in Fishing Economies*, 211–29. St John's: ISER, Memorial University.

– 1993. 'When Men Become "Women": Gender Antagonism and the Changing Sexual Geography of Work in Newfoundland.' *Sex Roles* 29, no. 7/8: 457–75.

– 1995. 'Women in an Uncertain Age: Crisis and Change in a Newfoundland Community.' In C. McGrath, B. Neis, and M. Porter, eds., *Their Lives and Times: Women in Newfoundland and Labrador, a Collage*, 279–95. St John's: Creative.

Faris, James. 1972. *Cat Harbour: A Newfoundland Fishing Settlement*. St John's: ISER, Memorial University.

Firestone, Melvin. 1967. *Brothers and Rivals: Patrolocality in Savage Cove*. St John's: ISER, Memorial University.

Gillespie, Bill. 1986. *A Class Act: An Illustrated History of the Labour Movement in Newfoundland and Labrador*. St John's: Newfoundland and Labrador Federation of Labour.

Gregory, Ann. 1987. *Study of the Impact of a Household Based Support Programme upon Women in Newfoundland and Labrador*. St John's: Newfoundland and Labrador, Women's Policy Office.

Hay, Douglas, and Craven, Paul. 1993. 'Master and Servant in England and the Empire: A Comparative Study.' *Labour/Le Travail* 31: 175–84.

Historical Statistics of Newfoundland and Labrador. 1985. Vol. 2. St John's: Newfoundland and Labrador, Department of Public Works and Services.

Inglis, Gordon. 1985. *More than Just a Union: The Story of the NFFAWU*. St John's: Jesperson.

Iverson, Noel, and Matthews, D.R. 1968. *Communities in Decline: An Examination of Household Resettlement in Newfoundland*. St John's: ISER, Memorial University.

Kearney, John. 1993. 'Diversity of Labour Process, Household Forms, and Political Practice: A Social Approach to the Inshore Fishing Communities of Clare, Digby Neck, and the Islands.' PhD dissertation, Laval University.

Larkin, Maureen. 1990. 'State Policy and Survival Strategies in P.E.I. Lobster Fishing Households.' In J. Burns, G. Pool, and C. McCormick, eds., *From the Margin to the Centre: Proceedings of the 25th Anniversary Meeting of the Atlantic Association of Sociologists and Anthropologists*, 58–76. Saint John: University of New Brunswick.

McCay, Bonnie J. 1976. '"Appropriate Technology" and Coastal Fishermen of Newfoundland.' PhD dissertation, Columbia University.

– 1988. 'Fish Guts, Hair Nets and Unemployment Stamps: Women and Work in Co-operative Fish Plants.' In Peter Sinclair, ed., *A Question of Survival: The Fisheries and Newfoundland Society*, 105–32. St John's: ISER, Memorial University.

McCay, Bonnie, and Creed, C. 1990. 'Social Structure and Debates on Fisheries Management in the Atlantic Surf Clam Fishery.' *Ocean and Shoreline Management* 13: 199–229.

McGoodwin, James R. 1990. *Crisis in the World's Fisheries: People, Problems, and Policies*. Palo Alto, Calif.: Stanford University Press.

McGrath, Carmelita. 1990. 'Ethnography of Southeast Bight.' In M. Porter, with B. Brown, E. Dettmer, and C. McGrath, 'Women and Economic Life in Newfoundland: Three Case Studies,' 280–321, Project Report on no. 482-87-0005 funded by the Social Sciences and Humanities Research Council (SSHRC), Women and Work Strategic Grants.

Martin, Kent. 1979. 'Play by the Rules or Don't Play at All: Space Division and Resource Allocation in a Rural Newfoundland Fishing Community.' In R. Anderson, ed., *North Atlantic Maritime Cultures*, 277–98. The Hague: Mouton.

Memorial University Folklore and Language Archives (MUNFLA), St John's. Tapes 91-461-C14604, C14605, C14593.

Morgan, Bernice. 1992. *Random Passage*. St John's: Breakwater Books.

Murray, Hilda. 1979. *More Than 50%: Woman's Life in a Newfoundland Outport, 1900–1950*. St John's: Breakwater Books.

Nadel-Klein, Jane, and Davis, D.L. 1988. *To Work and to Weep: Women in Fishing Economies*. St John's: ISER, Memorial University.

National Archives of Canada. RG 27, vol. 3458, files 4–11

Neis, Barbara. 1988a. 'Doin' Time on the Protest Line: Women's Political Culture, Politics and Collective Action in Outport Newfoundland.' In Peter Sinclair, ed., *A Question of Survival*, 133–53. St John's: ISER, Memorial University.

– 1988b. 'From Cod Block to Fish Food: The Crisis and Restructuring in the Newfoundland Fishing Industry, 1968–1986.' PhD dissertation, University of Toronto

– 1991. 'Flexible Specialization: What's That Got To Do with the Price of Fish?' *Studies in Political Economy* 36: 145–76.

– 1992. 'Fishers' Ecological Knowledge and Stock Assessment in Newfoundland.' *Newfoundland Studies* 8: 155–78.

– 1994. 'Female Fish Processing Workers' Occupational Health and the Fishery Crisis in Newfoundland and Labrador.' *Chronic Diseases in Canada* 15, no. 1: 12–16

Neis, Barbara, and Susan Williams. 1996. 'Women and Children First: the Impacts of Fishery Collapse on Women in Newfoundland and Labrador.' *Cultural Survival Quarterly* 20, no. 1: 67–71.

Neis, Barbara, L. Felt, D. Schneider, R. Haedrich, J. Hutchings, and J. Fischer. 1996. 'Northern Cod Stock Assessment: What Can Be Learned from Interviewing Resource Users?' NAFO SCR Doc. 96/45, 22.

Newfoundland and Labrador. 1991. 'Plant Employment Survey Final Results for the Province of Newfoundland and Labrador.' St John's: Planning Services Division.

Newfoundland Fisheries Commission. 1963. *Report and Recommendations of the Newfoundland Fisheries Commission to the Government of Newfoundland*, Vols. 1 and 2. St John's: Centre for Newfoundland Studies, Memorial University.

Newfoundland Fisheries Conference. 1962. 'Transcript.' St John's: Centre for Newfoundland Studies, Memorial University.

Ommer, Rosemary E. 1990. 'Merchant Credit and the Informal Economy: Newfoundland. 1919–1929.' *Historical Papers* [Canadian Historical Association]: 167–89.

Patey, Nina. n.d. 'Perceptions of Women in the Labrador Fishery.' Unpublished paper, Maritime History Group, Memorial University, St John's

Pierson, Ruth Roach. 1990. 'Gender and the Unemployment Insurance Debates in Canada. 1934–1940.' *Labour/Le Travail* 25: 77–103.

Porter, Ann. 1993. 'Women and Income Security in the Post-War Period: The Case of Unemployment Insurance, 1945–1962.' *Labour/Le Travail* 31: 111–43.

Porter, Marilyn. 1983. 'Women and Old Boats: The Sexual Division of Labour in a Newfoundland Outpost.' In E. Gamarnikow et al., eds., *Public and Private: Gender and Society*, 91–105. London: Heinemann and BSA.

– 1985. '"She Was Skipper of the Shore Crew": Notes on the History of the Sexual Division of Labour.' *Labour/Le Travail* 15: 105–23.

– 1988. 'Mothers and Daughters: Working Women's Life Stories in Grand Falls, Newfoundland.' *Women's Studies International Forum* 11, no. 6: 545–58.

– 1993. *Place and Persistence in the Lives of Newfoundland Women*. Aldershot, England: Avebury.

Rowe, Andy. 1991. *Effect of the Crisis in the Newfoundland Fishery on Women Who Work in the Industry*. St John's: *Newfoundland and Labrador*, Women's Policy Office.

Silk, Victoria. 1995. 'Women and the Fishery.' In C. McGrath, B. Neis, and M. Porter, eds., *The Lives and Times of Women in Newfoundland and Labrador: A Collage*, 264–9. St John's: Creative.

Sinclair, Peter. 1985. *From Traps to Draggers: Domestic Commodity Production in Northwest Newfoundland, 1850–1982*. St John's: ISER, Memorial University.

Sinclair, Peter, and L. Felt. 1993. 'Coming Back: Return Migration to Newfoundland's Great Northern Peninsula.' *Newfoundland Studies* 9, no. 1: 1–25.

Sissenwine, Michael P., and P. Mace. 1992. 'ITQs in New Zealand: The Era of Fixed Quota in Perpetuity.' *Fishery Bulletin* 90: 147–60.

Smallwood Papers. 1952. Memorial University, St John's. 1.06.006 Burin.

Snell, James G. 1993. 'The Newfoundland Old Age Pension Programme, 1911–1949.' *Acadiensis* 23, no. 1: 86–109.

Statistics Canada. 1991. Census of Canada, 1991. *Dwellings and Households: The Nation*. Cat. No. 93-311. Ottawa: Industry, Science, and Technology.

Story, George M., W.J. Kirwin, and J.D.A. Widdowson. 1990. *Dictionary of Newfoundland English*. Toronto: University of Toronto Press.

Task Force on the Atlantic Fisheries. 1983. *Navigating Troubled Waters: A New Policy for the Atlantic Fisheries*. Ottawa: Supply and Services Canada.

Tenth Census of Newfoundland and Labrador, 1935. 1937. Vols. 1 and 2. St John's: Newfoundland and Labrador, Department of Public Health and Welfare.

Tripartite Committee–Newfoundland and Labrador Fish Processing Sector. 1992. 'Discussion Paper: Implications of Resource Crisis for Newfoundland's Fish Processing Sector.' Ottawa: Department of Fisheries and Oceans.

Ursel, Jane. 1992. *Private Lives, Public Policy: 100 Years of State Intervention in the Family*. Toronto: Women's Press.

Valverde, Mariana. 1991. *The Age of Light, Soap and Water: Moral Reform in English Canada, 1825–1925*. Toronto: McClelland and Stewart.

Walsh, Jane. 1993. 'Women and the Moratorium – Reasonable Doubts.' Paper presented at the Annual Meeting of the Atlantic Association of Sociologists and Anthropologists, Antigonish, NS.

Women's Unemployment Study Group. 1983. *Not for Nothing: Women, Work and Unemployment in Newfoundland and Labrador.* St John's: Centre for Newfoundland Studies, Memorial University.

Woodrow, Helen M., and F. Ennis. 1994. *Women of the Fishery: Interviews with 87 Women across Newfoundland and Labrador.* St John's: Educational Planning and Design Associates.

Wright, Miriam. 1995. 'Women, Men and the Modern Fishery: Images of Gender in Government Plans for the Canadian Atlantic Fisheries.' In C. McGrath, B. Neis, and M. Porter, eds., *The Lives and Times of Women in Newfoundland and Labrador: A Collage*, 129–43. St John's: Creative.

4

'The Water and the Life': Family, Work, and Trade in the Commercial Poundnet Fisheries of Grand Bend, Ontario, 1890–1955

JEAN L. MANORE AND JOHN J. VAN WEST

Lake Huron from Bayfield to Sarnia (Figure 4.1) once supported flourishing commercial fisheries that were differentiated in time and place by their capturing technologies – poundnets, gillnets, seines, and trapnets (Figure 4.2).[1] Yet from the perspective of social history, our knowledge of these once regionally significant fisheries remains obscure. For one thing, the historic administrative records of the federal and provincial fisheries departments are fragmentary and discontinuous.[2] For another, while existing finding aids and reference guides for important records concerning fisheries have survived, they are either incomplete or dedicated to selected archival holdings, making access difficult. With some notable exceptions (Gearhart 1987; Lloyd and Mullen 1990), scholars have not recorded the oral testimonies of Great Lakes fishers, effectively denying them a central place in the history of their industry (see Thompson 1988: 2). Finally, while scholars have directed considerable effort over the last decade towards studying Ontario's large and well-established lower Great Lakes fisheries, particularly in Lake Erie (Cox 1992; Prothero 1980; Van West 1983, 1986, 1989), and Ontario's Aboriginal fisheries throughout the Great Lakes (Doherty 1990; Hansen 1991; Lytwyn 1990, 1992; MacDonald 1978; Schmalz 1992), they have not given similar attention to the non-Aboriginal commercial fisheries of the central and upper Great Lakes (excepting Peters 1981; Goodier 1984, 1989, and, for lower Lake Huron, Orr 1993: 61–72).

The Grand Bend fisheries' traditional and relatively substantial operations have been inexplicably neglected. Situated approximately halfway between Sarnia and Bayfield, on the lower Lake Huron coast in southwestern Ontario, the community is well known today for its wide and white sandy beaches on which large crowds converge each summer. The beaches there were not always used for amusement purposes, however. Until the mid-1940s, Grand Bend was a small hinterland fishing village. Its fishers staked their poundnets to the sandy

Figure 4.1   Location of the Manore family fishing grounds and community of Grand Bend, Ontario. *Source*: original draft provided by Victor P. Lytwyn

shoreline from which they harvested Lake Huron's rich and productive fish resources.

John Manore spent most of his life on Grand Bend's marine frontier, and his father spent most of his. The younger Manore, though now retired, continues to reside in Grand Bend and is one of its last surviving poundnet fishers. What follows is his story: an oral history, now recorded, of family, work, and trade in a once-thriving commercial fishing industry. His comments on gender are especially timely because the fairly recent addition of feminist analysis to the study of fishing communities has started a lively debate about the nature of gender relations in general and patriarchy in particular (Allison, Jacobs, and Porter 1989; Davis and Nadel-Klein 1988; Porter 1985; Neis, this volume). Such an approach rounds out written history, which, despite the constraints of fragmentary and discontinuous records, none the less reflects what oral historian Paul Thompson (1988: 6) refers to as the 'standpoint of authority.' John Manore's oral account joins the voices of Canadian Atlantic coast fishing families, thus bringing recognition to ordinary working people and making their history immeasurably more realistic (Andersen 1980; Candow 1990).

The Familial and Entrepreneurial Context

The history of the Manore family's poundnet fishery of Grand Bend, Ontario, begins with John's father, Christopher Columbus Manore, or Cub. Cub was born in Sandwich, Ontario (now part of Windsor), in 1865 and raised in Frenchtown (today, Monroe), Michigan. His father, Gilbert Menard (before Cub's day, the family surname had been spelled variously as Menard, Manard, Menare, and Munore) had been a Great Lakes sailor before joining the Michigan 9th Calvary to fight in the American Civil War. Cub fished for an uncle who owned and operated a small family enterprise named the J.B. Dewey Company, then operated Dewey's tugs on Lake Erie and Lake Huron, and later fished in various waters all the way from Lac St Jean, Que., to Sault Ste Marie, Ont., and across Canada to Prince Rupert, BC. He learned of Grand Bend's productive fishing grounds around 1886, at the Dewey Company's Port Huron fish house. As non-residents, neither he nor his cousin, who was also interested, could obtain a Canadian fishing licence. So they began fishing at Grand Bend using a Canadian licence taken out by a 'little Frenchman' named Demos Stebbins, who had first told them about the place.

Soon after he had established a seasonally resident fishery there, Cub met his future wife, Mary Kennedy, who had been born into a fishing family in Stephen Township, Ont. They married in Grand Bend, probably in 1887 or 1888, and had two daughters and seven sons, the last of whom was John, born in 1907

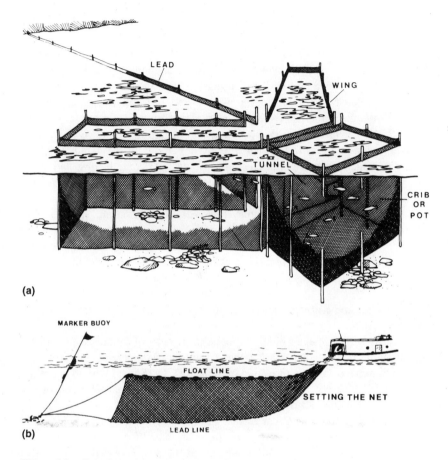

Figure 4.2   Great Lakes fishing gear: (a) poundnet, (b) gillnet, (c) seine, and (d) trapnet. *Source*: Adams and Kolenosky 1974. Images are reproduced with authority of the Ontario Ministry of Natural Resources.

(Figures 4.3, 4.4). The family followed a pattern of seasonal migration. It wintered in Point Edward, a small village bordering the larger community of Sarnia, Ont. During the spring, summer, and fall fishing seasons, Cub and his sons, as each came of age, went north to Grand Bend, then a hinterland outpost, to work their whitefish (*Coregonus clupeaformis*), yellow pickerel (*Stizostedion vitreum vitreum*), and sturgeon (*Acipenser fulvescens*) fishing grounds.

John Manore remembers his early experiences with his father and older

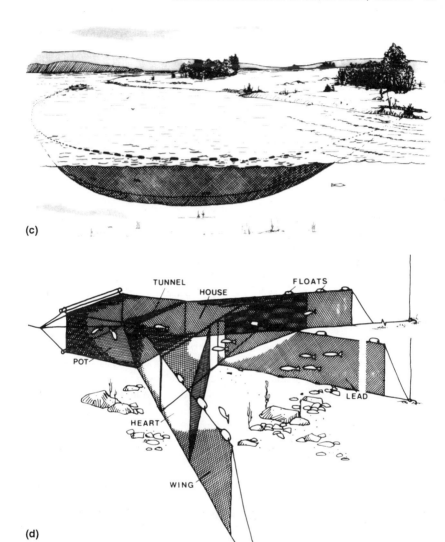

(c)

(d)

brothers at Grand Bend. The boys would, in his words, 'batch it' during most of the year, and the numerous domestic chores would be divided among them: 'I think I come to the Bend when I was about four years old and stayed here and my sister brought me up and I was here for two or three weeks ... And they were batching it [at Grand Bend] you see so my brother [Gladstone] ... and Lloyd,

Figure 4.3   Cub Manore, 1887.   Figure 4.4   Mary Kennedy Manore, 1887.
Source: Manore Family   *Source*: Manore Family

they were older and they'd be here permanent but Carl, Stan, and Ted and I, we weren't old enough so ... Carl and Stan would maybe come [for] a couple of weeks or maybe a month or maybe two months. So outta my life I only spent maybe five years at the Point [Point Edward] in the summer the rest of the time was on the fishery.'

Mary inherited the fishery and kept the family and its fishery going after Cub died in 1929. Both Lloyd and John continued to work the family fishing grounds for shares, as they had done with their father, until Lloyd's death in 1934. By this time Ivan had died, and brothers Stan, Carl, Gladstone, and Ted had left for other work elsewhere. This left John, the youngest brother, to ponder the future of the family fishery. John sought his mother's counsel: 'I talked to her and she said "You'll never get rich [from fishing] but you'll never starve." So that was her advice.' His decision to remain in fishing was, however, more fundamentally determined by his strong and lasting attachment to the water. Fishing was a way of life (see Gilmore 1990); in John's words, he simply 'liked the water and I just liked the life.'

The Manore fishery flourished during the Second World War, struggled for a decade thereafter, and finally ended altogether in 1955. John had inherited ownership of the family fishery when his mother died in 1943, and he continued to fish with poundnets at Grand Bend and vicinity until 1945. At this point, plunging demand, high labour and material costs, and the decline of the higher-valued migratory species began to undermine the relative advantage offered by the fixed, shore-based capturing technology that had sustained the local fishing economy since the late nineteenth century. Like other poundnet fishers in the area, John had to turn to more versatile, efficient trapnets and gillnets for his principal capturing apparatus.

In 1955, John Manore was forced to stop fishing because of declining fish stocks. Additionally, he says that he had difficulties finding men to work as crew on his boat. Since the men who used to look for seasonal employment could now get year-round employment in industry with regular hours and weekends off, they sought higher wages and better hours than John could offer.

John Manore had not intended to abandon the fishery completely and purchased a gas station in Grand Bend only to make ends meet until the fish stocks recovered, but there was no recovery. He eventually converted the gas station to a marina, which is now operated by his son, Richard. John still lives in 'The Bend,' but when we talked in August 1993 to him and his wife, Edith (who has since died), they said that they felt like outsiders in a community now supported by a seasonal recreational resort economy and a transient summer tourist trade. He conceded that Grand Bend had become 'a different place altogether ... Now we're strangers here. We don't know people.'

The Development of the Great Lakes Poundnet Fisheries

Poundnets originated in Scotland and were first used by Great Lakes fishers in the early 1830s (McCullough 1989). Fished from the shore, these nets incorporated four interdependent components. The leader directed the migratory fish into the pound and consisted of netting secured to a succession of stakes (usually tamarack) securely driven into the lake bottom in a line perpendicular to the shore; the heart was located at the outer end of the leader, in deeper water; fish entered the crib through a funnelled opening; and beyond the heart lay the pound or crib itself. John Manore's leads 'would run from 60 rods to 100 rods and the hearts, they were 5 rods, and the tunnels were 30 to 35 ft wide and 35 ft deep and the crib was 30 to 35 ft square [one rod measuring 16.5 ft or 5.03 m].'

Commercial fishers were quick to adopt the poundnet, preferring it over the less efficient, beach-based haul seine. With sufficient stake length and suitable lake bottom, a poundnet could be fished to a maximum of fifteen to twenty-four metres of water, compared to only four metres for the seine, thus offering a four-fold increase in range (Gordon 1965: 6). By the 1850s, American fishers were using poundnets as their preferred fishing gear at the western end of Lake

Erie, around Sandusky, Ohio (MacLaren 1965: 145). Canada's federal government first licensed Great Lakes fishers to use them in 1869, probably facili-
tating distribution of the net northward from Lake Erie to the more distant fisheries of the Upper Great Lakes (Adams and Kolenosky 1974: 8). Between 1914
and 1920, Cub Manore also fished as many as twenty-four poundnets from
Colchester to the Detroit River on Lake Erie. The Manores used fifteen-metre
lengths of white ash timber for their net stakes and paid $5 a piece for them,
compared to $10 to $12 for the more commonly used tamarack.

John's seven poundnets generally required, 'well, fifteen stakes [for the
lead]. We would generally have from twenty-seven to twenty-nine stakes to
thirty stakes to the net. We'd have fourteen stakes in the crib and hearts and that
whatever for the lead, depending on the length of the lead ... And twine was
dear. It was very expensive.' The lead directed spawning or feeding migratory
fish into the wings and from there into the crib. The crib, with its smaller-
meshed netting, prevented the escape of all but the larger fish, which could be
identified and selectively harvested. These fixed or stationary characteristics of
an ostensibly shallow-water, beach-based net contributed to the development
of local management regimes and exclusive usufructuary rights to fishing
grounds in the commercial fisheries of Grand Bend and vicinity.

Finding a Place

Maritime anthropologist James Acheson, among other scholars, argues that
commercial fishing exposes fishers to considerable financial risk and uncertainty because the resource itself – at least notionally – is shared in common
(Acheson 1981: 275–316). Commercial fishers shoulder greater risk and uncertainty than hunters because they 'are operating on a flat, undifferentiated surface and are exploiting animals that are difficult to see ... For commercial
fishermen, locating one's position is always problematic' (Acheson 1981: 276,
281). There is widespread evidence that fishers of *all types* develop strategies of
exclusive use or ownership of defined fishing areas for themselves, particularly
where fixed or stationary harvesting devices are employed, in order to reduce
uncertainty and competition and reserve the fish found within these areas for
their own use and benefit.

While Great Lakes fisheries records, cited by McCullough (1989) and Van
West (1983), sketch the pre-conditions favourable to poundnet fishing, they do
not specify which factors influenced poundnet fishers' selection of certain fishing grounds. John Manore informs us that government regulations, fishing
ethics, environmental conditions, and onshore services were the determining
factors in the use and selection of poundnet grounds and led to the fishers'

obtaining usufructuary (but not property) rights to those grounds over time. He notes that fishing families along the lower Lake Huron coast fished their pound-nets and gillnets and, in an earlier period, their beach seines, in mutually exclusive areas. He suggests that Grand Bend's poundnet fishermen consisted of 'the Greens and ourselves for quite a few years. In the late 1920s, the Gills started gillnetting. And the Desjardins. And there was two or three others started gill-netting then. It just picked up the gillnetting. So there maybe got to be six or seven gillnetters here.'

It is likely that these gillnet fishermen operated in Huron County waters, where, unlike in the Grand Bend area, gillnetting was permitted. Provincial reg-ulations demarcated the two fisheries, preventing the type of conflicts endemic to the Lake Erie fisheries from the 1920s into the early 1940s. John Manore remembered: 'Years ago, see this part from the River [in Grand Bend] down ... is Lambton County. North is Huron County. Well, you couldn't get a gillnet licence out of Lambton County. Your licence had to come out of Huron County. And then there is a five-mile limit on gillnets off Lambton's shore. So, we didn't have too much trouble [with gillnet fishermen] because the gillnetters stayed outside of us and the poundnets only went up a mile and three-quarters into Huron [County].'

The issuing of licences also influenced selection of poundnet fishing grounds. Poundnet fishing licences in Ontario were originally issued by the fed-eral government, and subsequently by the province, following a decision made by the Privy Council in 1898 (*Attorney General* 1898). The crown thereby con-veyed to individual fishers the right to fish within a specified area of water, oth-erwise known as a poundnet lot. These crown lots typically measured some eighty rods at beach frontage. The *length* of the lot remained boundless. It was customary for poundnet fishers to select their own individual lots and request a licence from the government to fish them.[3] If the chosen lots had already been spoken for, then the interested fisher would buy the owner out, or go elsewhere, or wait until those lots were available again.

These circumstances probably governed Cub's selection and (through Demos Stebbins's fishing licence) acquisition of his fishing lots on arrival in Grand Bend in the 1890s. As a matter of custom, a fisher would not harvest from, or even encroach on, the allocated fishing lots of other fishers. John affirmed that poundnet fishers 'pretty well worked together. Yes. The only thing maybe if they got too close to each other, the fisherman might say, "well, you are getting too close to me, stay over more to the centre of your lot or don't crowd me." See, if you had [adjoining] lots you could set right next to the fisherman beside you. But generally, you stayed pretty well in the centre of your lot ... You didn't encroach on another fisherman.'

While government regulations and fishers' respect for one another's 'territory' thus established the external boundaries of each fisher's poundnet grounds, the fishers made their own decisions on how they managed the fisheries within them. The Manore family's fishing grounds covered twenty-nine beach lots, lots 2 to 30 in Bosanquet Township. These lots were adjacent to each other and covered approximately seven miles (11.3 km) of Lake Huron shoreline, from Grand Bend to Port Franks (Figure 4.1). Only a handful of fishers shared the area, quite unlike the fisheries from Kettle Point to Sarnia to the south, which during their peak period in the 1930s comprised as many as seventy-five poundnets.

John Manore staked his seven poundnets each spring, selecting sites from among his father's lots. He could have fished up to twenty-nine nets, since he was licenced to fish one net on each of the twenty-nine lots, but, like his father before him, he decided not to do so. It would have made his small fishery unmanageable, and he therefore opted to keep working on a smaller scale: self-imposed, economically rational limitation on his use of the 'commons.' As he says, 'Seven nets was enough. We could [have] fished more, but then you would have to hire bigger crews, bigger outfits and then you get a bad year ... I asked my dad about the seven nets when I was younger. I said, "Why don't we fish more nets?" He said, "We got enough nets ... Fish seven good nets, fish them right and we got things to ourselves."'

Other factors also influenced the selection and use of these grounds, the most notable being environmental conditions. Even though a poundnet fisher could theoretically place his nets anywhere within his lots, certain areas were not practical for staking: 'You got a rocky bottom. Now you see, a mile and three-quarters north of here [north of Grand Bend] you can't drive [your poundnet stakes]. Its like a shale [bottom]. So you can't drive ... Or you may get big boulders. Or it may be such a rough spot that you can't hold a net.' Therefore, when a fisher wanted to move a poundnet from one of his lots to another or to a different location entirely, he would 'look at the bottom ... The whitefish don't go on the clay or the rock ... The sand is for the whitefish. If you get a sandy beach, you get a sandy bottom.'

Finally, proximity to a road, access to a team of horses, and necessary staples for the seasonally resident fishers on the beach also determined the acceptability of poundnet grounds, in John's experience. The fishers would investigate the availability of these onshore services when selecting their fishing sites. The beach was 'free ground,' and the presence of farms crucial to poundnet fishers: 'We could land anywhere on the beach. We could put our shanty on the beach, because it was government property. We had the right to land our boat anywhere ... but if this was your farm up here or your fence, we couldn't go through unless that fella gave us the right. But when we were fishing on the beach, we tried to locate on [adjacent to] somebody's farm because he had a driveway to the main road and he had eggs, meat, vegetables and he had a post [box] and he had a team of horses. So he had everything there.'

John Manore's narrative tells us much about how poundnet grounds were chosen and managed. Formal regulation and informal practice, from stationary gear and established poundnet boundaries to fishing ethics and connections with specific farmers, were instrumental in establishing the usufructuary rights of these fishers to specific fishing territories and their access to the resource.

Women and Men at Work in the Poundnet Fisheries

Maritime anthropologist James McGoodwin (1990: 24) argues that within the overarching social fabric of the world's fisheries circumstances, a sharp gender-based division of labour 'has the men working aboard fishing vessels while women typically serve as the mainstay of home and community life.' This analysis of gender relations in fishing communities has been subjected to heavy criticism from female anthropologists and other social scientists of late (see Neis, Sinclair, Squires, and Downton, and for Native fisheries, Ray, Newell, this volume).

McGoodwin himself recognizes that in some fisheries women participate in the marketing and distribution of the catch, and in others they actually labour as primary producers. Regrettably, the role of men and women in the Great Lakes fisheries, and the issue of how their roles reflected gender relations, have received only scant attention in the literature. Significantly, we know even less about these matters in the context of poundnet fishing, and we were not able to interview the women of this family (McCullough 1989: 65).

So John Manore's anecdotal impressions about the work performed by men and women in Grand Bend's poundnet fisheries is a first hint at gender relations in a Great Lakes fishery community. The spatial and temporal nature of men and women's participation in this fishery undermines the notions of naturalized 'separate spheres' and patriarchal dominance, since the division and organization of labour were conditioned in large measure by the material means of production. Poundnet fishing crews consisted of kin and of contractually employed individuals who were recruited on a seasonal basis. In sharp contrast to other types of fishing, a poundnet operation ideally worked as many as six or seven men to a boat to remove and reset the nets in late fall and early spring and only half that number to fish these nets in the intervening late spring, summer, and early fall fisheries. The seasonal demands on labour might have been problematic had fishing crews been difficult to recruit, but this was not the case, until after the Second World War. The Manores' crews usually consisted of 'men we picked up ... They just seemed to come. And they, well there was always men around ... Some would stay and others circulate.' Some 'stayed' (remained loyal) for as long as thirty years. When periodic shortages of labour occurred, crews were pooled.

Poundnet fishers relied on unattached or 'free' labour as required and also, significantly, on the support of farmers whose lands abutted the remote beaches whence the poundnets were fished:

Now, here at [Grand Bend], [my brother] Lloyd's wife, her family lived up on a farm where Grand Coves is. They had horses ... We used the Hamiltons.' But, when haying time come, our whole fishing crew went up to that farm. Well, Lloyd and I would get up and harness their horses and feed them – we'd go and start in the hay and the men that were working for us, we would take all their hay and grain off. Then, when it come to thrashing, then we'd go up and help them thrash. Every year we did that. So, then, in the spring, when we wanted the [poundnet] stakes pulled down, Alec come down with the [horse] team pulled the stakes down. In the fall, when we wanted them loaded off the scow, when we wanted them pulled up before the high water, Alec come down with the team and pulled the stakes up.

Poundnets, as distinct from gillnets, for example, demanded considerable preparatory and ongoing maintenance work year round if they were to be fished at their maximum efficiency. The crews were therefore retained through the summer to repair nets and remove bark from new stakes to replace those that had been lost or broken: 'There was the odd bad year ... In other years, you'd get broken stakes and you'd get deadheads through the twine. Some years, we'd be setting and pulling every day. Big seas.' Crews were required to haul stakes to remote beach locations, and they sharpened and, barring storms, set and pulled them in the spring and fall. The nets themselves required repeated coatings of tar to extend their serviceable life. Crews were also recruited in winter to cut ice.

For all this work, the poundnet crews received wages. This form of remuneration was unusual, because in fisheries throughout the world, crews have traditionally worked for shares. According to John Manore, 'Crews were paid $30 per month. Then they said they wanted to get paid for every day in the month, so we had to pay them for another day generally. And we paid them whether they worked or not, because some days they might be working 8 hours, 14 hours, and some days they may not be working at all ... Yes. A dollar a day. See we were only getting 5 or 6 cents a pound for fish.'

There is no definitive explanation for these poundnet crews' receiving wages as remuneration for their labour. Perhaps it was because their work involved a variety of tasks external to the sphere of production, or it may have been because they had no proprietary interest in the family businesses that employed them, since John Manore and his brothers, by contrast, did receive shares when Cub and Mary were in charge.

Men's work in the poundnet fisheries of Lake Huron differed from conditions elsewhere on the Great Lakes. The interdependent relationship with local farmers and the disbursement of wages rather than shares meant that the Manore fishery remained a family business that relied on the wider community for essential services and labour. Therefore, if this operation was typical, it is clearly inappropriate to conceptualize the operations of the poundnet fisheries in Grand Bend as an offshore male domain and an onshore women's domain. Does this conclusion also hold up when we examine the role of women in poundnet fisheries?

Women in Grand Bend's poundnet fisheries seldom if ever worked as harvesters, nor did they participate in processing and marketing. This latter point illustrates a distinctive difference between poundnet and most other fisheries, including the gillnet fisheries of Lake Huron, which typically employed women in onshore processing. If John Manore's fishing practices were typical, then women in the poundnet fisheries did not accompany their men to remote fishing sites, nor did they cut or fillet fish, because fish captured with poundnets were, at least in the earlier years, shipped to market directly from the beach locations, in 'the round' (i.e., whole) rather than dressed. In contrast, fish caught offshore in deeper water with gillnets were landed at the community dock for processing, usually by women (see Lloyd and Mullen 1990: 75–7; McCullough 1989: 69).

The exclusion of women in the poundnet fisheries from harvesting and processing left them free to work in other spheres, both within and external to the fisheries. For instance, many family fisheries along the lower Lake Huron coast were in fact owned by women. John Manore lists a few such women, all of whom, like his mother, succeeded to this position through inheritance. He recalls: 'George Harrow died ... and I think Mrs. Harrow went on and the two boys run the fishery but she was still the head of it. And I think Purdy, when Bill Purdy died, his wife run the fishery and the boys went on but she was still the head of it.'

Ownership of fisheries by women is not common worldwide, though field research conducted among Portuguese and Scottish fisherfolk by anthropologists Cole (1991) and Davis and Nadel-Klein (1988) suggests that a fisher's property, such as fishing equipment or possibly a house, was often inherited by his widow, though this practice was not the tradition in the Newfoundland inshore cod fishery (see Neis, this volume). The fishing enterprise itself, however, would not pass to her.

It thus seems that the Great Lakes poundnet fisheries are a rare example of customary rules not excluding women from inheriting their family fishing firms. This 'anomaly' may have existed because fishing areas were privatized by virtue of exclusive usufructuary rights granted by licences and therefore legally transferable. For his part, John Manore simply suggested that his mother inherited the family fishery because it was the 'natural' thing to happen.

Women contributed substantially to the family economy in other ways as well. It was customary for women to manage and tend independently the small parcels of land that many commercial fishers had acquired in addition to their fisheries. John Manore noted, for example, that 'a lot of these fishermen had ten to fifteen acres of land. They'd have a cow, a couple cows, a couple pigs and chickens, and a garden. Well, the women would look after all that.' As well, revenue from 'odd jobs' done by women whose husbands worked as members of poundnet fishing crews was apparently needed to supplement household income, though not necessarily in families such as the Manores, who owned and operated their own enterprises. This practice is known in fishing communities outside the Great Lakes area (Sinclair, Squires, and Downton, and Nels, this volume; Murray 1979).

Significantly, women's exclusion from, or only marginal participation in, the harvesting, processing, and marketing of fish did not necessarily compromise or diminish their authority in decision-making. John Manore remembers that his mother regarded the annual period of absence of the men (i.e., her husband and sons) as somewhat of a 'holiday' for her. He often stated that fishers were very independent in outlook and action, and so were their wives: 'Well I asked my Mother: "Why don't you come [to Grand Bend], Mother?" "No way," she said, "this is my holiday. I've got everything here." So she said, "If you boys are up there, this is my holiday." She was right because [at Grand Bend] we had outdoor plumbing, [we had to] carry water, [we had] wood stoves. Nothing like we had at home – convenience. So at home, she had everything right there – she didn't have to carry water or cook on a wood stove or go out to the bathroom.'

The geographical and seasonal separation from family encountered by Cub Manore and his sons ceased after John Manore took over the family operations in 1943 and he and his wife Edith relocated from Point Edward to Grand Bend. For Edith, the decision to move did not come easily, since she had lived her entire life in Point Edward and was reluctant to live in what to her was a strange community. At first they summered at the Bend and wintered at Port Edward, but when their son Richard reached school age, Edith saw that Grand Bend was the place to settle. As John Manore recalls: '[Richard] had to go to school ... Were we going back to the Point and then move to the Bend again in the spring and back to Sarnia in the fall? Or were we going to stay? ... [Edith] thought well, she had been here in the summer, she'd take Dick up town everyday in a buggy ... And the older ladies along the street stopped to talk to her and she got acquainted and she taught Sunday school while she was here ... I come home, around the first of July, I said, "Well, we have to make a decision, what are we going to do?" ... She said, "We are staying [in Grand Bend]. This is a good place to raise the kids."'

This narrative suggests a traditional voice for women in decision-making about family and the fishing enterprise, in addition to a role in supplementing incomes. John Manore's mother adopted the role of matriarch, providing advice to her sons. She managed the home in Point Edward and directed the children through school, but 'never bothered interfering in the fishing,' says her son John. She ran the household: 'She was boss.' In large measure her being 'boss' was characterized by the long periods of separation from her husband. John remarked: 'A lot of mothers say, "Wait till your dad comes home." ... If we got into trouble, mother settled it. She didn't wait "till dad come home." He might be away for three months.' Subsequently, Mary Kennedy Manore ran the family fishery for fourteen years after Cub's death. In this manner, women crossed over from the domestic realm to the fishery, though this was not always the case for women married to members of the crew who were labourers. Mary Kennedy Manore may not have pulled fish from the nets, but she had a significant say in how the fishery was managed.

Thus gender roles and responsibilities for women were not limited to onshore, domestic activities, nor were gender roles and responsibilities for men limited to the offshore. Both men and women carried out significant parental, if not domestic duties, even during the fishing season. Overall, both parents were figures of authority to their children, and both made decisions related to community life, work, and business.

Fish to Market

An essential aspect of small-scale commercial fisheries is marketing the harvest. Sociologist Wallace Clement argues that capitalism has all but eliminated (or subsumed) small-scale units of production (i.e., petty commodity production) from the social, political, and economic landscape (Clement 1984: 5). Many fisheries involving wild stocks have not been affected by this historical process because, among other more general factors, the characteristics of the resource generate high risk and uncertain returns. Existing petty commodity forms of productive activities therefore remain relatively undisturbed in many of the world's commercial fisheries, while merchant, and more recently industrial, capital has sought to control them through market domination and contractual relations with the fishers (Clement 1984: 7; Ommer 1990) in an attempt to minimize risk. Control over the poundnet harvesters of lower Lake Huron took this form; a brief discussion of it here illustrates how this process worked in fishing operations such as the Manores'.

John Manore suggests that small-scale fishers along Lake Huron's maritime frontier sold their fish to American mercantile interests that controlled them. Ontario's Great Lakes fishers were ruled by these foreign interests in large measure because in Canada consumption of fish was low and that of freshwater fish even lower. For example, of the fish consumed in southwestern Ontario, freshwater species comprised only 35 per cent of the total (McKenzie 1931). John Manore said that the market for the Canadian freshwater fish varieties, including his own harvests, 'was all in the States. It was all American market ... Cause you couldn't sell enough fish here to keep going.' Approximately 90 per cent of all the fish harvested from the Great Lakes and Ontario's northern waters went to the United States for sale. This export trade was usually managed on a wholesale basis from the larger lake ports, where merchants' agents negotiated prices with the fishers directly. Fishers in the smaller remote ports had to ship their fish on consignment, as agents were generally not being assigned to these ports.

The American market was indeed indispensable to the economic survival of Canada's freshwater fisheries (see also Tough, this volume). It was controlled by 'commission traders' on Peck Slip, centrally located in New York City's Lower East side, within the famous Fulton Street fish market area (Cornell 1916) and in Chicago, controlled by the Booth Fish Company (see anonymous 1965; 'Robert Poulson Fletcher, Jr.' 1964). Through horizontal integration, Booth also dominated the fish trade in Buffalo and Detroit. Booth initially occupied a dominant middleman role in Grand Bend's community-based fisheries and in many of the other Ontario Great Lakes and interior lakes fisheries (i.e., Lake Nipigon and Lake of the Woods). The company used freighter steamers to collect fish from the isolated fisheries along the Lake Huron coast, including that of the Manores. About 1918, however, the Manores stopped selling their fish to Booth because its marine-based fish-collection service had 'gradually died down' and began trucking their fish 'to the railhead at Park Hill [Ontario] ... The fish went mostly to New York [to the Peck Slip market]. It was a big market.' This marked the beginning of the Manores' long involvement with the New York City trade (Figure 4.5).

Figure 4.5   Loading and unloading space in front of Peck Slip fish market, New York. At left: several groups of buyers bidding on fish still on the waiting motor trucks (Fiedler and Matthews 1926)

About 1926, just before Cub died, the Manores began selling their fish to the Peck Slip-based Lakeside Fish Company; later they sold to Sol Broome, after he left Lakeside to set up his own operation. The family initially dealt with both firms on a price-contracted basis. Contracts were negotiated annually in Grand Bend, before the fishing season commenced: 'We dealt with the individual company, and then [later] they had a representative [the resident agent], and we dealt with him. But to market the contract, the President of the fish company would come himself and talk to you. And you would make a deal and then you would work through the [local] representative. He'd call you every day to tell you what the market was, or how many fish you'd had – we were under contract, it didn't matter. He just wanted to know how many you had.'

Their contracts with Lakeside and Broome doubled fish prices (to nine cents a pound, or twenty cents a kilogram) over those paid by Booth in earlier years. Participating fish merchants probably made such contracts in order to reserve for themselves an allocation of the harvest in a market characterized by demand volatility, price elasticity, and intense inter-merchant competition (Van West 1983: 65–81).

The good prices did not last after John Manore inherited the family fishery. Several years into the Second World War, the merchants on Peck Slip 'clammed up and come together,' he said, maintaining that Hitler and the rise of the Third

Reich caused the Jewish Americans on Peck Slip to consolidate their operations: 'Before they were individual [merchants]. You could call up Lakeside or
you could call Sol Broome up or anyone of them, you dealt with them individually. They wouldn't tell the other [merchants]. They were dealing with you. But
after that [the beginning of the Second World War], they knew what the price
was gonna be. Apparently they sat there every morning and they come up with
a price.'

However, price consolidation by buyers had begun much earlier, shortly after
the First World War, driven largely by the bankruptcies and mergers of many
small trading operations on Peck Slip. Under a unified Peck Slip, merchants
started to establish dockside prices through questionable if not unprincipled
practices, and fishers began to refer to them as a 'fish trust' or 'combine' (see
anonymous 1918, 1921, 1922, 1940; Conlon 1922; Cornell 1916; Dancey 1938;
Hinricks 1918; Hudd 1930).

The merchants' information networks helped them to control the terms of
trade with the fishers. Merchants were fully acquainted with the sometimes vol-
atile demand cycles for freshwater fish, many of which were linked to, and governed by, religious and secular holidays in Jewish social and cultural life. The
merchants were also able, through their dealings with the fishery, to follow the
movements of fish as they migrated along Huron's shoreline sinuosities (in the
case of whitefish, southward from north of Grand Bend down to Sarnia). This
knowledge allowed merchants to forecast the quantity of fish that each fishing
village along the coast could expect to land. As John Manore explained to us:
'Fish were caught at different times. Now, like here, in the spring, whitefish was
our catch here. Below Kettle Point, it was yellow pickerel. Well, we would
watch when the fish hit Sauble Beach north of Southampton; we would hear,
well it would only be ten days or so they [the fish] would hit us and the fish
would just keep on working down the shore ... Well, then, when they leave here,
I asked them in New York, ... those fish cut across the American shore, the
whitefish, and they would get them up Harbour Beach and on up the American
shore. And they would go right around. So [the merchants] just follow the fish
around.'

The merchants were thus in the enviable position of being able to assess landings on Lake Huron relative to those of other Great Lakes ports and beyond,
evaluate discrete community harvests in total against demand, and arrive at
notional dockside and retail prices for the freshwater species: 'They knew when
our fish come on the market in June. We would look – we could fish earlier –
but our heavy run would start the 10th of June, and then we would get a heavy
run of whitefish until the 15th of July. They knew that. And on down the shore,

pickerel below Kettle Point would come on at the end of April and May, they'd get a heavy run, Saginaw Bay would come on with "yellows," some place else, as I said Sauble Beach would come on with whitefish, up the American shore ... Well they knew that because they were buying them. They had it all ... yes.'

Under a unified Peck Slip, negotiations with the fishers were often leveraged. Many of the merchants conducted their operations with 'a big name but it'd be one man doing business out of a telephone booth.' For example, merchants would unilaterally decrease the dockside prices for fish in Grand Bend, based on their assessment of the local, regional, and national freshwater fish landings, relative to demand. John Manore explained that the merchants would 'tell ya, because you'd ask about the [price], you'd say "what price," and he'd say, "Well, they are hitting them [fish] here, or they are hitting them there, and we are getting a lot of fish from here and that is what is knocking the price."' Unorganized and lacking freezers to hold back their catches as a countervailing measure, commercial fishers had to dispose of their fish 'no matter what the price was.' John Manore indicated that he would either have to 'get rid of the fish, or dump them in the lake or bury them in the ground. You had no alternative. You either fished or you didn't fish. If the price didn't suit you, that was up to you. You had no control over the price.'

Fishers resisted the low dockside prices, but to no avail. Even in times of increased demand, during the Jewish religious and secular holidays, when merchants were willing to pay producers higher prices, merchants actually remitted little, or sometimes nothing, citing as reasons 'quality deterioration or non-delivery of fish shipments ... non-settlement or delayed settlement for fish received, and ... a number of other artful practices including the device of contrived bankruptcy' (Frick 1965: 48). John Manore remembers another tactic:

New York buyers would start to call you ... and you'd ask them what's their price? Well you may as well have not asked them their price because it was all set up. They knew the price and then they'd say well how many fish you got and ... how much ... well maybe they were up around fifteen to eighteen cents then and you'd set the highest price and they'd say OK, so maybe after talking to three or four of them saying 'OK, I'll give you maybe 1 cent more than you'd been offered,' you'd say 'OK I'll send you maybe a hundred boxes and you'd send another [buyer] fifty boxes, and another, a hundred.' You'd split the shipment up. But you'd get your cheque back it'd be the lowest price, the lowest price that was quoted amongst those three or four buyers no matter what buyer you sold 'em to.

Lake Huron fishers were thus unable to control the conditions of sale under most circumstances.

John Manore pointed out that poundnetters had an advantage over gillnetters in the district because poundnets produced better-quality fish. 'The gillnet fish are good if they're alive but most of the gillnet fish are just [dead] ... Some are [alive] if they lift [remove the fish] every night but they can't lift every night ... [Well], as soon as you set [the gillnets], you got fish so you're gonna get some dead fish ... They've taken water ... [but] they're too good a fish to throw away. Ya look at 'em and boy you're not gonna throw that fish away. There's nothing wrong with it ... but the flavour's not there.' John Manore's poundnet harvest did not necessarily return higher prices relative to fish caught with gillnets, but he appears to have been protected from claims of quality deterioration by the merchants. He indicated that the merchants gave him 'no kick [complaints] ... No complaints, they paid us in American money, our cheques come every week. Never had one bounce ... We always got our money.'

John Manore's experiences with the Peck Slip merchants mirror the circumstances under which many small-scale fishers laboured. Despite the relative disadvantages of his class, however, he was able to promote himself as a reliable and trustworthy supplier, which, he believed, explained his fair dealings with the Peck Slip merchants. He did not perceive himself to be a member of an alienated class of producers or of a class controlled by outside economic interests – a perspective that, if typical, might help to explain why Great Lakes fishers have apparently not been successful in their struggle to organize as a class for themselves.

Conclusion

John Manore's narrative has enriched our understanding of the rather elusive and relatively unknown history of the lower Lake Huron poundnet fisheries. He has shared stories about the allocation and control of fishing rights and space, social organization of production, external trade with merchants, and the interconnecting domains of male and female work.

This tightly focused history of a single individual operating a family-based fishery within a community-specific setting for a New York market also illuminates the established institutional practices for fishing and marketing that may have been common to many (perhaps all) poundnet fisheries situated throughout the Great Lakes littoral, if we allow for specific geographical and historical variation. This history sheds some light on the community-based fisheries of Ontario's 'inland seas,' making the twentieth-century history of lower Lake Huron accessible and comprehensible.

NOTES

1 We presented a version of this paper to the 20th Annual Congress of the Canadian
   Anthropology Society, York University, Toronto, 7 May 1993. We conducted three
   lengthy interviews with John Manore at his home in Grand Bend in September 1989,
   June 1990, and August 1993, which were taped, transcribed, and analysed in the intervening periods. The transcripts were circulated to John Manore and members of his
   extended family, including his son Richard. We raised with John Manore any new
   questions that resulted. The original transcripts remain in our possession. John Manore has read this paper and agrees with its storyline and analysis.
      Our thanks to John and Edith, Charles G., and Richard and Rhonda Manore, and
   other family members, for their generous assistance and support. Jean Manore's presence in this project helped to break down the barriers of field entry normally encountered by historians and anthropologists in a new setting. Appreciated is the help of
   A.B. McCullough, Christine Hughes, and Ben Osemeke (Ontario Native Affairs
   Secretariat), and Ed Cox.
2 An accidental fire in Parliament's West Block in 1894 destroyed almost all of Canada's pre-1892 manuscript fisheries records, and Ontario's fisheries records were
   intentionally expunged early this century, seemingly in the absence of a regulatory
   system of records retention that would have protected their integrity and because they
   were believed to be of little historical value. For a short history of fisheries administration in Ontario, consult Payne (1967).
3 Underlying title to, or ownership of, the lake bed, however, remained vested with the
   provincial crown, pursuant to Ontario's Bed of Navigable Waters Act, 1911.

WORKS CITED

Acheson, James M. 1981. 'Anthropology of Fishing.' *Annual Review of Anthropology*
   10: 275–315.
Adams, G.F., and D.P. Kolenosky. 1974. *Out of the Water: Ontario's Freshwater Fish
   Industry*. Toronto: Ontario Ministry of Natural Resources, Division of Fish and
   Wildlife, Commercial Fish and Fur Branch.
Allison, Charlene J., Sue-Ellen Jacobs, and Mary A. Porter, eds. 1989. *Winds of Change:
   Women in Northwest Commercial Fishing*. Seattle: University of Washington Press.
Andersen, Raoul, 1972. 'Hunt and Deceive: Information Management in Newfoundland
   Deep Sea Trawler Fishing.' In R. Andersen and Cato Wadel, eds., *North Atlantic
   Fishermen*, Newfoundland Social and Economic Papers no. 5, 120–4. St John's:
   ISER, Memorial University.

– 1973. 'Those Fishermen Lies: Custom and Competition in North Atlantic Fishermen Communication.' *Ethnos* 38: 154–64.
– 1980. 'Millions of Fish.' *Canadian Issues* (special issue on Canada and the sea) 3, no. 1: 127–39.
Andersen, Raoul, and G. Stiles. 1973. 'Resources Management and Spatial Competition in Newfoundland Fishing: An Exploratory Essay.' In Peter Fricke, ed., *Seafarer and Community*, 44–66. London: Croom Helm.
Anonymous. 1918. 'Annual Convention of the Lake Erie Fisheries Association.' *Canadian Fisherman* 5, no. 3: 640–50.
– 1921. 'Lake Erie Fishermen's Convention.' *Canadian Fisherman* 8, no. 2: 36–9.
– 1922. 'Erie Fishermen to Sell Co-operatively.' *Canadian Fisherman* 9, no. 4: 76–7.
– 1925. 'Peck Slip – Past and Present.' *Fishing Gazette* (annual), 75–8.
– 1930. 'Freshwater Fisheries.' *Canadian Fisherman* 17, no. 2: 39.
– 1940. 'Lake Erie Convention: Dishonest Dealers Still Said to Be Preying on Freshwater Fishermen.' *Canadian Fisherman* 27, no. 2: 71.
– 1965. 'The Booth Story.' *Quick Frozen Foods* 27: 110.
Apostle, Richard, and Barrett, Gene. 1992. *Emptying Their Nets: Small Capital and Rural Industrialization in the Nova Scotia Fishing Industry*. Toronto: University of Toronto Press.
Attorney General for the Dominion of Canada v. Attorneys General for the Provinces of Ontario, Quebec, and Nova Scotia. 1898. *Law Reports* (Judicial Committee of the Privy Council) 700.
Bed of Navigable Waters Act. Statutes of Ontario, 1911, c. 6.
Bluestone, Daniel M. 1991. '"The Pushcart Evil": Peddlers, Merchants and New York's City Streets, 1890–1940.' *Journal of Urban History* 18, no. 11: 68–92.
Candow, James E. 1990. '"They Done Alright": A History of the Mudge Family Fishery at Broom Point, Newfoundland, 1941–1975.' Microfiche Report Series, no. 427. Ottawa: Environment Canada, Parks Service.
Clement, Wallace. 1984. 'Canada's Coastal Fisheries: Formation of Unions, Cooperatives and Associations.' *Journal of Canadian Studies* 19, no. 1: 5–32.
Cole, Sally. 1991. *Women of the Praia: Work and Lives in a Portuguese Coastal Community*. Princeton: Princeton, NJ: University Press.
Conlon, J.H. 1922. 'Let's Tighten Up.' *Canadian Fisherman* 9, no. 3: 43.
Cornell, H.S. 1916. 'Co-operation in the Fishing Industry.' *Canadian Fisherman* 3, no. 11: 356–8.
Cox, E.T. 1992. *An Indexed Chronology of Some Events in the Development and Administration of Commercial Fishing on Lake Erie*. Wheatley, Ont.: Lake Erie Fisheries Assessment Unit, Report 1992–8.
Dancey, A.W. 1938. 'Lake Erie's Fishing Industry.' *Canadian Fisherman* 25, no. 4: 8–10.

Davis, Donna Lee, and Jane Nadel-Klein, eds. 1988. *To Weep and to Work: Women in Fishing Economies*. St John's: ISER, Memorial University.

Doherty, Robert. 1990. *Disputed Waters: Native Americans and the Great Lakes Fishery*. Lexington: University of Kentucky Press.

Fiedler, R.H., and J.H. Matthews. 1926. *Wholesale Trde in Fresh and Frozen Fishery Products and Related Marketing Considerations in New York City*. Appendix VI to the Report of the U.S. Commissioner of Fisheries for 1925. Washington, DC: Government Printing Bureau.

Frick, Harold C. 1965. *Economic Aspects of the Great Lakes Fisheries of Ontario*. Bulletin No. 149. Ottawa: Fisheries Research Board of Canada.

Gearhart, Clifford Ross. 1987. *Pity the Poor Fish Then Man*. Grand Haven, Mich.: Marine Publishing Company.

Gilmore, Janet. 1990. 'Fisherman Stereotypes: Sources and Symbols.' *Canadian Folklore* 12, no. 2: 26–7.

Goodier, John L. 1984. 'The Nineteenth Century Fisheries of the Hudson's Bay Company Trading Posts on Lake Superior: A Biographical Study.' *Canadian Geographer* 28: 341–51.

– 1989. 'Fishermen and Their Trade on Canadian Lake Superior: One Hundred Years.' *Inland Seas* 45, no. 4: 284–306.

Gordon, William G. 1965. *Haul Seine in the Great Lakes*. Fishery Leaflet no. 577. Washington, DC: United States Department of the Interior, Fish and Wildlife Service, Bureau of Commercial Fisheries.

Hansen, Lise C. 1991. 'Treaty Fishing Rights and the Development of Fisheries Legislation in Ontario: A Primer.' *Native Studies Review* 7, no. 1: 1–21.

Hinricks, H. 1918. 'The South-Shore Association.' *Canadian Fisherman* 5, no. 3: 652–5.

Hudd, F. 1930. 'Marketing Canadian Fish in New York City.' *Canadian Fisherman* 17, no. 5: 16–17.

Lloyd, Timothy C., and Patrick B. Mullen. 1990. *Lake Erie Fishermen: Work, Tradition, and Identity*. Chicago: University of Illinois Press.

Lytwyn, Victor P. 1990. 'Ojibwa and Ottawa Fisheries around Manitoulin Island: Historical and Geographical Perspectives on Aboriginal and Treaty Fishing Rights.' *Native Studies Review* 6, no. 1: 1–30.

– 1992. 'The Usurpation of Aboriginal Fishing Rights: A Study of the Saugeen Nation's Fishing Islands in Lake Huron.' In Bruce W. Hodgins, Shawn Heard, and John S. Milloy, eds., *Co-Existence?: Studies in Ontario–First Nations Relations*, 81–103. Peterborough, Ont.: Frost Centre for Canadian Heritage and Development Studies.

McCullough, A.B. 1989. *The Commercial Fishery of the Canadian Great Lakes*. Studies in Archaeology, Architecture and History, National Historic Parks and Sites, Canadian

Parks Service, Environment Canada. Ottawa: Ministry of Supply and Services Canada.

MacDonald, Graham Alexander. 1978. 'The Saulteur-Ojibwa Fishery at Sault Ste. Marie, 1640–1920.' MA thesis, University of Waterloo.

McGoodwin, James R. 1990. *Crisis in the World's Fisheries*. Palo Alto, Calif.: Stanford University Press.

McKenzie, R.A. 1931. *The Fish Trade of Southern Ontario*. Bulletin No. 23. Ottawa: The Biological Board of Canada.

MacLaren, A. 1965. 'A Brief History of Fisheries Development on Each of the Great Lakes.' Appendix to H.C. Frick, *Economic Aspects of the Great Lakes Fisheries of Ontario*, Bulletin No. 149. Ottawa: Fisheries Research Board of Canada.

Murray, Hilda. 1979. *More Than Fifty Per Cent: Women's Life in a Newfoundland Outport, 1900–1950*. St John's: Breakwater Books.

Ommer, Rosemary E., ed. 1990. *Merchant Credit and Labour Strategies in Historical Perspective*. Fredericton: Acadiensis Press.

Orr, Sandra. 1993. *Huron: Grand Bend to Southampton*. Erin, Ont.: Boston Mills Press.

Payne, Robert. 1967. 'A Century of Commercial Fishery Administration in Ontario.' *Ontario Fish and Wildlife Review* 6, nos. 1–2: 7–15.

Peters, John. 1981. 'Commercial Fishing in Lake Huron, 1880–1915: The Exploitation and Decline of the Whitefish and Lake Trout.' MA thesis, University of Western Ontario.

Porter, Marilyn. 1985. 'She Was Skipper of the Shore Crew: Notes on the Historiography of the Sexual Division of Labour in Newfoundland.' *Labour/Le Travail* 15: 105–23.

Prothero, Frank. 1980. *The Good Years: A History of the Commercial Fishing Industry on Lake Erie*. Port Stanley, Ont.: Nan-Sea Publications.

'Robert Poulson Fletcher, Jr. [President of Booth Fish Company].' 1964. In *National Encyclopedia of American Biography*, 81–2. New York: J.T. White and Company.

Schmalz, Peter. 1992. 'The European Challenge to the First Nations' Great Lakes Fisheries.' Paper presented at the 48th Annual Meeting of the Canadian Historical Association, Charlottetown, PEI.

Thompson, Paul. 1985. 'Women in the Fishing: Roots of Power between the Sexes.' *Comparative Studies in Society and History* 27, no. 1: 3–32.

– 1988. *The Voice of the Past: Oral History*. 2nd ed. First pub. 1978. Oxford: Oxford University Press.

Van West, John J. 1983. 'The Independent Fishermen in the Port Dover Fishing Industry: A Case Study of Their Production and Market Relations.' PhD dissertation, University of Toronto.

– 1986. 'The Need for a Social Science Perspective in Fisheries Management: A Socio-

Historic Case Study of the Independent Fishermen in Port Dover, Ontario.' *Environments* 18: 43–54.

– 1989. 'Ecological and Economic Dependence in a Great Lakes Community-Based Fishery: Fishermen in the Smelt Fisheries of Port Dover, Ontario.' *Journal of Canadian Studies* 24: 95–115.

White, C.H. 1905. 'Fulton Street Market.' *Harper's Magazine* 111: 616–23.

5

'Ould Betsy and Her Daughter': Fur Trade Fisheries in Northern Ontario

ARTHUR J. RAY

It is safe to assert that all down the centuries of the fur trade in Canada fish has been the principal sustaining food for the fur trader and his dog-team. It is true, of course that most trading posts had one or more of the various kinds of game ... But the weight of evidence indicates that, when all else failed, even the buffalo of the plains, the trader as well as the native had to fall back on the humble fish.

ANDERSON 1961: 63

Fishing was so important to most Aboriginal communities in central and western Canada that Aboriginal people sought protection for this aspect of their economies when negotiating land surrender treaties with Canada during the late nineteenth and early twentieth centuries. Each major numbered treaty contains clauses that granted Native people the right to fish in their customary ways, subject only to such regulations that the crown might impose from time to time for conservation purposes. As Euro-Canadian settlement spread into treaty areas, provincial and federal governments sought, in the name of conservation, to limit Aboriginal fishing and hunting rights to subsistence use. Aboriginal people assert that this limitation is an infringement of the treaty rights negotiated by their ancestors and that the action fails to take into account Indian economic history. Many First Nations argue that fishing has always been important in their societies and that it included a commercial dimension.

The historically important Treaty 9 territory of northern Ontario is one of the regions where recurring disputes occur about this issue. To resolve them it is essential to obtain a clear understanding of the histories of the Aboriginal economies of this sprawling region during the pre-treaty era. The central legal question is this: was commercial fishing an established custom for Native groups living south of the Albany River in 1905, when the agreement was first signed,

Figure 5.1   Map of the central subarctic showing relevant fur-trade posts

and for those living farther north in 1929, when the accord was extended into that area? In other words, did Native people occupying this vast region customarily trade the commercial supplies of 'humble fish' of which trader Anderson speaks and that he, his colleagues, and their dog teams required?

The Contact Setting

Treaty 9 encompasses what anthropologists call the central subarctic culture area of Aboriginal Canada, and it was the traditional heartland of the Hudson's Bay Company (HBC) for the first two centuries of its operation in Canada (Figure 5.1).

The London-based merchants built two of their first three posts – Fort Albany (1672) and Moose Factory (1673) – there, on the shores of James Bay, which afforded a cheap waterway into one of the best fur-producing regions in North America. When the company began operations, the Swampy Cree occupied the marshy James Bay lowlands, and the Upland Cree the rugged Canadian Shield portion farther inland.

These two groups were nomadic fishers, hunters, and gatherers. During the long winter they lived in small groups of closely related families (winter bands), pursued big game and fur-bearing animals, and conducted some ice fishing. During the warmer months of the year, when freshwater fish were one of the main-stays of their diet, the winter bands gathered into larger, regional bands at productive fishing places on the countless lakes and numerous rivers of the shield. During the eighteenth century, major population migrations took place within the vast area that would come under Treaty 9 when the Upland Cree moved farther west, into Manitoba and beyond, in response to the fur trade. After that migration, the Northern Ojibwa, who had previously lived in the country bordering eastern Lake Superior and northern Lake Huron, replaced them. Apparently Ojibwa people traditionally depended more heavily on fish than did their Cree neighbours, for Charles Bishop's research shows that for half the year – from early spring until late autumn – the Northern Ojibwa had lived in large fishing villages located along the north shore of Lake Huron (Bishop 1974: 374).

Native Provisioning in the Treaty 9 Area

After a brief attempt to conduct their Canadian operations from a single depot on Charlton Island in James Bay, the HBC's directors abandoned the idea in favour of building trading posts at the outlets of the major rivers flowing into James Bay and Hudson Bay – most notably the Albany, Moose, Nelson, Rupert, Severn, and Eastmain rivers. In the early 1670s it was too risky for company officials to concentrate all their assets at one location because of the threat of French attack. Equally important, and of relevance here, the company's men quickly learned that they needed to be situated on the mainland where they could turn to the Native people whenever they needed help – particularly with food supplies. The risk of starvation was especially great on isolated Charlton Island. Senior HBC officials also knew that importing European foodstuffs into its chartered northern territory, which they had named Rupert's Land, was uneconomical. Thus from an early date the directors encouraged their officers and men to obtain substantial provisions from their Aboriginal 'clients.' When the company launched a major inland expansion a century later, its demand for Native food products, known as 'country produce,' grew rapidly.\

Country produce was essential to the success of all fur-trading operations: the French (before 1760) and the Canadian trading rivals (1765–1821) of the HBC were equally dependent on Indian produce. From the outset, the so-called fur trade was in fact a commercial enterprise involving the exchange not just of fur but of myriad Aboriginal supplies and foodstuffs, including fish and fish products. There is no question that fur trading would not have been a profitable venture for Euro-Canadians had Aboriginal people refused to sell them food or to work for them at very moderate wages as fishers, hunters, and collectors.

When English traders first arrived, large herds of caribou (probably the barren ground variety) frequented Swampy Cree territory from beyond the Severn River in the northwest to Akimiski Island in the east (Lytwyn 1993). Fur-bearing animals, especially beaver (which all Cree prized for meat and pelts) abounded, as did waterfowl during the warmer months of the year. Less well recognized is the fact that a wide variety of fish also were plentiful in 'contact' Cree territory. Sturgeon (*Acipenser fulvescens*, or 'Nemew' in Cree), whitefish (*Coregonus clupeaformis*, or 'Tickomeg,' 'Tichameg,' or 'Tickemeg' in Cree), lake trout (*Salvelinus namaycush*, or 'Namaycush' in Cree), jackfish (*Esox lucius,* or 'Keneshue' in Cree), and several species of suckers (*Catostomus,* which the Cree referred to as 'Namepith') seem to have been the most important species, judging from the company's eighteenth- and nineteenth-century post records (Williams 1969: 118–24).

At the time of contact and throughout the early fur trade, Cree men of the district specialized in hunting big game, and both men and women pursued fur-bearing animals and small game. Likewise, both sexes engaged in fishing, though it was primarily women's domain. These fishers used a variety of gear, including hook and line, spears, nets, and brush and stone weirs, and conducted a series of major seasonal fisheries every year. The most significant of these took place in spring (for sturgeon) with spears and weirs and in early autumn (for whitefish) with dip nets and brush and stone weirs. According to HBC trader Andrew Graham, daily catches of whitefish commonly amounted to between 500 and 600 fish per day (Williams 1969: 118–24). Lake trout and suckers were popular in summer. Jackfish, available most times of the year, were a type of 'back-up' food. Graham observed that jackfish 'are very numerous and are much valued by the Lake Indians (of the upland shield), as they are a supply for them at all seasons, when their gun and ammunition fails, or other food fails' (Williams 1969: 118–19). Once the Cree and Ojibwa were drawn into the fur trade, European barbed metal fish hooks, net lines, and twine, as well as seine nets, became useful additions to their personal fishing gear.

The Cree and Ojibwa women of the region preserved all types of fish by freezing their winter catches and by custom drying and smoking them on racks in temporary bark smokehouses during the warmer months. These traditional practices continued into the modern fur trade, as witnessed in the early twentieth century by HBC trader James W. Anderson:

There were two methods of smoking – cold and hot. Cold smoking was the simplest and consisted of drying in the sun. This was the method mainly used for preserving fish for dog food. Hot smoking was by far the best method of preserving fish for human consumption. The fish were not only smoked and dried over a fire but also partly cooked. The selection of firewood for use in the fish-smoking wigwam was of prime importance, for certain woods gave certain flavours, and in general hardwoods were preferred. Indian women developed great skill in curing ... Smoked fish were light and easily transported, with high food value relative to weight, a small bundle being sufficient for a relatively long journey. (Anderson 1961: 66)

In their records, traders often referred to smoked fish as 'hung fish.' Clearly, and of significance here, the Cree and Ojibwa of northern Ontario offered the newcomers and their dog teams quite an impressive variety of fish and other country foods.

Initially, venison had dominated the country foods that Swampy Cree supplied to the HBC. Historical geographer Victor Lytwyn's detailed analysis of York Factory's records reveals that barren ground caribou meat, commonly referred to in the HBC records as 'deer flesh,' was paramount in that major post's lively provision trade until overhunting decimated the herds late in the century. This overkill was the direct result of the burgeoning market for country foods that the inland expansion of the fur trade generated during the era of HBC and Nor' Wester rivalries between about 1765 and 1821 (Lytwyn 1993: 453–4). Account book data from Fort Albany, which is the post with the longest documentary record in the Treaty 9 region, reveal a similar pattern of provisioning for the lower Albany River area. It was in the expenses sections of these books that the traders recorded their purchases of country food. The account books for the period 1693–1797 disclose that in the beginning the company's men mostly purchased meat (venison) and fat or, alternatively, contracted Native men to hunt for the post (Hudson's Bay Company Archives [HBCA], Fort Albany Account Books 1692–1805: 1–118). Normally the post managers gave Native hunters on contract a small advance of goods, then paid them for their meat in goods at an agreed Made Beaver (the currency of trade) price. Before 1744, there is only 1 recorded fish purchase, and there are only three references to Natives being paid to fish.

A more diversified purchasing pattern is evident after the mid-eighteenth century, however, even though post managers continued to buy substantial

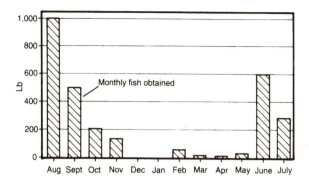

Figure 5.2   HBC Fort Albany Post, monthly fish harvests for 1810. *Source*: HBCA, B 3/D, 1810

quantities of venison. Foreshadowing nineteenth-century developments, purchases of fish, rabbits, and geese appear more often. The references to fish purchases increased substantially after 1760, though the species that the traders were buying at Fort Albany are seldom noted. When they are mentioned, one species – whitefish, listed as 'tickameg' or 'tickomeg' – appears most often.

By the early nineteenth century the situation changed dramatically throughout the James Bay and Hudson Bay lowlands. The decimation of the local caribou herds forced the Swampy Cree and the European traders to turn more heavily to fish, migratory waterfowl, and small game for their food. The rapid expansion of inland operations after the 1770s reinforced this trend – especially the heightened and expanded seasonal demand for fish. A key reason for this new demand for fish was the increasing role that dog teams played in winter transport. Whitefish was the predominant food for dog teams. The increased importance of fisheries to provision procurement at Fort Albany by the early nineteenth century is evident from Figure 5.2, which shows the monthly fish harvest at the post in 1810. The crucial role of the big autumn whitefishery is also readily apparent.

So significant was fish to the inland expansion of the fur trade that when the traders built new posts in the interior one of their first concerns was the establishment of a local fishery. The traders usually hired Natives to show them where to fish and, similar to the bayside posts, they also employed Native fishers on contract or as seasonal labourers. Typical, for example, is the journal entry of 23 October 1794 for Fort Kenogamissi, which the HBC set up that summer on the upper headwaters of the Kenogamissi River: 'Paid two Indians for showing us a fishing place at the head of yr Lake [Kenogamissi] and assisting us in catching fish' (HBCA, Kenogamissi Post Journals [KPJ] B 99/a/1).

Retaining Native fishers was essential for the continuing operation of the posts. During competitive trading periods, these Natives were in a strong position, as the KPJ entry for 23 October 1799 indicates: 'the Canadians has sent Brandy and tobacco to the Indians to entice them to get fish for them, in consequence of that I was obliged to send them a small keg of Brandy and Oatmeal because without their assistance we shall not be able to get any for the want of hands' (B 99/a/6).

The Native practice of selling fish, fishing seasonally under contract, and working as seasonal or permanent salaried fishers for trading posts continued into the twentieth century. The relative weight of these various kinds of arrangements within the HBC's fish procurement operations varied considerably from post to post by the time Treaty 9 was negotiated, as the following discussion demonstrates.

At Fort Albany, the only surviving post journals for the late nineteenth century (covering the years 1890–5) reveal that here the company obtained most of its fish – predominantly whitefish – from company-operated seines located at three harvesting sites: Fishing (the most important location), Yellow, and Devils creeks. Company servant James Linklater supervised seining activities, and other employees helped him. Male and female Native workers made net floats, hauled in the fish, and dried the catch (HBCA, Fort Albany Post Journals [FAPJ], B 3/a/198: 2–3 and 13–14).

Fort Albany obtained some of its fish through trade with local bands, though the significance of these supplies to this post is not clear. What is apparent is that most of the exchanges took place during the big autumn whitefish harvest. Unfortunately, the journal references rarely indicate the quantities of fish that the Aboriginal fishers sold. Typically, the journals merely state 'a number of Indians in this morning with fish and ducks,' or simply, 'fish coming in' (FAPJ, 1891–2: B 3/a/198: 2–3 and 13–14). For a small outpost of Fort Albany, Henley House, only a very few journals survive (HBCA, Henley House Post Journals [HHPJ] 1873–5, B86/a/69–71), and these offer little insight into provisioning activities, except to mention that the officer in charge of the post was assisted by two Aboriginal men – Big Head and Mekeekwies – who helped him make a 'fish basket trap' (HHPJ 1874–5: B 86/a/70: 2).

Farther inland, the Martin Falls Post relied on a local seine fishery. Here, however, the Ojibwa did most of the fishing for the post on contract. The post managers often referred to these Natives in general terms as 'our fishers' or 'the fishermen.' Important Native contract fishers who operated from 1869 to 1899 included women as well as men: Lame Man, 'Ned and his sister,' 'Ould Betsy and her daughter,' Old Sturgeon, Moss Sturgeon, Sandy Sturgeon, 'Osshkapay and his brother,' and Patrick and Jack Wich ee capay (HBCA, Martin Falls Post

Journals [MFPJ], B 123/a/85–93). Members of the Sturgeon family (clan) were among those Ojibwa who had moved into northern Ontario after the commencement of the fur trade, and they have long played a major role in the commercial fishery of the upper Albany River area (Bishop 1974: 305–40;. Lytwyn 1993: 110–40; Ray 1998: 94–114). The Martin Falls post supplied the contract fishers with salt, kegs, nets, and sometimes tents (MFPJ, B 123/a/86: 28, 41; 88: 33; 89: 15). The Indians under contract also usually obtained small advances and received final payments for their catch on delivery of it to the post. Occasionally, the post manager would pay other Natives to pack the fish in kegs and haul them to the post on behalf of the fishers (B 123/a/89: 20).

The story of two women who were under contract with the Martin Falls post, 'Ould Betsy and her daughter,' offers a rare glimpse into the crucial role of Aboriginal women in the local economy after contact. According to post journals, these two women served the post's fisheries during the late 1860s and the 1870s, and their duties extended well beyond harvesting. Ould Betsy and her daughter carried messages from the post to the fishing stations, hauled nets back and forth between the post and these sites, and did most of the drying and smoking of fish. During the time of the spring sturgeon fishery of 1870, the post journal records: 'Lame Man came in with 10 Sturgeon this forenoon. Betsy at the smoking of them, also cleaning those she smoked a few days ago [on the 27th of May], and giving them some more smoking' (MFPJ, B 123/a/86: 19). The two women also smoked some of the autumn whitefish catch. The post's journal entry for 15 October 1871 notes: 'Old Betsy & her daughter got up a bark tent and are smoking the whitefish we got in yesterday.' Other Native women made and repaired fish nets for the post (B 123/a/89: 9 and 90: 45).

Fish trading also was commonplace at Martin Falls post throughout this period, though the journals for the 1890s make few specific references to the kinds of fish bartered (B 123/a/86: 18–25, 28, 42; 87: 20, 44; 88: 33, 36, 44; 89: 5, 9, 14; 90: 1, 4–7, 10, 12, 23–9; 91: 6, 9, 13, 62, 82, 87; and 93: 14, 24, 46–7). Similar to those for Fort Albany, the entries for Martin Falls typically state: 'Indians arrived with some fish,' or with 'a little fish.' References to specific species indicate that whitefish and suckers dominated the trade, but sturgeon also were included in the traffic. Most of the fish that Aboriginal people harvested on their own and sold to the post were fresh or smoked, whereas most of the 'contract' fish obtained from Natives was salted. Salting, a European introduction, gave the fish a longer 'shelf life.' Some of the contract fishers, 'Ould Betsy' among them, made independent sales of fish to the post during the off-season.

The surviving late-nineteenth-century journal records for the nearby post of Fort Hope are much scantier than those for Martin Falls, covering only the

Figure 5.3   Indian woman making nets at New Brunswick House, n.d. *Source*: Ontario Archives, S7607

years 1890–2. None the less, the record indicates that the fishery here was very similar to that of the Martin Falls post. The post manager contracted to have Aboriginal people operate his camps. These fishers included Ayumeawumie, Drake 'with sons,' Kachang, Lame man (perhaps same individual who appears in the Martin Falls post records), Shabayorkishick, Shatagamy, Sugar Head, and Chief Yesno (HBCA, Fort Hope Post Journals [FHPJ], B 291/a/1: 7; 2: 6; and 5: 6). The manager also hired at least one of these contract fishers – Sugar Head – to fish for the post (FHPJ, B 291/a/1: 1). Various other Ojibwa arrived at Fort Hope from time to time to sell several varieties of fish, including sturgeon (B 291/a/2: 16–26; 4: 15; and 2: 1–5).

The late-nineteenth-century records for Osnaburgh House, which span the years 1871–7, identify eight fishing stations. Three stations were located at an undetermined place called simply 'the narrows' (probably located on the Albany River); another three on Lac Seul, on the Albany; a seventh at the 'Doghole'; and the eighth somewhere in the 'Fir Hills' (HBCA, Osnaburgh House Journals [OHPJ], B 155/a/80: 2 and 5; and 81: 4). The harvesters at Osnaburgh House included both company servants and Native men and their families, whom the company retained in the same fashion as at Martin Falls. Listed among these Aboriginal fishers were David and John Skunk, and 'Fanny and his wife,' who operated the fish stations at Fir Hills and at Doghole (B 155/a/ 81: 1, 3). Whitefish dominated the catch at this post, and for this species the fisheries at 'the narrows' and on Lac Seul were the most important. During the autumn of 1872 'the narrows' stations yielded an impressive 6,500 whitefish, and those of the lake, just over six thousand (B 155/a/81: 1, 4). The journals mentioned sturgeon only once (B 155/a/83: 13).

At Fort Severn, the journals for the years 1873–97 indicate a post heavily dependent on seines operated by 'the men' – a reference that normally implies regular, non-Native, HBC servants. It is also clear, however, that local Aboriginal people assisted. The entry for 6 June 1874, for example, mentions: 'Cromarty [a company servant] down at the beacon with 3 Indians to look for fish with the Seine got 10 fish' (HBCA, Fort Severn Post Journals [FSPJ], B 198/a/ 123: 8). Eighteen days later, the journal reports: 'Indians employed at the Seine and getting no fish' (B198/a/123: 9). They were eventually successful in getting fish.

Significant bartering of fish by men and women also occurred at this post. A single Native woman, referred to both as 'the Fish Wife' and 'Fisher Wife,' dominated fish selling at Fort Severn for four years, from 1876 to 1879. It is possible to pull together her story from the post journal. In the summer of 1876 she made twelve visits, from 22 June to 11 August, hauling in six to 130 fish each time. Her season total was an impressive 336. Normally the Fort Severn journalist failed to indicate what types of fish she sold, though the entry for 29 June did mention suckers (FSPJ, B 198/a/123: 35–7). The following summer 'the Fish Wife' made thirteen visits and brought with her 262 suckers and 'other' fish (B 198/a/124: 2–4). Her visits in 1878 began much earlier in the season, on 12 June, and she made her twenty-eighth and final call of the year on 28 August. All in all, she delivered a grand total of 722 fish that year, which was a record for her and probably a particularly good year for fishing in the region. Trout were mentioned as being among her sales (15–18). In 1879 she commenced her visits on 16 June, and that season it took her seventeen visits to sell 547 fish by 23 August (29–31). During the summer of 1880, the 'Fisher Wife' made only two visits, both at the end of June, and she bartered a mere thirty-eight fish (42–3). It is not known what species she brought in this year, but it is unlikely that whitefish figured prominently, because she concluded her transactions well before the autumn fishery would have begun. Thereafter the woman disappears from the Fort Severn record. Because she generally visited the post every two or three days during her selling periods, the fish she sold presumably were being caught by her family rather than by herself: she was a trader.

In subsequent years, most of the independent fish sales mentioned at Fort Severn involved Aboriginal men, suggesting that the men generally dominated this aspect of the business. Specifically named are 'Snipper' and 'Benjamin Blue Coat,' both of whom also fished for the HBC on contract (FSPJ, B 198/a/ 125: 25–6; 127: 16; and 128: 27).

For the twentieth century, even fewer HBC post records are available for the James Bay District. Yet, scant as that record is, its contents do at least suggest that little to do with fish had changed from the previous century. The 1929–30 journal for Attawapiskat, for example, mentions that Natives were bringing in fish to trade (HBC 1929: 20–47).

The surviving records from the various HBC posts thus make it clear that the Cree and Ojibwa of the Treaty 9 area were accustomed to selling fish to fur traders for local consumption. They did so through barter, under contract, and, to a lesser extent, as seasonal or permanent servants of the HBC. From the outset Aboriginal women, both young and old, took an active part, either as members of fishing camps or as traders and labourers in their own right. As the fishing sector of the commercial economy became more important, partly because of the depletion of key fur-bearer populations – especially in the southern reaches of the Treaty 9 area – men became increasingly active in this sphere of the commercial economy, which initially had been primarily women's domain.

Women's Isinglass Trade

Besides selling fish to feed both people and dog teams, the Aboriginal women of the Treaty 9 area produced a highly specialized sturgeon product for sale – isinglass (see also Tough, this volume). Traditionally, women manufactured this glue-like substance from the swim bladders of sturgeon during the spring fishery. HBC fur trader James Isham noted this practice in 1743, when he remarked: 'The glue the Natives saves out of the Sturgeon is very strong and good, they use it in mixing with their paint, which fixes the Colours so they never Rub Out' (Holzkamm et al. 1988: 98). Europeans also valued isinglass for glue and as a fining, or clarifying, agent in the production of beers and wines (Holzkamm and McCarthy 1988: 200).

The HBC directors first showed an interest in obtaining the product in the late seventeenth century, when a glut of beaver pelts on the London market led them to search for alternative Native products to send to England. Accordingly, on 30 May 1693, they wrote to James Knight, their commander at Fort Albany, and instructed him: 'Wee should be glad you could *procure us some isinglass* being only the *sound of sturgeon dryed*, of wch. wee are Informed great

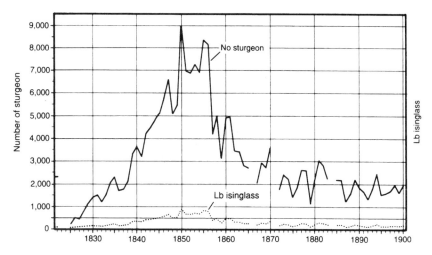

Figure 5.4   Isinglass sales data for HBC Southern Department, with estimated sturgeon harvest, 1822–1900. *Source*: HBCA, B3/D, 1810

quantitys may be had, that Comodity is alsoe very currant here/ for our *Bissiness is to find out new Comotitys etc to Improve the small Furr Trade* nothing being more vendable then small Furrs &c, for since we have Imported great quantitys of beavr. it is become a drugg & sells at a very low rate' (Rich 1957: 231–2). This early effort to stimulate an isinglass trade from the Swampy Cree failed, however, because these people thought the company's price too low to compensate them adequately for the time and effort needed to produce a saleable product (Lytwyn 1993: 345–6). By the early nineteenth century the company had established a lively trade in this product, and in the Treaty 9 area Ojibwa women were the major producers.

Figure 5.4 displays isinglass sales data from HBC's 'fur sales' and auction catalogues pertaining to its Southern Department from 1822 to 1900. This historically important department encompassed an area that became part of Treaty 9. It also included most of the territory covered by the Robinson-Superior and Robinson-Huron treaties of 1850 – the lands lying between lakes Superior and Huron and the height of land marking the southern limit of the James Bay–Hudson Bay drainage basin. Isinglass sales rose sharply between the early 1820s and 1855 and declined abruptly thereafter. Research shows that it took about twenty-five adult sturgeon to produce one kilogram of isinglass (or ten sturgeon for one pound), and this makes it possible to estimate the minimum annual sturgeon harvest that the sales data represent (Holzkamm and McCarthy 1988: 921–3). This is also shown on Figure 5.4. The erratic downward trend in the data after 1855 is mostly the result of the HBC's reorganization of its Southern Department in 1856, which led to the elimination of the area included in the Robinson treaties (the company's Huron and Superior districts). The effect of this particular administrative adjustment on isinglass trade figures is readily apparent from Figure 5.4, which shows that from 1825 to 1856 the returns for the Lake Huron and Albany River districts accounted for most of the Southern Department's isinglass production. It was the elimination of the Lake Huron district from the Southern Department after 1856 that caused the sharp drop in the production figures.

The returns of the Albany District, which accounted for the bulk of production in the Southern Department after 1856, indicate no major decline in the sturgeon harvest. Even though returns for the Albany River district are unavailable after 1870, Figure 5.4 suggests that sturgeon production levels in the district probably remained fairly steady, given that this district yielded the bulk of the department's output. Especially surprising, the company's isinglass trade did not decrease sharply, even though there were growing numbers of competitors in the Southern Department after 1870 and London prices declined by about 50 per cent between the late 1870s and 1900 (Ray 1990; HBCA, Account of Fur Sales A 51/41–53).

Holzkamm and McCarthy's (1988) research suggests that the isinglass returns of the Lac la Pluie area could be used to determine the minimum potential for a sturgeon fishery along the Rainy River (121–3). Perhaps the Albany District returns could have similar implications for an Albany River fishery. The HBC records, however, give no post-by-post breakdown of the isinglass trade. Nevertheless, scattered references suggest that most of the district's production came from the area between Martin Falls and Osnaburgh House posts. The HBC's isinglass sales data suggest that sturgeon harvesting had not yet reached sufficient levels to threaten stocks, though historical geographer Frank Tough (this volume) cautions that drawing such conclusions about the resilience of lake sturgeon stocks is risky.

Another problem with the record on isinglass is that the company's traders made very few observations about the character of this commerce. Isinglass simply was always a minor concern to them. Yet there are some insights into the importance of the isinglass trade to Natives that can be pieced together from the record. Take, for example, the entry for 5 August 1869 in the Martin Falls Post Journals; it indicates that the 'Moonee ass' family took part in the isinglass trade (MFPJ, B 123/a/85: 26). We have already seen that the 'Sturgeon family' (clan) took an active part in all aspects of the commercial provision fishery.

The Osnaburgh House Account Book, 1853–54 Indian Debt Book (HBCA), provides a rare picture of the general nature of this commerce. It reveals that most individual transactions (twenty-five out of a total of thirty-nine) ranged between one-half pound and one pound (0.24 kg and 0.49 kg) of isinglass. It would have taken approximately five-and-a-half to eleven sturgeon to yield these amounts for the Indian traders in question. The debt book also suggests that most sales probably represented a family's (wife's) annual production fur trade, since the named individuals appear only once per year. Quite striking, a mere handful of families outproduced most of the other suppliers of isinglass – the top four accounted for nearly one-third of the total. One-third of the total represents a catch of at least 225 sturgeon, suggesting, among other things, that the leading producers were fishing specialists.

Supplying Fish to Government Agents

New economic opportunities developed for Aboriginal people soon after Confederation, when the fur trade began to decline and to diversify. Agents of the Canadian government spearheaded this change. Geologists working for the newly formed Geological Survey of Canada were among the first of these outsiders to arrive on the scene. Like the fur traders before them, they depended on the Aboriginal men and women to guide them and supply food.

The first reference in the HBC records to Native men providing such services to survey parties in the Treaty 9 territory is that of 15 September 1871 at Martin Falls, when Old Sturgeon (who, we have seen, was one of the key fishers in the area) guided the celebrated Robert Bell and his party (MFPJ, B 123/a/87: 14). Bell, who had joined the Canadian Geological Survey in a full-time capacity in 1869, led a reconnaissance of the entire Hudson Bay drainage system and reconnoitred the route for a transcontinental railway. In offering his services to Bell, Old Sturgeon was continuing a traditional way of dealing with newcomers to his territory.

When the Treaty 9 commissioners themselves visited the area in 1905, they too bought fish and other country foods as they travelled, though Native suppliers are not specifically indicated. The government agents included the record of their purchases in their expense reports to the Department of Indian Affairs in Ottawa, which show that the commissioners made a number of small purchases of fish as they travelled and during their visits at various HBC posts (NA 1906). There is little doubt that Natives were the ones who sold them fish in the field.

Conclusion

A number of general conclusions follow from this analysis. HBC fur-trade records show that the ancestors of the Cree and Ojibwa currently inhabiting the

Figure 5.5   Indian fish weir, northern Ontario, n.d. *Source*: National Archives of Canada, PA43316

Treaty 9 and Robinson Treaty territories engaged in commercial fisheries well before these treaties were signed in the mid-nineteenth century. Sturgeon and whitefish were the key species harvested, and Aboriginal people bartered them fresh, frozen (during the winter), or dried and smoked (all varieties hot smoked for human consumption; whitefish cold smoked for dog teams).

In the eighteenth century the commercial harvest was on a comparatively small scale. During the nineteenth century this commerce grew through barter sales and contracts. The HBC also hired Aboriginal men and women to make and repair fishing gear, to locate and operate company fishing sites, and to pack, transport, and process (dry, smoke, or salt) fish. When government agents and others began arriving later in the century, Aboriginal people sold them fish as well. Besides making and repairing fish nets and selling and processing fish for food and dog teams, Aboriginal women of the district specialized in manufacturing isinglass from the swim bladders of sturgeon for an expanding export market. All of this leads to the conclusion that the commercial use of fish by Aboriginal peoples was a well-established tradition in the areas throughout present-day northern Ontario before treaties were signed and that the women as well as the men played significant and diverse roles in those fisheries.

NOTE

1 This term also included a wide variety of non-food items, such as canoes and canoe supplies (birch rind, spruce root, and wattap), snowshoes, hides, and babiche (leather strips used for cord).

WORKS CITED

Anderson, J.W. 1961. *Fur Trader's Story*. Toronto: Ryerson.
Bishop, Charles A. 1974. *The Northern Ojibwa and the Fur Trade*. Toronto: Holt, Rinehart and Winston of Canada.
Holzkamm, Tim, Victor Lytwyn, and Leo Waisberg. 1988. 'Rainy River Sturgeon: An Ojibway Resource in the Fur Trade Economy.' *Canadian Geographer* 33, no. 3: 194–205.
Holzkamm, Tim, and M. McCarthy. 1988. 'Potential Fishery of Lake Sturgeon (*Asipenser fulvescens*) as Indicated by the Returns of the HBC Lac la Pluie District.' *Canadian Journal of Fisheries and Aquatics Science* 45: 921–3.
Hudson's Bay Company (HBC). 1929. Attawapiskat Post Journal, 1929. Outift 260. Ms. in possession of the author.
Hudson's Bay Company Archives (HBCA), Provincial Archives of Manitoba (PAM):
– Abstract of Fur Returns [London sales], 1840–1850, A 50/1–13.
– Account of Fur Sales, 1822–1904, A 51/41–53.
– Accounts of Fur Sales: Fur Sale Books, 1822–1907, A 51/1–54.
– Accounts of Fur Sales: Fur Sale Books, 1872–1923, A 48/13–26.
– Fort Albany Account Books, 1692–1805, B 3/d/1–118.
– Fort Albany Post Journals (FAPJ), 1890–95, B 3/a/197–201.
– Fort Hope Post Journals (FHPJ), 1890–1903, B 291/a/1–5.
– Fort Severn Post Journals (FSPJ), 1873–97, B 198/a/123–8.
– Henley House Post Journals (HHPJ), 1873–75, B 86/a/69–71.
– Kenogamissi Post Journals (KPJ), 1794–95 B 99/a/1.
– Martin Falls Post Journals (MFPJ), 1869–99, B 123/a/85–93.
– Osnaburgh House Account Book, 1853–54 Indian Debt Book, B 155/d/1–9.
– Osnaburgh House Journals (OHPJ), 1871–77, B 155/a/80–3.
Lytwyn, Victor. 1993. 'The Hudson Bay Lowland Cree in the Fur Trade to 1821: A Study in Historical Geography.' PhD dissertation, University of Manitoba.
National Archives of Canada (NA). 1906. RG 10. Department of Indian Affairs, 1906, vol. 3033, file 235.225, pt 2.
Ray, Arthur J. 1998. *Indians in the Fur Trade*. 2nd ed. First pub. 1974. Toronto: University of Toronto Press.

– 1990. *The Canadian Fur Trade in the Industrial Age.* Toronto: University of Toronto Press.

Rich, E.E., ed. 1957. *Hudson's Bay Copy Booke of Letters Commissions Instructions Outward, 1688–1696.* London: Hudson's Bay Record Society.

Williams, Glyndwr, ed. 1969. *Andrew Graham's Observations on Hudson Bay.* London: Hudson's Bay Record Society.

6

Depletion by the Market: Commercialization and Resource Management of Manitoba's Lake Sturgeon (*Acipenser fulvescens*), 1885–1935

FRANK J. TOUGH

Freshwater fisheries were crucial to Native societies in the subarctic before and during the fur trade (Holzkamm, Lytwyn, and Waisberg 1988; Lytwyn 1990; Ray, this volume; Tough 1984; Van West 1990). Lake sturgeon was one of the most important species harvested before and after European contact, and Manitoba's fisheries were among the most productive in the subarctic. Cree and Ojibwa people conducted their main sturgeon fisheries near the spawning grounds in early summer and during the autumn river runs, and there is evidence that Cree at the north end of Lake Winnipeg also fished sturgeon in open water rapids in late winter (Hudson's Bay Company Archives [HBCA], Provincial Archives of Manitoba [PAM], Norway House Post Journal [NHPJ], B 154/a/71, 26). Sturgeon served remarkable multi-purpose use for Indians. Its flesh, oil, and eggs provided a high-calorie food and a source of protein and essential vitamins. Indians made sturgeon-skin jars for storing oil, produced a type of fish pemmican from fire-dried sturgeon mixed with fish oil, and created a paint base and glue from the inner membrane of the swim bladder. After contact, subarctic fur traders depended on sturgeon and whitefish (*Coregonus clupeaformis*). Sturgeon was important to Natives and European traders for food; isinglass, a significant by-product made by Native women from sturgeon swim bladders, became an article of trade at fur-trade posts (see Ray, this volume). Quite clearly, sturgeon was a significant resource during the long occupation of the subarctic, but the demise of sturgeon stocks after treaties were signed requires elaboration and explanation.

The demise of many sturgeon fisheries, associated with the advent of commercial fishing, has been noted by fisheries biologists. Like sturgeon fisheries in Ontario, Lake Winnipeg's yields collapsed within a few decades after the commencement of commercial exploitation. Fisheries biologists working for the Ontario government, Harkness and Dymond (1961: 81), had ample evidence to declare: 'There is no more striking example of the reckless and wasteful exploitation of the natural resource than that of the sturgeon.' Accounts of population dynamics, and descriptions of the demise of fish stocks, by fisheries biologists seldom consider the political and economic forces associated with the collapse of sturgeon fisheries. Despite the short duration of the high-yield commercial fishery, sturgeon played a significant role in the history of the freshwater industry. The geographical spread of the market-oriented fishery occurred because of the pursuit of the high-priced sturgeon (Figure 6.1). In contrast to the notion that Native people played no part in the Canadian economy following the fur trade, the economic history of the freshwater fishery documents Native involvement in a capitalist fishery.

Even though sturgeon has long ceased to be a major commercial catch in Manitoba, its demise raises issues of resource management and offers useful lessons. I argue that the particular failure of resource management to sustain sturgeon populations was the result of two main factors: the inability of sturgeon to respond reproductively to overfishing, and the relatively higher prices for sturgeon and caviar. None the less, we must look at these management problems within the context of the political economy of a capitalist industry and an ineffective regulatory system. This chapter concludes with an examination, in light of the historical evidence, of some of the assumptions of models employed by fisheries economics.

Sturgeon Population Dynamics

The biological literature has long noted the strikingly negative impact of fishing pressures on sturgeon populations (Ecologists Limited 1987; Houston 1987; Oliver 1987; Patalas 1988; Sopuck 1987; Sunde 1961; Threader 1981). For Ontario fisheries, Harkness and Dymond (1961: 67) observed that 'a distressing feature of sturgeon fisheries is the rapid decline in yield which occurs when populations are fished. The result has always been the same – relatively high initial yield, followed by a sudden and permanent decline to very low levels.' Presumably with respect to freshwater fish, they observed: 'Sturgeon fisheries are unique in showing a sudden and permanent decline to very low levels' (Harkness and Dymond 1961: 72). Although the pattern of the collapse of these fisheries is well known, the precise population dynamics is obscure because historical production records do not include subsistence yields.

The inability of sturgeon fisheries to support extensive and ongoing commercial exploitation can be explained in part by the relative ease of capture and this species' limited reproductivity. Sturgeon reach sexual maturity much later than

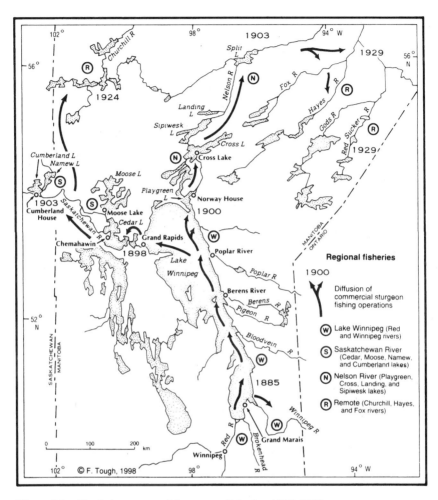

Figure 6.1   Manitoba commercial sturgeon fisheries, 1885–1935

any other freshwater fish. For the Nelson River sturgeon, males mature in fifteen to twenty years, but females take twenty-five to thirty-three years (Scott and Crossman 1973: 86). Moreover, spawning does not always occur immediately once sturgeon have matured. As well, sturgeon do not spawn every year. Spawning intervals are two to three years for males and four to six for females. During May and June, sturgeon leave the lakes to spawn in swift river waters at depths of two to fifteen feet (0.61 to 4.6 metres). Congregating in shallow waters during the spawning season, these fish are easily exploited. While sturgeon are quick to leave the spawning grounds at the sign of danger, their eggs quickly deteriorate unless deposited. Intensive exploitation of spawning grounds would have to continue for only several years before much of the population would be captured and future propagation threatened. Nevertheless, female sturgeon produce eggs in great numbers; mature fish register about five thousand eggs for each pound of body weigh. Lake sturgeon are large. A 275-lb (125-kg) sturgeon has been recorded for Lake Winnipeg, and some grow to six feet (1.8 metres) (Harkness and Dymond 1961: 8). The size and tendency of lake sturgeon to keep to shallow water make it both desirable and relatively easily captured; hatcheries and artificial propagation have not successfully provided any compensation for losses incurred by fishing.

Thus sturgeon fisheries are easily conducted, but late maturity and infrequent spawning set severe limits on the level of sustainability. The biological constraints on reproductivity are vital aspects of sturgeon fisheries, but a full consideration of management possibilities must be informed by the economic history of the exploitation of this species. The broader economic context of fisheries production, not just a narrow, technical, biological consideration of yields, is also required.

Commercialization of Lake Winnipeg Fisheries

In the 1880s, strong American demand for whitefish led to development of the Manitoba commercial fishery (Tough 1984). Lake Winnipeg was the main fishery. In contrast to the traditional fishery, in which Natives fished for their own consumption and for exchange in response to the provision needs of the fur trade, the commercial operation that developed after 1880 had the characteristics of frontier capitalism: dependence on export market (in this case, the large American market), increasing intensity of capital, paternalistic labour relations, and reckless rates of resource extraction.

The efficiency of capital-intensive companies, operating initially in an unregulated environment, raised the concerns of many residents of the lakes. Many Natives, vital to the progress of the industry, were alarmed at the scale of commercial exploitation, and Lieutenant-Governor John C. Schultz obtained independent evidence that fishing companies threatened traditional Native livelihoods. The new commercial fishery did not offer a viable alternative to traditional fisheries because companies provided wage labour mostly to young males, not to older men, women, and children, who had difficulty obtaining subsistence from a near-shore fishery that had previously been reliable. From the start, these companies encroached on this essential resource of Native communities. Some members of the Berens River Indian band, in turn, had threatened to cut the sturgeon nets that had been set on Indian near-shore fisheries. In 1890, this situation forced Schultz to visit the band to listen to its problems with commercial fishing companies. Chief Jacob Berens explained to Schultz:

We still go down to set out nets but the larger nets outside of the river have caught so many fish that little remains for us, and sometimes our children cry for food. Why does the white man with his large nets and miles of nets, and his steam boats to tow his other boats, not go out north into the deep waters of the lake where we cannot go with our small boats and nets, and catch fish there? Why does he come to spread his nets just at our feet, and take away the food from our children's mouths? Our people's hearts are sore for the last two years. We have complained and complained and still the big fishermen comes and we see only starvation for our children in front of us. When we made this Treaty, it was given us to understand that although we sold the Government these lands, yet we might still hunt in the woods as before, and the fish and the waters should be ours as it was in our Grandfathers' time ('Notes on Indian Council,' *Native Studies Review* 1987: 119)

Clearly the right to fish, expected from the treaty, was encroached on by the well-equipped commercial companies.

Continuing pressure from Indians and Indian agents, and lobbying from Schultz, resulted in fisheries official Samuel Wilmot's investigation of the Lake Winnipeg fishery in 1890. Wilmot agreed with Indians that the prevalent method of fishing was resulting in 'a gradual but steady depletion of the white-fish product of Lake Winnipeg going on, from the effects of the present system of fishing' (Canada, *Annual Report of the Department of Fisheries, Sessional Papers [SP]* 1891, 8: 61). He suggested closing whitefish spawning grounds to commercial fishing, thus protecting the fish from the companies, which were known to fish the approaches to spawning runs. His report recommended a more regulated industry, with restrictions on fishing effort and a distinction made between commercial and domestic licences (see Newell 1993: 73 for similar recommendations on the Pacific coast in the same decade).

Wilmot's concern for the conservation problems caused by unregulated fishing companies was not matched with an interest in promoting Indian commercial fisheries. When asked to respond to the advisability of providing Indian bands with more capital, he articulated a position that restricted Indian involvement in the capitalist frontier: 'It would be undesirable that Indians should be supplied with large boats and longer nets in order to fish in open or deeper parts of the lake. If the Indians desire to fish in waters outside their reserves, or other waters set apart for them, they place themselves in competition with other fishermen, and should therefore make their own provision for such outside fishing'

(*SP* 1891, 8: 62). In the early stages of the Manitoba commercial fishery Indians lost the opportunity to accumulate wealth from this resource. Their involvement would be restricted to selling fish and labour to the market.

The resulting management regulations, designed to sustain biological yields, reflected a shift from Native to non-Native political and legal control over fisheries wealth. The 1894 dominion fisheries regulations for Manitoba reflected Wilmot's effort to accommodate both an export-oriented commercial industry and a local, traditional fishery. These regulations provided separate annual commercial and domestic licences, closed seasons for commercial licences during spawning (15 May to 15 July was closed to sturgeon fishing), penalties for traders obtaining fish during closed seasons, limitations on the amount of gillnet allowed for each type of licence, and exclusion of commercial companies from near-shore spawning areas. However, the highly efficient poundnets, a destructive technique when used for sturgeon, were not banned. These regulations recognized the general right to fish for subsistence purposes: 'Every farmer, settler, or *bona fide* fisherman, or half-breed, who is an actual resident of the locality where he proposes to fish, shall be entitled to a "domestic license"' (*Canada Gazette* 1894). Commercial and subsistence fisheries were now formally recognized and regulated. The disruptions brought by commercialization led to creation of a legal regime that would have general application: 'Indians and half-breeds, as well as settlers and all other persons' (*Canada Gazette* 1894). In these regulations, Indian fishing rights were obscured: the fisheries minister 'may from time to time set apart for the exclusive use of the Indians, such waters as he may deem necessary, and may grant to Indians or their bands, free licenses to fish during the close season, for themselves or their bands, for the purposes of providing food for themselves, but not for the purpose of sale, barter or traffic' (*Canada Gazette* 1894). Regrettably, exclusive fishing reserves were not established. Given the importance of fish to these bands, recognition of exclusive reserves protecting Indian fisheries would have been as important as the treaty provision for surveying land reserves. Significantly, the imposition of restrictive regulations was not considered during treaty negotiations (Tough 1996: 234–39). However, holders of domestic licences could sell fish.

Management efforts to protect fish reproduction during spawning runs under a 'market regime' would be pointless as long as companies could obtain fish through trade with Indians. Thus in the late 1890s pressure was put on Indians to stop traditional fishing during the various closed seasons. For several decades, Indian requests for reserves were ignored by the dominion fisheries department. Considerable wrangling went on between the departments of Indian Affairs and of fisheries over a variety of Indian fishing issues. Despite their authority to set aside reserves, fisheries officials held the view that Indians did not need fishing reserves in areas lacking significant white settlement, whereas in areas of white settlement an Indian fishing reserve would interfere with the 'public' right to fish. The outcome was that the traditional mode of fishing was marginalized without Indians receiving the compensation of exclusive fishing reserves or the means to engage in large-scale commercial fishing. Though Indian political pressure had forced the state to acknowledge the problems cause by wide-open commercial fishing, restrictions on the traditional fishery were created. A new market for fish created an incentive for some Indians to harvest during the spawning season, thereby expanding traditional harvest levels.

Wilmot's recommendations and subsequent efforts at enforcement curtailed Native fisheries while commercialization of Lake Winnipeg fisheries continued unabated. Commercial production expanded in the period from 1890 to 1904 (Tough 1980). A record harvest of 7.5 million lb. (3.4 million kg) of whitefish was achieved in 1904. By 1907, whitefish harvests had fallen to 2.35 million lb. (1.1 million kg). The 'fishing-up' effect – the removal of the natural stock or resource 'capital' – was very marked for the Lake Winnipeg fishery. Expanded production in the 1890s was achieved by a similarly rapid capitalization of the industry. After 1893, the investment in steam tugs and shore stations (wharves, ice-houses, and freezers) surpassed the investment in fish boats and nets. The trend towards greater investment in large vessels and plant indicates the increasingly capitalistic nature of the industry. The shift of production to fisheries more remote from shipping centres contributed to the need for more capital. Also, the alleged overexploitation of the southern fishing grounds necessitated the shift of harvesting and shore facilities to the north end of the lake. Overall, the industry required more capital because yields were declining relative to effort. In 1892, each dollar of investment yielded about seventy pounds (thirty-one kilograms) of whitefish; by 1910, yields had dropped to ten pounds (4.5 kilograms) per dollar (Tough 1980: 72).

The need for more capital led to business concentration in the late 1890s. Captain William Robinson formed the Dominion Fish Company in 1898 through a forced merger of the other major Lake Winnipeg fishing companies. According to F.W. Colcleugh, fisheries inspector for Manitoba, the smaller firms had to submit to the inevitable monopoly: 'If they refused to join the syndicate and sell the plant and business ... they would be "frozen out,"' and moreover, once they sold out, they 'were compelled to sign a bond not to' enter the fish business again for a period of ten years' (NA, RG 23, vol. 112, file 110 (2), 2 May 1899). It turned out that Dominion Fish was owned by the American firm A. Booth Packing Company of Chicago. Reorganization of the industry in the late 1890s increased American control over this major local resource and guaranteed the vast American market supplies of freshwater fish. The merging and absorption of the Lake Winnipeg companies coincided with expansion of the Booth interests in the United States. Thus, in this period of development, business concentration on Lake Winnipeg had the backing of the powerful Booth corporation (Tough 1980: 59–63).

The merging of firms not only focused the capital on Lake Winnipeg but also resulted in lower fish prices and declining incomes for fish producers. In the case of sturgeon, the organization of the Dominion Fish Company allowed it to cut production costs by slashing the prices paid to producers for fish. Prior to the merger, companies paid $1.50 to $1.75 for a dressed sturgeon and $1.00 a pail (containing twenty pounds or nine kilograms) for sturgeon eggs or caviar. The companies sold caviar for seventy-five cents a pound (thirty-five cents a kilogram), or $15.00 per pail, which represented an enormous mark-up. (Fish producers were, however, allowed to keep the swim bladders and oil.) With the oligopolistic situation created by the Dominion Fish and the Fryer and Ewing companies, sturgeon were purchased in the round (undressed) for $1.25. The two companies were able both to lower the price of sturgeon and to obtain the valuable eggs at no extra cost (NA, RG 23, vol. 112, file 110 [2], 2 May 1899). The essential structure of the industry, based on a dependent relationship among fish producers, foreign capital, and an external market, was consolidated after Wilmot's investigation. Regulations proved to be a timid response to the changes wrought by rapid commercialization. This region was not a backwater of a stagnant fur trade: the bountiful fishing resources had attracted foreign capital, which further increased the market orientation of local economies.

Commercial Sturgeon Fisheries

During the late-nineteenth-century consolidation of commercial interests, spatial expansion (largely a response to diminishing returns from discrete fisheries that made up the Lake Winnipeg fishery) and intensification of sturgeon fishing occurred. Figure 6.1 reconstructs Manitoba's commercial sturgeon fisheries for 1885–1935; for Lake Winnipeg, sturgeon were confined to the east shore and were generally caught at river mouths. Commercial sturgeon fishing had begun at Grand Marais and Pigeon River in the 1880s, but the pursuit of sturgeon quickly spread north to the lower Saskatchewan and the Nelson rivers. Figure 6.2 charts the production of the larger regional sturgeon fisheries for 1885–1935, but the published data often aggregated information from distinct fisheries. In the early years, estimates for the export market and home consumption were combined. In 1900–1 and 1906, the Nelson River catch was included in the Lake Winnipeg yields. None the less, these data provide a reconstruction of

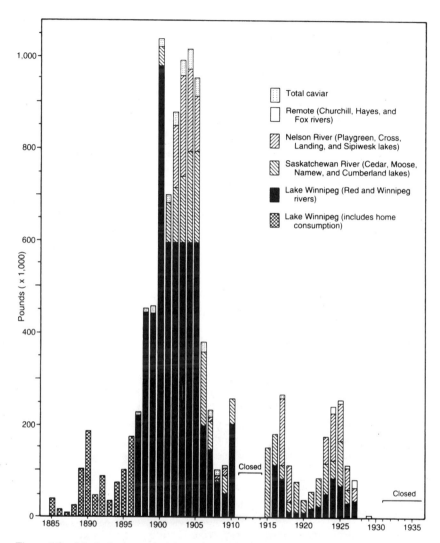

Figure 6.2   Manitoba commercial sturgeon yields, 1885–1935. *Sources*: Canada, *Annual Report for the Department of Fisheries*, 1885–1916, *Sessional Papers*; Canada, Dominion Bureau of Statistics, *Fisheries Statistics*, 1917–35; Canada, *Annual Report for the Department of Fisheries*, 1924–9.

Note: In 1900, 1901, and 1906, Nelson River catch was probably included with Lake Winnipeg. For 1919–23 Nelson River catch was included in the The Pas district (Saskatchewan River). The 'remote' fishery for 1929 includes Landing Lake (Nelson River). These data do not include Buffalo Bay (Lake of the Woods).

overall production trends, and inferences can legitimately be made about the spatial dynamics of the commercial fishery.

Fisheries Inspector La Touche Tupper noted in 1897 that 'there was a large increase of sturgeon fishing' and that 'the industry is gradually creeping up the east shore' of Lake Winnipeg (*SP* 1899, 11A: 208). The rapid increase in production can be understood from the relatively higher prices for sturgeon and caviar. Recorded caviar production reached four thousand pounds (eighteen hundred and fourteen kilograms) by 1897 (*SP* 1899, 11A: 207). In that year, the value of sturgeon and its products had doubled and tripled, respectively, while the price of whitefish had declined. In 1900, for example, Indians received $1 for each sturgeon, but companies paid only two-and-a-half cents for a whitefish. The per-pound value of sturgeon and whitefish did not vary greatly, but sturgeon were much larger than whitefish and sturgeon eggs added value (fifty cents per pound, or twenty-three cents per kilogram). By 1905, the price spread had increased: whitefish was worth seven cents a pound (three cents a kilogram), sturgeon was priced at ten cents a pound (four and one-half cents a kilogram), and caviar doubled to one dollar per pound (forty-five cents per kilogram) (*SP* 1901, 27: 109; 1902, 22: 166; 1906, 22: 194). Harkness and Dymond (1961: 56) noted the escalation of caviar prices: 'from $9 to $12 per keg of 135 lbs., in 1885; was worth $20, five years later; $40, in 1894; and before the end of that decade it had risen above $100 per keg.' Indicative of the comparative high value of sturgeon, Houston (1987: 182) reported 1987 prices for lake sturgeon caviar as $250 per kilogram and for smoked sturgeon, $39 per kilogram. With price increases, commercial production of sturgeon expanded and dwarfed the estimated home consumption. Even if the published data for home consumption understated actual harvests, commercial production of sturgeon for export had come to dominate the regional fishery.

When, at the end of the nineteenth century, sturgeon and caviar production spread beyond Lake Winnipeg, Native people in the affected areas did not outright oppose the commercial fishery. Nevertheless, the fisheries inspector, E.W. Miller, reported on the industry in the lower Saskatchewan River basin: 'The fishermen themselves petitioned for the closing of the fishery for the summer fearing the intrusion of outside men,' which resulted in a 'feeling leading to somewhat exaggerated statements being made as to the rapid depletion of the lake' (*SP* 1901, 22: 153). He placated them: 'Licenses were subsequently issued to permanent residents, only permitting them to take sturgeon during the winter season, when no fish are wasted and a far better price can be obtained by the fishermen' (153). Because Miller now planned to enforce a closed season on Indians, participation in the industry came as trade-off for Native people. Opposition re-emerged in 1908, when a large company moved to monopolize the local fisheries. Cumberland House Natives, assisted by Native clergyman J.R. Settee, protested 'the wholesale slaughter of our fish' and 'found it very strange that an American [Captain Coffee] can be allowed to deplete our waters of fish' (NA, RG 23, vol. 363, file 3216, 1, 16 March 1908). They could not oppose commercial harvesting as such, indeed, they wanted the benefits of commercial production. Rather they sought to protect their long-term interests by requesting 'a lease or power to have the full control to fish and sell fish to [for] the half-breed [Métis] & [and] Indians inhabit by this part of the country' (16 March 1908). Fishing sturgeon for the external market thus took hold of the Saskatchewan River, but not without sufficient involvement of local Natives.

To facilitate the spatial expansion of the industry, the dominion fisheries department as a matter of policy favoured local residents. It licensed Natives for sturgeon fishing for the export market, and, because domestic-licence holders could sell fish to commercial companies, the distinction between the two types of licences was not significant. In this respect, the domestic licence provided for a continuation of the traditional fishery. La Touche Tupper explained how Native labour was incorporated into a capitalist fishery and management regime: 'I only licensed residents, and near the Indian reserves at Berens River, Bloodvien, I only issued to the Indians of the reserve, much to their benefit and satisfaction. They all took out licenses and strictly observe the law. The Berens River chief personally sees that all the nets are taken up Saturday; all offal disposed of, and only the proper number of yards of twine used' (SP 1899, 11A: 208). Compliance with regulations occurred with the support of the chief, and consequently Indians got more involved with the industry. Of the 180 sturgeon licences issued for the Nelson River in 1903, all but four went to Natives. Similarly, the department granted fishing licences for the lower Saskatchewan in 1903 only to local residents. Because of skills and knowledge, Native involvement benefited the industry. For example, in 1923, Fisheries Department records reported of the Nelson River fishery: 'Nearly all Indians fishing in the upper river. The white men do not seem to do very good fishing in the swift waters and the Indians seem to know the very few places where nets can be set. Hooks are used very little and the Indians say that in about two weeks nets will be too dirty to fish' (NA, RG 23, vol. 526, file 711-12-3, 10 July 1923).

By the turn of the century a number of Ojibwa and Swampy Cree communities were sturgeon fishing for export. Poplar River, Norway House, and Cross Lake bands were among the communities who had also became engaged in sturgeon fishing by 1900. The inspector of Indian agencies, E. McColl, reported for 1900: 'I found the Indians of the Norway House Band scattered for seventy miles around Playgreen Lake busily engaged in fishing for sturgeon. They also make a comfortable livelihood at the fisheries, having caught about ten thousand sturgeon and over a hundred thousand whitefish' (*SP* 1901, 27: 109). Further north, McColl noted, 'The Indians of Cross Lake Band are doing a thriving business of fishing for sturgeon and whitefish this season' (108). The market-oriented fisheries were one of the attractions that had caused Natives from northeastern Manitoba to migrate to the Norway House–Cross Lake area (Tough 1996: 223). In 1900, the Moose Lake and Chemawawin bands, for example, were clearly involved in the commercial fishery and made 'a good deal of money from sturgeon fishing and the sale of caviar' (*SP* 1901, 27: 104). When describing the industry, the dominion commissioner of fisheries, E.E. Prince, mentioned the companies, settlers, Icelanders, and 'Indians and half-breeds who fish from their reserves, largely for food, but also for sale, especially sturgeon' (*SP* 1910, 22: lix). The incorporation of Natives in the market-oriented fishery by 1900 seems to have significantly reduced the type of opposition expressed at Berens River in 1890.

Additional fisheries of the Lake Winnipeg drainage system were brought under the sphere of the market because of the need to exploit new, unsullied sturgeon fisheries. The introduction of commercial fishing to the north often began with selective fishing for the high-priced sturgeon. Miller observed that 'the great demand for this fish [sturgeon] has caused the fish companies to push their operations' into the lower Saskatchewan and that the sturgeon was 'the only fish, which under existing conditions of transportation, could be profitably marketed' (*SP* 1904, 22: 202). By 1903, sturgeon fishing had moved west from Cedar Lake to Cumberland Lake. The firm of Ewing and Fryer found sturgeon fishing on the Nelson profitable enough to develop a transportation system employing York boats, a gasoline boat, and a steam tug to transship over rapids all the way upstream, from Sipiwesk Lake to a station at Spider Island in Lake Winnipeg. As with other profitable staples in Canada, frontier capitalists found the means to deal with the geographical challenge of moving commodities. Sturgeon fishing rapidly moved down the Nelson River.

Eventually the industry moved out of the Lake Winnipeg drainage system altogether, because depletion led to the closing of Lake Winnipeg to sturgeon fishing in 1912. In 1924, the Churchill River near the Manitoba–Saskatchewan border was commercially fished for sturgeon. Sturgeon prices were high in the 1920s, when the average price per pound was fifty cents (per kilogram, twenty-three cents) (Canada 1924–5, 1926–7, 1928–9). By 1929, the commercial 'frontier' for sturgeon had reached the northeast corner of Manitoba – to the Fox and Gods rivers and Red Sucker Creek (Figure 6.1). The rapid exhaustion of sturgeon fisheries and high prices continued to encourage steady diffusion of sturgeon fishing into ever more remote areas for a large, open-ended market. In effect, the commercial frontier for sturgeon stopped at the northern limit of the range of this species. And, because sturgeon initially had been harvested selectively on a much larger 'fish frontier,' transportation and infrastructure developments had the spin-off effect of stimulating exploitation of other freshwater fish. Thus the economic significance of sturgeon for the industry outlasted the local collapse of its yields.

Given the population dynamics of the sturgeon and the huge market values for sturgeon and caviar, the rise and fall of production, as displayed in Figures 6.1 and 6.2, can easily be grasped. Despite the data problems, it is evident that the boom for sturgeon fisheries occurred between 1897 and 1906. Yields were dominated by Lake Winnipeg, though significant production was obtained from the Saskatchewan and Nelson rivers. In later years, a major part of Lake Winnipeg's production was obtained from the Winnipeg River fishery. Sometimes high transportation costs meant that northern fisheries were at the profit margin, and consequently, low yields for the 'remote' fisheries occasionally indicate that the cost of effort was too high relative to return to encourage producers to pursue the fisheries of these areas. Production was not sustained over the long term. Not surprisingly, one of the major sturgeon producers, Ewing and Fryer, went out of business. By 1910, Ewing had torn down his buildings on the Brokenhead Reserve and moved to St Paul, Minnesota (NA, RG 10, vol. 8046, file 501/32-3-4-3, n.d.).

Government fisheries managers were aware both of the importance of sturgeon to Indians and of the legacy of depleted sturgeon fisheries elsewhere in Canada. E.E. Prince identified the essential problem: 'An American came to Manitoba a few years ago and bought sturgeon and other fish, buying chiefly from Indians. He strictly kept the law in all his dealings, but he was able to clear $4,000 in a single year on sturgeon. No doubt this money is lost to Canada, and our Indians got no proper share of it, and if the system continues the country will have to feed the Indians and their families' (*SP* 1905, 22: lxii). Prince thus predicted that Native participation in a system of unequal trade would lead to a loss of economic self-sufficiency. Indians were becoming involved with the market and producing for export, but short-term increases in cash income led to an increased market orientation for the local economy. Despite the relatively higher prices for sturgeon, Prince argued, Indians did not get a fair share of the value of the resource. Low incomes for producers were a defining characteristic of this industry (Tough 1980).

Problems of Resource Management

Regulation schemes and local fisheries officials were incapable of protecting lake sturgeon. Their approaches towards fishing for the export market were inconsistent. In 1898, for instance, Inspector of Fisheries Miller had suggested that for Cedar Lake 'the development of the fishery at this point for export purposes is considered to be prejudicial to the interests of the resident population' (*SP* 1899, 11A: 215). Yet shortly thereafter he licensed commercial producers and permitted use of poundnets. Officials were overly optimistic about the sustainability of sturgeon under commercial pressures and about immediate economic returns. Similarly, with respect to Nelson River, Miller stated in 1904: 'A feeling exists in some quarters that allowing fish to be sold from these waters will endanger the food supply of the Indians, but the extent of water is so great and the resident population so small, that this fear is groundless, and from a material point of view, the existence of a market for fish cannot fail to be of great benefit to the Indians and encourage industry among them' (*SP* 1905, 22: 208). The next year Miller stated: 'A visit to the Nelson river country is sufficient to dispel any fear that it is possible for these waters to become depleted until the number of available fishermen is immensely greater than at present' (*SP* 1906, 22: 196). Miller's response to those concerned about resource depletion was typical of frontier capitalism – a belief that the resource was too bountiful to be exhausted. The expression 'to encourage industry' among Natives was an old cliché: 'Industry' meant 'work ethic.' The licensing of Natives to fish commercially for sturgeon not only facilitated the geographical expansion of the market but could also be rationalized in terms of cultural assimilation and short-term material prosperity.

Local fisheries officials had inadequate knowledge of the sturgeon, and they were unable to perceive the signs that populations were under stress. At the turn of the century they optimistically reported that the Winnipeg River was 'teeming with young sturgeon' (*SP* 1906, 22: 192). In 1910, the overseer was delighted that very little caviar had been manufactured out of the 173,800 lb. of sturgeon caught from the Winnipeg River, which 'would go to show that fish were not parent fish' (*SP* 1912, 22: 283). This 'very large catch of small sturgeon' did not alarm W.S. Young, inspector of fisheries, who apparently thought that catching young, immature sturgeon did not threaten future propagation (*SP* 1912, 22: 282). Similarly, Overseer D. McEwen reported: 'There was produced and exported to the United States, one thousand seven hundred and thirty-eight cwt. of sturgeon which, weighed in a dressed condition, averaged about ten or twelve pounds to the fish, and the fact that out of so large a production only twenty-four cwt. of caviare [*sic*] were manufactured, would go to show that the fish were not parent fish. While it is true some large sturgeon were taken, yet they were the exception and not the rule' (282–3). These observations strongly indicate that the age structure of sturgeon population had been disrupted and that smaller, immature fish had become the major component of the catch, because most of the older, sexually mature ones had already been caught. Given the late reproductivity of the sturgeon and the infrequent spawning, a population of young fish would be highly vulnerable. The inability of managers to foresee the consequence of a fishery dominated by young sturgeon is indicative of a failure of management.

If officials lacked knowledge to manage the resource scientifically, the problem of sustainabilty was further complicated by lack of enforcement. Despite the existence of a set of regulations going back to Wilmot's 1890 investigation, effective enforcement on the lake cannot be assumed. The sturgeon closed season could be circumvented. Sturgeon and the valuable eggs caught in the closed season could be kept alive in pounds until the open season (SP 1905, 22: 208). Overseer W. McEwen reported that Indians on the Pigeon River caught sturgeon during the closed season and kept the fish alive until they were purchased by dealers (SP 1897, 11A: 214). A candid comment made in the 1911 report of the dominion's 1909–10 Commission for Manitoba and the North West (Manitoba Fisheries Commission) suggests that enforcement efforts may have been somewhat incidental to the development of the industry: 'Moreover, many of the fishery officers have no aptitude for effective patrol duties, while the poorly paid fishery guardians, in many cases, do practically no duty at all, some of them being, apparently, not acquainted with the regulations, and, of course, make no attempt to enforce them, often for the reason that they are ignorant of what the law is' ('Report of the Manitoba Fisheries Commission' 1911: 38–9).

The absence of effective state management meant that the commercial companies controlled the fishery. The commissioners claimed that the decline of sturgeon was worse than the published data would suggest ('Report' 1911: 34A). Unlike some of the local fisheries managers, fisheries officials in Ottawa were aware that commercial exploitation elsewhere had proven harmful to sturgeon stocks. In 1905 E.E. Prince's research on lake sturgeon, which addressed issues of commercialization and depletion, was published in the *Sessional Papers*, but some ten years earlier he had warned that 'no fish is so easily depleted by reckless methods, as the exhausted United States sturgeon fisheries demonstrate' (SP 1905, 22: lv). To account for the situation of Manitoba sturgeon fisheries in 1905, Prince cited an observation made several years earlier by 'an exceptionally able and well-informed,' but unnamed dominion official: 'It will be hard to keep outsiders from the sturgeon waters to the north. Commercially the sturgeon are in demand and will be sought after more and more every day. The Fisheries Department in Ottawa would be in a better position to resist the greed of outsiders, who wish to clean out the sturgeon and make money fast in these northern waters, by taking into account the just claims of the Indians, and their future, both as regards food and appropriate employment' (SP 1905, 22: lxii). In this instance, the regulation of outside forces, in order to maintain the interests and claims of Indians, was advocated.

The *Interim Report* (1909: 5) of the Manitoba Fisheries Commission recorded that 'sturgeon, the most valuable fish found in these waters, is on the point of extinction and can only be preserved by stringent measures' (5). The *Interim Report* wisely recommended: 'No sturgeon or caviar shall be exported from the province of Manitoba on and after January 1, 1910' (10). Had the government closed off the external market at that time, sturgeon would probably have reverted back to a traditional resource for local consumption. No export prohibition was ever implemented, however. Sturgeon yields continued to decline, and Lake Winnipeg finally had to be closed to sturgeon fishing in 1912. Fishing effort shifted to the smaller northern fisheries after the closing of Lake Winnipeg, but production soon peaked at 146,800 lb. (66,576 kg) in 1924. The frontier beyond the Lake Winnipeg fishery was not as bountiful.

Rather than come to terms with the economic impulses that encouraged overfishing, resource management practices following the dominion royal commission of 1909–10 included annual yields limits for each fishery and the periodic closing of a lake to commercial fishing. In 1917, the government reopened Lake Winnipeg as a wartime measure, but so depleted were its stocks that the fishery produced only 555,900 lb. of sturgeon over the next twelve years – about equal to one year's harvests during the 1885–1910 period (Figure 6.2). Management officials did not seem to pay close attention to the harvest data they collected, reporting in one year that depletion was not occurring, only to 'discover' that, in the next, 'There seems reason to fear that here also this valuable fish may be facing the fate it has so generally met with elsewhere' (Canada 1924–5: 46; 1926–7: 49–50). Closing the lake was contemplated in 1925, but it did not happen until 1928.

In the decades that followed the royal commission, management did not resolve the problem of sturgeon fisheries. In 1926, dominion Inspector of Fisheries J.B. Skaptason wrote: 'The rapid decrease and depletion of the sturgeon fisheries of North America during the past 40 or 50 years, or since its commercial value became pronounced, is one of the most serious problems that have faced culturalists and biologists for a quarter of a century or more. A great many attempts at artificial propagation of this valuable fish have met with almost total failure' (Skaptason 1926: 34). The inability to propagate sturgeon in a hatchery made this species different from the whitefish and pickerel (*Stizostedion vitreum*), the other mainstays of the industry (Tough 1980). Skaptason also argued that management had been effective: 'A period of extensive pound net fishing made heavy inroads into the sturgeon' of the Saskatchewan River, but 'had it not been for the timely intervention of the Fisheries administration, we would no doubt have witnessed the depletion of another of our great sturgeon waters' (Skaptason 1926: 17).

Given that fisheries officials had originally approved use of poundnets, rationalizing them as the nucleus of a valuable fishery, despite the opposition of Natives, the praise seems somewhat misplaced. Effective management would not have permitted poundnets at all. As it was, effort was restricted by catch limits: 65,000 lb. (29,478 kg) for the Saskatchewan River, 140,000 (63,492) for the Nelson, and 100,000 (45,351) for the Churchill (Skaptason 1926: 17–19). For the Saskatchewan River fishery, the catch limit was below the yields during peak production and was therefore too little, too late. The provincial deputy minister of mines and natural resources recalled in 1935: 'When the natural resources were transferred to the province [in 1930] the Director of Fish Culture at Ottawa informed us that the last of the sturgeon in Canada were to be found in Manitoba and requested that we do everything possible to preserve the species' (PAM, RG 17, B1, box 50, file 44.5.4, 4 May 1935). Despite Ottawa's warnings, Manitoba officials permitted commercial sturgeon fishing in Lake Winnipeg and northern waters in the 1930s and 1940s. Not surprisingly, yields were not impressive. In 1943, provincial biologist David Hinks reflected: 'The sturgeon [had been] so intensively fished that by 1927 it was nearing extinction. In this year, sturgeon was closed, and remained so for ten years, but since the reopening, the very poor catches which have been made indicate that even this period of rest has been insufficient to permit the sturgeon to recover' (Hinks 1943: 17–18).

Fisheries management practices of the day simply could not accommodate the economic and biological causes of the depletion of sturgeon fisheries. In the decades that followed the peak in sturgeon production and the ineffective royal commission of 1909–10, dominion and provincial managers were unable to sustain the commercial production of sturgeon.

Conclusion: Explaining Depletion

Several assumptions made by mainstream fisheries economics do not hold for the lake sturgeon. Essentially, fisheries economics calls for the 'privatization' of common or collective property tenures (Crutchfield 1979; Munro 1982; Scott 1955, 1979; Smith 1968; Turvey 1964). In effect, marginal analysis argues that under certain property conditions a bio-economic equilibrium can be achieved. Over the years, economists have developed models to show that a sole owner will better manage a fishery as private property than the inefficiency and redundancy of common property. Guided by the tenets of marginal analysis, a sole owner will operate a fishery so as to capture a rent but will not deplete the fishery. Depletion of a fishery will not occur because a sole owner will be guided by present-value discounting (in which future net revenues encourage private owners to conserve and produce at sustained levels, or may even reduce landings as compared to a common property system of management). Scott argued that a sole owner could plan so as to maximize the present value (future net returns discounted to the present) and would employ alternative fishing techniques in order to reduce the investment required for a given level of output (Scott 1955: 120–1). In the long run, according to his analysis, a rising user cost curve – a reconceptualization of the effect of overfishing on fish populations – would cause the sole owner to reduce production to as to optimize future incomes (123). Because Scott argued that a sole owner could take advantage of integration and scale, he shifted the problem from simply 'rent dissipation' because of common property rights to a problem of productivity of capital. Or more crudely, the long-term interest of a monopolist coincides with a sustained yield from fisheries.

This line of analysis began with H. Scott Gordon's famous and highly influential 1954 essay 'The Economic Theory of a Common-Property Resource: The Fishery.' Gordon's model argued that the common property regime of fisheries resulted in a dissipation of the rent of a bountiful resource. An integral part of his argument was the assertion that fish populations are 'entirely unaffected by the activity of man,' and he sincerely believed that fish populations are indestructible (Gordon 1954: 126). Gordon cited A.G. Huntsman as an authority who claimed in 1944 that 'we have as yet been unable to learn of a clear, documented case of under-replacement through overfishing for this continent [North America]' (Huntsman 1944: 534). These curious claims by both Gordon and Huntsman were made after nearly every major commercial sturgeon fishery had been depleted – a situation acknowledged by those involved in resource management (Hinks, 1943: 17–18). More important, these unsound assertions indicate that the difficulties in recognizing the biological limitations to achieving and sustaining a bio-economic equilibrium did not have a good foundation. The indestructibility of fish was one of the crucial premises of the model advanced by Gordon; but the economic history of the lake sturgeon, in particular, exposes the fallacy of an assumption-laden diagramatic model too far removed from reality.

The regulation of the productivity of a fishery held by a monopolistic owner could be problematic, since the possibility exists that maximization of present value can result in greater net revenues through accelerated overexploitation than can conservation for future returns (net revenues). If the maximum biological reproductive potential of species is less than the discount rate, then overfishing, not conservation, becomes the rational choice. The limited reproductive response characteristic of lake sturgeon makes this fishery economically similar to whaling. For whaling, Clark and Lamberson (1982) demonstrated that the net revenues of a total harvest of the whale population, reinvested at the going interest rate, would generate a higher return in the long run than revenue from an ongoing sustained harvest. Thus they suggested that a bio-economic equilibrium was not assured in all instances simply because a firm maximizes the present value of profits according to the opportunity cost of capital and the present value of an anticipated future profit based on discounting profit according to the expected rate of return.

Historically, sturgeon fisheries in Manitoba were oligopolistic, and operators did not have sole ownership in terms of long-term planning. Nevertheless, the high immediate return from the larger fisheries and the limited growth rate of sturgeon suggest that the sturgeon, like whales, would not achieve a bio-economic equilibrium through present-value discounting. Moreover, a sturgeon fishery regulated essentially by the rational self-interest of sole ownership, in keeping with Scott's reasoning, would have employed alternative fishing techniques, in this case poundnets, in order to reduce the investment required for a given level of output. In Manitoba, the North American market provided high market prices for sturgeon and caviar, in part because of a succession of depleted regional fisheries, which made the immediate net revenues very high. The combination of high value and slow intrinsic growth, or biological replenishment, makes the sturgeon an ideal candidate for severe depletion, not sustainability, according to the logic of present-value maximization. A long-term bio-economic equilibrium is unlikely under such a property system.

The case of lake sturgeon, along with other freshwater fish, conforms rather well to the pattern of a staple economy. In the 1880s, Indians argued in support of their subsistence rights and against commercial fishing. An old and unidentified Indian explained in 1888:

We have waited long for our guardians to do something for us. They must know surely that our food is being carried away to the [United] States. The fish are becoming fewer and fewer and more and more difficult to catch especially in our ways of taking them. Our little canoes and handful of net toiling along the shore in shallow water are but a drop in the lake compared with these companies with their steamers and enormous nets enclosing fish of all kinds. This state of things can go on for a little while. Just as the traders left us, stores and all, when the furs were gone and it did not pay them to stay, so the fishermen will go, but with this difference: in the former case we did all the catching and received an equivalent for our furs; but here these men catch all the fish, and we get nothing, and when they leave the mischief is done. (PAM, MG 12, E3, box 19, Robert Phair to J.C. Schultz, 1888)

This Elder had identified many of the problems of Canada's regional economies based on the export of staples – depletion of resources, demise of the local economy, and the inevitable departure of the outsiders once money could no longer be made.

The participation of Indians in commercial fishing proved to be similar to that in the fur trade. Native integration with the market for fish may have been motivated by the reality that the only choices were outside pillage or some level of local participation. Outside pillage, such as industrial megaprojects in the subarctic, brought untold havoc to traditional 'subsistence' economies (in reality, domestic economies supported by exchange activities). Yet the case of sturgeon would seem to show that instability was also created for Native economies when the traditional activity of fishing was first attracted, then dominated, by the external market.

Opposition to outside encroachment diminished when conditions favoured Indian participation in commercial production. The spread of the market system appeared unstoppable; it must have become clear to the Manitoba Cree and Ojibwa that they had to choose between immediate dispossession and involvement in the industry. Prince's report noted: 'A female sturgeon with roe is worth more than a beaver, for the roe is worth over 50 cents per pound, and a ripe fish may yield 20 pounds' (*SP* 1905, 22: lxii). Indian participation in the commercial fishery was the 'rational' choice, since in the short term the cash return for labour time was very favourable. There can be little doubt that the introduction and spread of what Karl Polanyi called the 'market pattern' are relevant to this region (Polanyi 1944). The price system, in existence since the fur trade, created the possibility of gain with the expanded commercialization of the fishery. None the less, given the biological characteristics of sturgeon and failure of government resource managers to conserve the resource in the face of such strong commercial demand, Indian benefits from the commercial sturgeon fishery were short-lived.

Understandably, there is a tendency to explain the history of Indian fisheries in terms of dichotomies such as commercial versus domestic, or white versus aboriginal. The failure of resource management in the case of the Rainy River sturgeon fishery, for example, has been blamed on governmental preference for non-Indian over Indian fishers. This preference resulted, Holtzkamm, Lytwyn, and Waisberg have argued, in 'a switch in harvesting processes from long-term sustained yields under Ojibwa management to short-term productivity by white entrepreneurs' (1988: 203). This type of interpretation fails to identify the historical role of the market in changing harvesting patterns. Holzkamm, Lytwyn, and Waisberg did not offer a comprehensive explanation for the depletion of sturgeon but only argued that the government favored white commercial producers over Indian 'subsistence' fishers. This argument does not consider the particular problem of the sturgeon, which was driven to near-extinction while other freshwater species recovered or experienced only periodic shortages, despite also being harvested by white entrepreneurs.

Along with biological limitations, two inter-related but distinct causal factors – the market and the preference of the state for white entrepreneurs – must be evaluated. To limit a historical argument about fisheries to simply the racial preferences of the state may overlook the economic forces at work. My evidence indicates that the state favoured involving Natives in the commercial exploitation of the Lake Winnipeg, Nelson River, and Saskatchewan River sturgeon fisheries, yet the fishery collapsed. The question arises: had the government favoured the Rainy River Ojibwa in the production of sturgeon, would the sturgeon fishery there have remained intact?

The economic history of sturgeon fisheries indicates that the tragedy for Manitoba Cree and Ojibwa was caused by the rapid and reckless penetration of the market and the uncontrolled geographical spread of harvesting. The investment of foreign capital into a large-scale fishery for an external, open-ended market, and particularly the market in sturgeon eggs, resulted in problems of resource allocation. The history of this fishery should provide some useful background for Aboriginal and treaty right negotiations over co-management. For Ojibwa people, the multi-purpose uses and cultural significance of sturgeon gave it a special position in the regional economy akin to that of the bison on the North American plains. And like the bison, the sturgeon was quickly driven to near-extinction.

NOTE

In accordance with the editorial style adopted in this collection, the publisher has provided metric conversions for all quantities originally given in imperial units only.

WORKS CITED

Canada. 1885–1916. *Annual Report of the Department of Fisheries. Sessional Papers (SP).* Ottawa.
- 1891, no. 8, *Fisheries.*
- 1897, no. 11A, *Fisheries.*
- 1899, no. 11A, *Fisheries.*
- 1901, no. 27, *Annual Report for the Department of Indian Affairs.*
- 1901, no. 22, *Fisheries.*
- 1902, no. 22, *Fisheries.*
- 1904, no. 22, *Fisheries.*
- 1905, no. 22, *Fisheries.*

– 1906, no. 22, *Fisheries.*
– 1910, no. 22, *Fisheries.*
– 1912, no. 22, *Fisheries.*
– 1909. Dominion Fisheries Commission for Manitoba and the North West, 1909–10. *Interim Report and Recommendations.* Ottawa: Government Printing Bureau.
– 1911. Dominion Fisheries Commission for Manitoba and the North West, 1909–10. 'Report of the Manitoba Fisheries Commission.' Unpublished. National Library of Canada, Canadian Federal Royal Commission Reports, microfiche 51.
– 1917–35. Dominion Bureau of Statistics. *Fisheries Statistics.* Ottawa.
– 1924–5, 1925–6, 1926–7, 1928–9. *Annual Report for the Fisheries Branch.* Ottawa: King's Printer.
Canada Gazette. 1894. 'Regulations Relating to Fishing in Manitoba and the North West Territories' 26 May.
Clark, C.W., and Lamberson, R. 1982. 'An Economic History and Analysis of Pelagic Whaling.' *Marine Policy* 103–20.
Crutchfield, J.A. 1979. 'Economic and Social Implications of the Main Policy Alternatives for Controlling Fishing Effort.' *Journal of the Fisheries Research Board of Canada* 36, no. 7: 742–52.
Ecologists Limited. 1987. 'A Lake Sturgeon Yield Study on the Kenogami River Year 3 – Phase 1 Report.' Prepared for Ontario Ministry of Natural Resources.
Gordon, H. Scott. 1954. 'The Economic Theory of a Common-Property Resource: The Fishery.' *Journal of Political Economy* 61, no. 2: 124–42.
Harkness, W.J.K., and J.R. Dymond. 1961. *The Lake Sturgeon: The History of Its Fishery and Problems of Conservation.* Toronto: Ontario Department of Lands and Forests.
Hinks, David. 1943. *The Fishes of Manitoba.* Winnipeg: Department of Mines and Natural Resources.
Holzkamm, Tim E., Victor P. Lytwyn, and Leo G. Waisberg. 1988. 'Rainy River Sturgeon: An Ojibway Resource in the Fur Trade.' *Canadian Geographer* 32, no. 3: 194–205.
Houston, J.J. 1987. 'Status of the Lake Sturgeon, *Acipenser fulvescens*, in Canada.' *Canadian Field-Naturalists* 101, no. 2: 171–85.
Hudson's Bay Company Archives (HBCA), Provincial Archives of Manitoba (PAM), Norway House Post Journal (NHPJ).
Huntsman, A.G. 1944. 'Fishery Depletion.' *Science* 99, no. 1583: 534.
Lywtyn, Victor P. 1990. 'Ojibwa and Ottawa Fisheries around Manitoulin Island: Historical and Geographical Perspectives on Aboriginal and Treaty Rights.' *Native Studies Review* 6, no. 1: 1–30.
Munro, Gordon R. 1982. 'Fisheries, Extended Jurisdiction and the Economics of Common Property Resources.' *Canadian Journal of Economics* 15, no. 3: 405–25.

National Archives of Canada (NA):
- RG 10, Department of Indian Affairs.
- RG 23, Department of Marine and Fisheries.

Newell, Dianne. 1993. *Tangled Webs of History: Indians and the Law in Canada's Pacific Coast Fisheries*. Toronto: University of Toronto Press.

'Notes on Indian Council at Treaty Rock, Beren's River, Lake Winnipeg, Man., 12 July 1890.' 1987. *Native Studies Review* 3, no. 1: 117–27.

Oliver, C.H. 1987. 'Proceedings of a Workshop on the Lake Sturgeon (*acipenser fulvescens*).' Unpublished report. Ontario Fisheries Technical Report, Series 23. Toronto: Ontario Ministry of Natural Resources.

Patalas, J.W. 1988. 'The Effects of Commercial Fishing on Lake Sturgeon (*Acipenser fulvescens*) Populations in the Sipiwesk Lake Area of the Nelson River, Manitoba, 1987–1988.' Ms. Report No. 88–14. Winnipeg: Department of Natural Resources, Fisheries Branch.

Polanyi, Karl. 1944. *The Great Transformation*. Reprint 1957. Boston: Beacon Press.

Provincial Archives of Manitoba (PAM):
- MG 12, E3, Lieutenant-Governor John Christian Schultz Papers.
- RG 17, Department of Mines and Natural Resources.

Scott, Anthony. 1955. 'The Fishery: The Objective of Sole Ownership.' *Journal of Political Economy* 62, no. 2: 116–24.

- 1979. 'Development of Economic Theory on Fisheries Regulation.' *Journal of the Fisheries Research Board of Canada* 36, no. 7: 725–41.

Scott, W.B., and E.J. Crossman. 1973. *Freshwater Fisheries of Canada*. Ottawa: Fisheries Research Board of Canada.

Skaptason, J.B. 1926. *The Fish Resources of Manitoba*. Winnipeg: Industrial Development Board of Manitoba.

Smith, Vernon L. 1968. 'Economics of Production from Natural Resources.' *American Economic Review* 53, no. 3: 409–31.

Sopuck, R.D. 1987. 'A Study of the Lake Sturgeon (*Acipenser fulvescens*) in the Sipiwesk Lake Area of the Nelson River, Manitoba, 1986–87.' Ms. Report no. 87–2. Winnipeg: Manitoba Department of Natural Resources, Fisheries Branch.

Sunde, L.A. 1961. 'Growth and Reproduction of the Lake Sturgeon (*Acipenser fulvescens Rafinesque*) of the Nelson River in Manitoba.' MSc thesis, University of British Columbia.

Threader, R.W. 1981. 'Age, Growth and Proposed Management of the Lake Sturgeon (*Acipenser fulvescens*) in the Hudson Bay Lowland.' Typescript, Ontario Ministry of Natural Resources.

Tough, Frank. 1980. 'Manitoba's Commercial Fisheries: A Study in Development.' MA thesis, McGill University.

- 1984. 'The Establishment of a Commercial Fishing Industry and the Demise of

Native Fisheries in Northern Manitoba.' *Canadian Journal of Native Studies* 4, no. 2: 303–19.

– 1987. 'Fisheries Economics and the Tragedy of the Commons: The Case of Manitoba's Inland Commercial Fisheries.' Discussion Paper no. 33. Toronto: Department of Geography, York University.

– 1996. *'As Their Natural Resources Fail': Native People and the Economic History of Northern Manitoba, 1870–1930.* Vancouver: University of British Columbia Press.

Turvey, Ralph. 1964. ' Optimization and Suboptimization in Fishery Regulation.' *American Economic Review* 54, no. 2: 64–76.

Van West, John. 1990. 'Ojibwa Fisheries, Commercial Fisheries Development and Fisheries Administration, 1873–1915: An Examination of Conflicting Interests and the Collapse of the Sturgeon Fisheries of the Lake of the Woods.' *Native Studies Review* 6, no. 1: 31–65.

7

'Overlapping Territories and Entwined Cultures': A Voyage into the Northern BC Spawn-on-Kelp Fishery

DIANNE NEWELL

Early in April 1992 I travelled to the Southern Tshimshian village of Kitkatla, British Columbia, to observe the annual commercial herring spawn-on-kelp harvest, whose essential requirements are laid down by the lucrative Japanese market for the product.[1] The Japanese are renowned for their preoccupation with, as one journalist put it, 'ceremonial gestures that incorporate aesthetically beautiful cuisine,' much of which is derived from seafoods such as herring spawn-on-kelp, which they consume as a traditional food called *Kazunoko Konbu* (Hanson 1992: 40). But for centuries, roe (eggs) or spawn (fertilized eggs) deposited on seaweed or other plant material on the northwestern coast of North America has also been a special food for coastal Indian societies that routinely harvested and processed it for personal consumption, elaborate ceremonial use, and trade.

A great delicacy, on a par with Russian caviar, Pacific herring (*Clupea pallasii*) spawn and kelp with spawn (the latter Kwakwala speakers called Qā'x·q!ɛlīs aɛ'nt, Boas 1921, 254–5, 422–8) qualify as traditional high-status food. Among the Northwest Coast nations, spawn-gathering and -harvesting, occurring in the early spring, usually launched the annual seasonal round in select areas of the coast, and spawn was a special food served to visitors and at feasts, as well as being a major item of trade. It also was one of several fisheries to involve women in every stage, from harvesting to processing and trading. The ancient harvests and uses continue along much of the coast today, but, in keeping with Western economic and legal traditions, federal government regulations for BC fisheries restrict these Indian harvests to household subsistence use and trade among Indians only and prohibit other sales of herring spawn except under commercial licences, which are extremely limited and operated almost exclusively by all-male crews.

Herring spawn-on-kelp is, however, the only commercial fishery in the Canadian Pacific region in which Indians have managed to secure a major share of licences, and commercial production (with its roots in ancient traditions) has large-scale cultural, economic, and legal implications for BC First Nations fishing communities.

I came to think of my trip to the north coast as an opportunity to witness an example of what Edward Said (1993: xxix) calls the complex state of overlapping territories and entwined cultures brought about by imperialism, and also as a voyage of discovery into the shifting environment of Pacific coast fisheries traditions.

'Digging Sticks and Fishing Lines'

A string of Kitkatla villagers, young and old, dropped in for a brief visit with Cecil Hill, the skipper-owner of the seiner *Western Spirit*, his wife, Karen Hill, who was in charge of life on board and assisted with packing the spawn-on-kelp product, and his all-male Native crew of three, before we headed south to the herring ponding operations. None of the guests came empty-handed; most of what they brought was food – jars of jam, boxes of plain cakes. One small boy arrived with a large spring salmon. What, if anything, had been or would be given in exchange for the items brought to the boat by villagers was never mentioned, though clearly connections and obligations were being formed or recognized. Pacific salmon (genus *Orcorhynchus*) was not the skipper's favourite type of fish for eating; he preferred lingcod (*Ophiodon elongatus*) and various other species of Pacific fishes commonly called cods, which he and the crew caught with rod and reel to eat during my five-day stay on the fishing grounds. The government's official designation for the various types of fish and seafoods it licences Indians to catch for personal consumption is 'food fish' (caught in the 'First Nation's fishery').

The Canadian government's 'invention' of an Indian food-fishing tradition in the late nineteenth century, which equated Indian fisheries strictly with subsistence harvesting, is a far cry from either the past or present reality of the commercial importance of traditional foods for Northwest Coast Native communities (Newell 1993). Pacific coast methods for what anthropologist Wayne Suttles (1987) calls 'coping with abundance' (which really meant accommodating to gluts and scarcities, minor fluctuations, and local failures in food supplies) included establishing elaborate systems of resource exploitation, co-use of harvesting sites among groups, food preservation and storage, patterns of specialization, and inter-village and -regional exchange. People could thus live at a level and population density above the average of the world's non-agricultural societies – and many of the agricultural ones.

Though the Aboriginal peoples of the Pacific Slope were known as the salmon people, for whom the staple of diet and trade was the anadromous Pacific salmon, they also exploited any other available type of fish, shellfish, and sea mammal, preserving most of the harvest. Accomplished, prolific traders, for whom seafoods were major trade items, they also possessed an elaborate ceremonial life in which seafood as symbol was crucial. This unity of economy and culture requires elaboration, because it was, and remains today, central to most BC First Nations.

Northwest Coast peoples developed as the great ranked societies of Aboriginal Canada. They were strongly hierarchical. Each group occupied specific territories, and there were formal individual or group rights and accompanying obligations to particular hunting territories and harvesting sites. Every group used fishing, gathering, and fish-processing places and technologies within the context of annual seasonal rounds of resource-gathering and trade, which lasted from early spring until late fall. Winter was the important ceremonial and feasting season.

Margaret Blackman (1990: 245) suggests that the foods that the Haida traditionally held in the highest esteem were those that were scarce, seasonal, and socially or geographically restricted – those such as salmon, for example, that came from lineage-controlled territories, and those that were offered to guests at feasts or to special individuals. The same was generally true for the Tsimshian, who, according to Marjorie Halpin and Margaret Seguin (1990: 271), additionally deeply valued foods that required intensive labour 'organised by a person of rank,' 'imported items' (meaning those obtained from non-Indians), and eulachon (*Thaleichthys pacificus*) oil, called grease, and 'anything preserved in grease.'

This pattern emphasizes the way in which wealth and status were inextricably entwined, since foods were, among other things, a source and measure of wealth in these feast-oriented societies. Women organized the processing and storage arrangements for fish and most types of seafoods and in many cases were solely, or at least partly, responsible for the harvests, especially for mass-gathering of shellfish and molluscs, marine plants, and herring spawn. They developed their own specialized techniques and equipment for harvesting, processing, and storing aquatic resources, often exploited their own harvesting grounds, and in some cases operated their own transportation and trading networks.

The Aboriginal practice of harvesting, processing, and using seafoods and other traditional foods continued alongside the state-regulated industrial fisheries after British Columbia became a province of Canada in 1871. Blackman tells a story about one of her field trips to Old Masset Village, Haida Gwaii (Haida peoples, Queen Charlotte Islands), in 1977, to undertake a life history project with the eighty-one-year-old, high-ranking Haida woman Florence Edenshaw Davidson. Blackman noticed a separate freezer on the back-porch pantry. It was brimming with 'Indian foods': venison, herring roe, berries, and fish. These had either been gathered from traditional sites and processed using a mixture of customary methods (such as smoking and air- or sun-drying, together with the more recently available techniques, such as salting, canning, or freezing) or received through local networks of sharing or through inter-village and -regional trade. In Masset in the 1970s it was still the case that 'every woman's got to have her digging stick and every man his fishing line and devilfish [octopus] stick' (Blackman 1990: 245).

Herring Spawn-on-Kelp

As the floatplane that shuttled me the short distance from Prince Rupert to Kit-katla in 1992 touched down at the village jetty, I, like Blackman nearly twenty years earlier, spotted many small boats filled with Indian food – in this instance, egg-laden seaweed. The women of the village had been busy with their annual spring harvest. Large schools of Pacific herring migrate inshore from their sum-mer feeding grounds on the continental shelf to spawn once a year in the quiet, shallow waters of sheltered inlets, bays, and kelp patches everywhere along the BC coast. The peak period of spawning activity of the adult herring occurs in March and early April. A number of distinct populations, differing in size and form, swim in separate schools and have diverse spawning grounds; spawning time varies in different localities and weather conditions, and spawning inten-sity fluctuates from year to year. For centuries, coastal Indians caught local pop-ulations efficiently with nets and special rakes in inshore waters, principally in spring, just before the schools of fish entered the spawning areas closer to shore. But a more popular, special food for subsistence, ceremony, and trade was herring eggs or spawn (Brown and Martin 1985; Lane 1990; Newell 1993: 189–92).

In Pacific waters, sexually mature females (those who are at least three years old) deliberately deposit sticky eggs, 20,000 to 40,000 each, on selected sur-faces, or substrate – usually marine plants growing in or below sheltered inter-tidal waters – and the egg masses remain attached to these surfaces. In some areas the herring schools are so dense that the fish spawn in successive waves, and so thickly do the herring congregate during spawning that the eggs are laid in many layers. In recent years annual BC herring-spawn production has amounted to over two trillion eggs, though fisheries scientists predict that the generally high levels of predation mean that on average possibly only one in 10,000 eggs laid will become a mature herring and return to spawn.

Men and women traditionally collected spawn-covered kelp growing naturally from the beach. They also found that pre-set material was easier to harvest (Stewart 1977: 124). So from ancient times, prior to the run, many Pacific coast groups have placed plant materials – whole trees (cedar, spruce, hemlock) or boughs and branches, or specially harvested varieties of kelp hung on lines or sometimes on tree boughs – in places where herring normally run, submerging and anchoring them in place with lines and sinker and anchor stones (see Stewart 1977: 125–7).

My days on the spawning grounds in 1992 gave me a first-hand picture of just how innovative and efficient was the ancient harvesting technology of collecting spawn on trees. Crew members studied the forested shoreline for several days. Eventually they spotted a perfectly shaped western hemlock to use: 'The overall shape and density is really important for the tree to work,' they explained. If they anchor a short, properly shaped tree right-side up in a spawning bay, the herring should deposit eggs fairly evenly both sides of the foliage. Initially, the tree's branches float towards the surface of the water, thus exposing their outer surfaces. Once spawning begins, the egg deposits adhere to the branches, slowly weighing them down, which action, together with the gentle action of the ocean tides and currents, exposes previously untouched areas. Thus 'animated,' the entire tree is eventually encrusted with spawn.

Popular traditional methods of harvesting on selected kelp or tree boughs were to bring them into spawning bays and suspend the blades of kelp or weighted boughs from floating log frames, placing and anchoring the frames close to the shore in areas with shoals of spawning herring, or to set individual lengths of weighted kelp or tree branches (set among eel grass) in the spawning bays (Stewart 1977: 125). The historical and ethnographical literature contains mentions of several specific varieties of kelp having been used for collecting spawn in kelp ponds. Giant kelp (*Macrocystis integrifolia*) was in general use among the northern and central peoples of the coast, including the Bella Bella (Heiltsuk) and the Hesquiat and Ahousat of Vancouver Island; boa kelp (*Egregia menziesii*) was also used by the Bella Bella; and common kelp (*Nereocystis leutkeana*) was popular with the Kwakiutl (Kwakwaka'wakw) and Clayoquot (Nootkan, now Nuu-chah-nulth) peoples (Lane 1990: 4). In all areas of production, herring spawn was usually eaten with the kelp – but never with the tree branches – on which it was deposited.

European stories about the persistence of the annual spawn harvests abound, beginning with the pioneer English explorer of the Pacific, Captain James Cook, thought to remark in the 1770s that the Kyuquot (Nuu-chah-nulth) of Vancouver Island 'make a very good caviare, preserved dry on small pine [probably hemlock] branches and seaweed.' The English fur trader and colonial official William Tolmie wrote on 5 April 1834: 'The Sound now abounds with herring, which are depositing their spawn plentifully – they eject it on branches of pine placed in convenient stations by the [Bella Bella] Indians by whom it is used as food & collected in great quantities. It has a slightly saline taste, not disagreeable' (Lane 1990: 9). One hundred years later, when Chief Charles Nowell of Alert Bay attended a feast at the village of Bella Bella that featured herring eggs, he would remark to his biographer: 'Those Indians don't eat much of white man's food' (Ford 1968: 236).

In the homeland of the Skidegate Haida, Indians continued to conduct spawn-on-kelp harvests in the 1970s. Hilary Stewart's sensitive eye and lively prose capture the drama of the harvest in the early spring of 1975, when there was an exceptionally heavy run of herring. From the dock at Skidegate Landing, she watched a small rowboat arriving, 'laden with piled-up lengths of spawn-covered kelp draped over the seats and in the bow.' By this time 'long lengths of seaweed, creamy amber with spawn, hung from nearly every porch and sun deck [in Skidegate]; racks and clothes lines in gardens and carports were festooned with it. Some of the kelp was draped over the lines, but much of it hung down full length, held at the top with clothes pins, blowing freely in the breeze. The owner of the large hardware store in [Queen] Charlotte City said he hadn't a clothes pin left in the place' (Stewart 1977: 147).

Traditionally, some herring spawn was eaten fresh locally, but most of it, tons of product, was cured for local consumption and inter-village and -regional trade (Figure 7.1). Franz Boas's Kwakiutl ethnographic materials on fishing and gathering, and food recipes recorded by George Hunt and mostly provided by his wife, Mary Hunt, offer rare early-twentieth-century details of the special harvesting and processing of herring spawn on hemlock and cedar branches and on kelp (Boas 1921: 184–5, 254–5; Lane 1990: 5). Among Kwakiutl societies, they noted, 'husbands' harvested spawn on hemlock branches, hanging up the poles bearing spawn-coated branchlets to dry and 'set' at a windy site (drying took about six days), then carrying the dried spawn to a rocky area where their 'wives' carefully 'wiped' the dried herring spawn from the hemlock, spread the spawn on mats to dry thoroughly, and then bundled it up for storage or transport, both requiring special containers. Various elaborate gendered rituals of reconstituting, cooking (usually steaming), serving, and eating spawn are mentioned. Hunt indicated that the Kwakiutl sometimes ate herring spawn dry, but only when mixed with salmon-berry shoots and fish oil; taken straight, dried spawn was 'too rough' (indigestible), she said (Boas 1921: 254, 428; a similar pattern existed for the Tlingit – see Stewart 1977: 148). The most important form of herring spawn for ceremonial use among the Kwakiutl, however, was 'kelp with spawn' (Boas 1921: 254–5, 422–8). It was out of this Aboriginal

Figure 7.1   Mrs Dorothy Gordon and Mrs Ruben Mason drying herring eggs, place unknown, n.d. *Source*: British Columbia Records and Archives Service (HP27028)

kelp-with-spawn fishery that the present, licensed spawn-on-kelp fishery would evolve.

When I visited Kitkatla in 1992, at least one elderly women in this Tsimshian village still practised the local traditional processing methods for rubbing or wiping off dried spawn collected on branches, using an ancient method not unlike the one described by Hunt for the Kwakiutl (Boas 1921: 422–3). Today, of course, most First Nations, including the Kitkatla, cure spawn either by salting down the eggs on branches or kelp and then storing them in brine in plastic containers, or by freezing them (see Bouchard and Kennedy 1989). Some raw egg-laden kelp is eaten fresh as snack food, which is especially popular with children, and much of it, fresh or frozen, as I witnessed, is diced and added to hot dishes such as scrambled eggs and chop suey.

Both differential availability and taste preference motivated the inter-village and intracoastal trade in distinctive varieties of herring spawn and other traditional foods. This explains why many groups – the Nootka, Kyuquot, and Kwakiutl, for example – traded for herring roes even though they had local supplies (Newell 1993: 31–2). Barbara Lane (1990: 4) discovered that people procured herring spawn on a variety of substrates because each one imparted some flavour to the product. The Bella Coola people (Nuxalk), for example, sometimes obtained spawn on boa kelp from the Bella Bella but did not use spawn on giant kelp, a variety that the Bella Bella also produced.

Mary Hunt's information suggests that some groups traded away what they considered to be the less-desirable spawn. Spawn collected on cedar branches, easily removed from this material, was also somewhat undesirable. The use of cedar was restricted to collecting raw spawn to be eaten fresh: '[Cedar-branch spawn] is bad when it is dried, and it quickly gets red; and it also tastes of cedar-branches when it gets dry' (Boas 1921: 422). The women would separate the dried 'white' (fertilized) herring spawn collected on hemlock branches from red spawn, placing the large pieces of white egg masses in air-tight cedar boxes for storage in a dry place in the house until winter. The red spawn, however, went into her medium-sized cedar-bark hampers for 'other tribes, for it is not good to keep long' (Boas 1921: 254). Thus the Kwakiutl took dried herring spawn (probably on kelp, which had important ceremonial use) in trade from the Bella Bella and traded out fresh, red herring spawn to other groups.

Herring spawn was also an item of external commerce with local merchants and Japanese-Canadian neighbours, even before the government introduced a regulated commercial spawn-on-kelp fishery in the 1970s. The renowned Pacific coast artist and writer Emily Carr, writing about an Indian friend in the Queen Charlottes about 1912, said of her annual harvest of dried herring eggs on kelp: 'Mrs. Green knew where the fish put their eggs in the beds of kelp, and she went out in her canoe and got them ... After she had dried them she sent them to the store in Prince Rupert and the store shipped them to Japan, giving Mrs. Green value in goods,' which she chose from a store catalogue (Carr 1965: 75). Heiltsuk women and men today say that they routinely traded dried herring spawn to Japanese Canadians for supplies of soya sauce – and still do (testimony of various Heiltsuk witnesses in *Heiltsuk Indian Band v. Canada* 1993).

The federal fisheries department paid little or no attention to herring-roe and spawn-with-kelp fisheries until 1955, when it amended the British Columbia Fishery Regulations to ban all harvesting and use of herring spawn except by Indians for their food (Statutory Orders and Regulations [SOR]/55-260, sec. 3 amended by the addition of section 21A), and 1974, when it decided that there was a commercial market for herring roe products and so allowed commercial harvests by permission of the regional director (SOR/74-50, sec. 9, 21A [1]).

During and immediately following the Second World War, there had been an international boom in herring products, such as oil, fish meal, and solubles, leading the federal department to permit unusually intense commercial fishing for herring in British Columbia, which was a major supplier. Under this kind of pressure, Canadian Pacific coast herring stocks quickly collapsed. Ottawa responded by closing the herring fishery permanently in the late 1960s. Out of that ruin, the licensed spawn-on-kelp and roe-herring fisheries struggled into existence in the early 1970s, in response to the apparent partial recovery of the herring stocks, the opening up of secure Japanese markets for herring-roe products, and Indian requests to sell spawn-on-kelp. The federal government has shown a willingness to encourage and protect Indians' involvement in these new commercial fisheries, in an attempt to compensate them for loss of halibut and salmon licences under the Davis Plan licence-limitation program for the Pacific region, introduced in the 1960s (Newell 1993).

Though Ottawa granted Indians special preference in the commercial spawn-on-kelp fishery, a brief glance at the *Commercial Licensing Handbook* ('Category J – Herring Spawn-On-Kelp' 1994: 43) reveals a fishery that is very different – impersonal, market-driven, and universal in application – from the ancient commerce. The department regulates the herring spawn-on-kelp fishery by determining the amount of herring stock that will be harvested each year, by allocating the herring stock to the different herring fisheries, by allocating the herring spawn-on-kelp fishery to commercial users and the Indian food fishery, and by allocating the commercial herring spawn-on-kelp licences.

The requirements of harvesting operations, licensing, and marketing arrangements initially were, and largely remain, those of Western industrial capitalism and science: they seek to limit, privatize, and domesticate the fishery in every possible way. In 1980, commercial permits gave way to new category ('J') of licences (the 1914 provision for commercial use was transferred to section 17 of the Pacific Herring Fishery Regulations and amended: SOR/80-876, sec. 8, 17[a], [b]). Permits, and later licensing regulations, set the conditions for this commercial fishery. The fisheries department restricted the commercial spawn-on-kelp fishery originally to closed ponds (impoundments) and, until recently, to one species of marine plant for the use of substrate – giant kelp – usually referred to in the fisheries literature by its Latin name, *Macrocystis* (the use of *Laminaria saccharia* has also been permitted).

J licences are issued for the herring ponding operations, and in the case of impoundments (whereby catcher vessels are used to impound live herring and draw them into an enclosure filled with kelp) they specify the area of operation of the catcher vessel (a commercial fishing vessel must be designated in the application as the catcher vessel) and ponds. Gear restrictions may apply to the catcher vessel 'as required,' as may other restrictions – on dates of operation for the catcher vessel, type and number of impoundments, and dates of dismantling of the impoundments. Whether for impoundments or open ponds (whereby herring are left free to enter and exit the pond during spawning, as in the traditional manner), restrictions may cover maximum harvested product weight ('quota') of herring spawn-on-kelp, standards of quality for the spawn-on-kelp product, and species of marine plants used in the fishery, as well as dates and method of harvesting and packing, type, marking, and dimensions, and even the colour of containers used to pack the product.

Licence fees match those of roe-herring (H) licences and in both cases are nominal for Indian licence-holders. J licences are renewable but not transferable – holders are required to operate the licences – and holders are not permitted to hold also a herring-roe licence or even to participate in the roe-herring fishery. A provincial harvesting licence allows each operator to harvest manually individual stipes, or lengths, of specified types of kelp from areas of his or her own choice, with no quotas set.

As a matter of long-standing policy, with the introduction of the new spawn-kelp-fishery, Ottawa classed Indian harvests as non-commercial and granted permission for them as a privilege, subject to conservation needs. Indian 'food fishery' licences authorize the harvesting from beaches of not more than five hundred pounds (two hundred and twenty-six kilograms) of herring spawn-on-kelp for food purposes, and the material so taken can be traded among Indian people (Dickson 1977: 31). Under Canada's Fisheries Act (RSC 1970, c. F-14, contrary to Pacific Herring Fishery Regulations (SOR/84-324, sec. 20[3]), however, it is unlawful to attempt to sell herring spawn-on-kelp unless it has been taken or collected under the authority of a J licence.

What has happened *in practice* on the harvesting grounds of the ancient homelands of BC First Nations is another story. As Michel Callon observed when looking at humans and objects symmetrically in studying the domestication of the scallops and the fishers of St Brieuc Bay, France, compliance in BC commercial spawn-on-kelp operations can be said to depend on a 'complex web of interrelations in which Society and Nature are intertwined,' and fish communities, like fishing communities, have needs and are 'stakeholders' (Callon 1986: 201). Canadian fisheries managers, who thought that they had a tight grip on this fishery, have found themselves dealing with a shifting fishing mode, in which the essential flow of change always inclines towards ancient practice.

Making Tradition

BC First Nations both initiated and improved the regulated herring spawn-on-kelp fishery and expected to be intimately involved in it. Commercial production of spawn-on-kelp, with markets for traditional seafoods in Japan, began in Prince William Sound, Alaska, in 1969. The Alaskan product was low in quality, for the Americans harvested spawn deposited naturally on inferior types of kelp: the product contained sand, and supplies were highly irregular.

The Skidegate Indian Band of the Queen Charlottes therefore approached the

Canadian fisheries department for permission to sell a portion of its spawn-on-kelp harvest, and in 1971 and 1972 federal fisheries service field staff closely studied the band's traditional techniques and materials, assessed the feasibility of impoundment (used to rear bait) and its impact on herring and kelp, and surveyed the market potential for the BC product, which had to compete with Alaskan production (Dickson 1977: 28; Dickson, Buxton, and Allen 1972). The department issued the Skidegate a spawn-on-kelp permit in 1974 for the purpose of conducting an impoundment operation. This proved successful and opened the way for a viable industrial fishery in which Native licence-holders could continue to innovate.

Ottawa based the original selection of permit holders in the 1970s on the criteria of remoteness of residence and of previous experience in catching, holding, and handling live herring in impoundments, which at the time were used to produce bait for other fisheries. One of the individuals selected for a permit was Cecil Hill, whose experience with impoundments in the bait fishery was invaluable, as was his membership in the Kitkatla Indian Band. The week I spent in early April 1992 with him and his crew on his seiner *Western Spirit* hinted at the rich texture of life on the water that was possible for First Nations' participants in this fishery, despite demanding restrictions and specifications.

Kitkatla licence-holders conduct harvests in their traditional territory around Porcher Island (for impoundments) and the waters south of it, into the southern end of Banks Island (see Figure 7.2). This portion of the coast south of the Skeena is a heavily forested, fiorded region of high precipitation. It includes both open and exposed waters on the outside island shores, which provide habitat for giant kelp, and more restricted and sheltered waters in narrow passages and inlets, which offer ideal spawning grounds for herring. As Hill navigated the labyrinth of channels to reach one of his harvesting grounds, I witnessed an evolving tradition of participation and innovation, which has woven together several methods for harvesting spawn.

Like most modern industrial fishers, Hill had invested heavily in his vessel and the latest electronic communications and fish-finding and navigation equipment, and he had more than one computer on board. But he was manoeuvring all of this 'high' technology through ancestral territory whose terrain and fishing grounds he knew intimately – in a three-dimensional way – and he put that local knowledge to good use. I saw him piloting from natural benchmarks taught to him by his grandfather: using his 'sea eyes' to spot the concentrations of seabirds that stick close to the schools of feeding herring, studying the intensity and configuration of surface bubbles caused by herring as they follow the natural flow of plankton to the surface of the water at dusk, and (using such clues) speculating about the size, age, and sexual maturity of the schools of her-

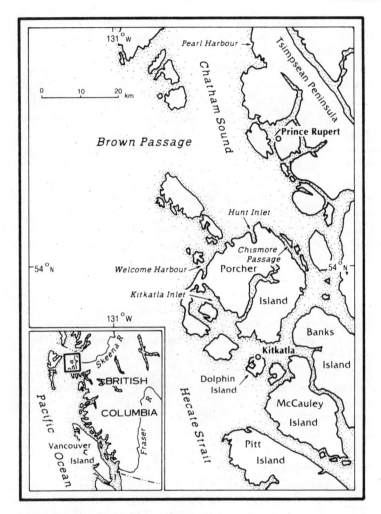

Figure 7.2   Locations (place names) of the herring spawn-on-kelp fishery
of the Porcher Island area, BC, 1985. *Source*: Adapted from Shields,
Jamieson, and Sprout 1985: 39

ring that lay below. He used his electronics mostly to confirm and 'fine-tune'
these more subjective findings.

When using impoundments, licence-holders such as Hill select a sheltered
area, usually deep-water bays, near a herring spawning area and kelp patches,

and then construct the impoundment from a net suspended from logs arranged in a rectangle or square and pursed at the bottom to form an enclosure, as shown in Figure 7.3(a), to suit local circumstances. The entire apparatus is anchored, and if in shallow water, tied to the beach, and the crew then heads out in a herring skiff to areas containing wild beds of *Macrocystis*. The crew harvests stipes, transports them to the impoundments, and hangs them individually on ropes ('kelp lines') strung across the pond in the water. Harvesters have found that the ideal kelp for the purpose are robust, dark, and with large symmetrical blades free of rips and holes, since entangled blades cause uneven or incomplete egg coverage.

In harvesting the kelp, operators cut just above the ragged basal portion, which is left growing. They discard the upper portion (apical) because the young fronds growing at the tip (meristem) of the plant exude substances (polysaccharides) that prevent the egg masses from sticking to the kelp (Shields, Jamieson, and Sprout 1985: 10). Operators tie the kelp to lines in specific intervals, in lengths ranging from one to three metres, as it is harvested.

Harvesting and hanging kelp require great skill and judgment: contact between the blades of kelp and the webbing or the ocean bottom at low tide, once the kelp lines become heavy and sag with the weight of eggs, ruins the blades. Kelp density in the ponds is left up to individual operators and varies from operation to operation. With the kelp in place, a seiner is used to catch a set of herring about to spawn: the annual amount per licence is now officially pegged at approximately 100 short tons (90.7 metric tonnes) of herring. The vessel, with its net full of herring, is slowly towed sideways to the impoundment site by a second vessel. The herring are then released through a gate into the impoundment. Ideally, over several days sexually mature females deposit eggs evenly and thickly on the blades of kelp, and males fertilize them. The herring are then released, and the spawn-on-kelp is harvested to the quota (currently averaging 8 short tons, or 7.6 metric tonnes) established by the annual licence.

This technical description does not begin to do justice to the skill, judgment, luck, hard work, and experimentation involved in a successful, maximum-allowable harvest of top-grade product. The number of egg layers and the dimensions and condition of the kelp determine the official grade of product: top quality (number-one grade) requires a minimum of three, evenly distributed layers of eggs on both sides of a healthy blade, with minimum dimension of 35.5 by 10.2 cm. Each permit-holder harvests approximately 14.5 large totes to fill the eight-ton quota. One large tote yields between 544 and 635 kg of product (Shields, Jamieson, and Sprout 1985: 10, 13). In the first few years of the industry, permit-holders for the spawn-on-kelp fishery rarely filled as much as

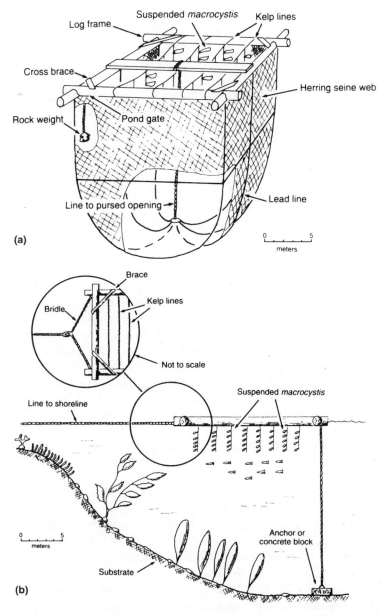

Figure 7.3   (a) A typical herring spawn-on-kelp deep-water impoundment, British Columbia, 1985. (b) A typical open pond, British Columbia, 1985. *Source*: Shields, Jamieson, and Sprout 1985: 41, 43

50 per cent of allowable quotas. In 1976, five of the twenty permit-holders produced no product at all (Dickson 1977: 31).

Coordinating the timing of spawning with the timing of fish impoundment and loading of kelp was, and remains, a major logistical problem: the *Macrocystis* has to arrive at the impoundments fresh and undamaged, and thin kelp is especially susceptible to damage from harvesting, which greatly reduces its pond life (Shields 1984: 31). Decisions about when to harvest the kelp require temporally fine judgment, since *Macrocystis* remains viable in impoundments and open ponds for only *six to ten days*, depending on the quality of the kelp and egg deposits and on the weather. Moreover, the natural rate of deterioration of the kelp increases once it is spawned on. Should the kelp in the impoundment deteriorate and have to be discarded, new kelp is needed, more herring must be added, and the operation has to be repeated, perhaps several times (Shields 1984: 4).

Along with these complexities, the spawning condition and 'loading densities' of herring in impoundments are critical for the survival of the fish and for both the timing and spatial pattern of spawning. Herring die as a result of stress, scale loss, or abrasions when fishers are catching, transporting, and especially impounding them. Impoundment may kill off some herring because of the high loading densities necessary to fill quotas before the kelp deteriorates; Shields and Kingston (1982) indicated significant herring mortality of 30 to 50 per cent in closed-pond systems. These rates are alarmingly high, given that (unlike in the roe-herring fishery) it is not necessary to kill whole female fish to obtain the roe.

It has also been found that a high percentage of impounded females do not deposit their eggs, or not fully, or not on the kelp blades themselves, thereby 'wasting' the large supplies of kelp that hang in the impoundments. What fisheries scientists refer to as the 'spatial selectivity demonstrated by spawning herring' (meaning the places where female herring choose to, or not to, deposit their eggs) often works against the interests of commercial operators. Scuba divers who were hired to scrutinize captured herring found the females depositing heavily against the impoundment webbing, attracted by free herring that were spawning *outside* the mesh – a case of nature outwitting commerce (Shields 1984: 30).

A perpetual problem is the differing size and degree of maturity of minor stocks (small schools) of herring, which results in a range of spawning times. If fish do not spawn within a few days of impoundment, the likelihood of their spawning at all decreases with time; and all the while, the harvested kelp in the impoundment is deteriorating (Shields 1984: 4). These problematic spawning 'behaviours' prevent the thick, even egg coverage of healthy kelp that fetches top prices with Japanese buyers (Dickson 1977: 29). Inducing the females to deposit their eggs in specific ways to conform to market requirements remains a critical technical problem, with implications for conserving herring and kelp. Licence-holders such as Cecil Hill work with these difficulties in innovative ways.

When I arrived in Kitkatla territory in early April 1992, Hill's impoundment fishery there was finished for the year. What I observed was a 'late fishery,' conducted later in the season *and* introduced later as an official type of spawn-on-kelp production – harvesting in open ponds. The open-pond principle differs from impoundment in that no webbing is hung from the log frame and no herring fishing is involved (see Figure 7.3[b], p. 134). The Department of Fisheries and Oceans (DFO) strongly resisted the use of open ponds. However, the determination of various First Nations to stay involved in commercial fisheries, together with the peculiarities of spawn-on-kelp production, meant that Aboriginal traditions and ecological knowledge became incorporated in post-1981 regulations that permitted use of open-pond methods. It was in the Queen Charlottes and on the northern mainland coast in 1982 and 1983 that DFO sponsored contract studies to test the effectiveness of the open-pond method (Shields and Kingston 1982).

The flexibility that comes with the use of ponds had been especially valuable to the Native communities in the Queen Charlottes, where impoundments were unsuccessful. With the open-pond method, log frames loaded with *Macrocystis* are pushed into areas of naturally spawning herring and positioned over the fish. This type of operation is relatively mobile: some operators have a number of ponds placed in traditional herring-spawning areas, while others have two to four ponds that they tow from area to area, depending on spawning activity (Shields, Jamieson, and Sprout 1985: 7).

Open ponds not only eliminate injury and mortality of herring but require less equipment and waste less herring reproductive potential. Nevertheless, problems do exist. Open-pond operations must use sites that are less than ideal for producing a top-grade commercial product. Fisheries scientists and spawn-on-kelp operators have identified the positioning of the frames as the 'most critical stage of the open-pond fishery ... Operators must place the ponds both close enough to shore so that the suspended kelp is near natural, attached vegetation yet so it remains off the bottom at low tide and in areas sheltered from prevailing weather,' but also away from mud, sand, and fresh-water runoff, all of which produce sediment that attaches to the spawn-on-kelp and ruins the product (Shields, Jamieson, and Sprout 1985: 11).

Some operators, Cecil Hill included, have been experimenting with improvements using both open ponds and impoundments. Hill has, for example, studied problems with spawn coverage and devised a special kelp-suspension system.

The experiments were sponsored by DFO and funded jointly by its Fisheries Development Branch and by Supply and Services Canada's Unsolicited Proposal Program, with impressive results. Consultants monitored and independently tested Hill's experiments (Shields 1984). Incomplete, uneven spawn coverage occurs because kelp has the natural tendency to float, and because the action of water currents causes individual plants to become entangled. Herring also tend to avoid spawning on kelp located at the surface of the water, where they are most vulnerable to birds and other activity around the pond. Thus the stipes of kelp must be properly weighted to keep them hanging straight and submerged.

The consultants found that compared to the industry's traditional nonweighted system, Hill's kelp-suspension system increased spawn-on-kelp production significantly as well as partially reducing the impact of the fishery on herring and kelp stock. Hill's system uses individual, refillable plastic containers, which he named 'Kelp Klips.' These are filled with gravel and attached to the bottom of each *Macrocystis* stipe in an impoundment or open pond (see Figure 7.4). These weights are emptied at the end of the fishery (usually they fall off and are easily recovered as the kelp ends deteriorate) and stored for future use. It was found that Hill's weighted kelp hung vertically as soon as it was suspended, did not become entangled, and hung at an equal depth, so herring have more spawning substrate available. The kelp produced a thicker coating of egg layers deposited on the kelp blades, had one-third more blades that met the topgrade egg-layer criteria (measured in thickness, evenness, and extent of surface coverage on both sides of the blade), and exhibited 7 per cent less waste in the form of trimmed and discarded product (Shields 1984). The amounts of herring and kelp required to fill the quota are reduced, as are the time, labour, and cost associated with the collection of additional kelp and the capture of additional herring. Hill and several other operators use the 'Kelp Klip' or a similar, weighted system to great advantage.

Entangled Interests and Agendas

Cecil Hill enjoys this industrial fishery and the flexibility that it affords. He prefers it to the roe-herring fishery, which, though more lucrative, is much more tightly controlled by government fisheries managers and, with its short openings (times to reach quota have typically been as short as fifteen minutes to two hours) and large number of licensed vessels, is more chaotic, dangerous, and stressful. Some fishers liken it to shooting fish in a barrel. Permitting holders of J licences the flexibility to harvest in open ponds, impoundments, or both, thereby extending the time and spatial area of legitimate harvests, has made it

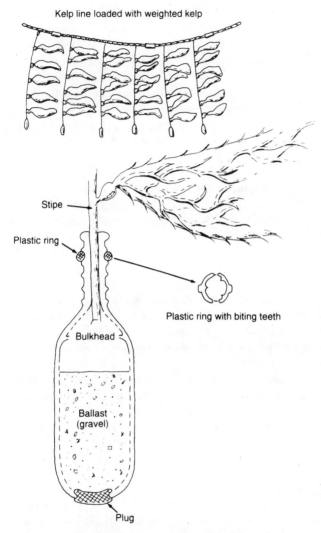

Figure 7.4   The 'Kelp Klip' (Cecil Hill's innovation) and its attachment to suspended *Macrocystis*, British Columbia, 1984 *Source*: Shields 1984: 7

easier for all operators to fill their annual quotas. Open ponds, in fact, are the only viable method of filling licence quotas in the Queen Charlottes (Shields and Watson 1983).

For Native licence-holders, flexibility of method enables them to use traditional kelp-harvesting and herring-fishing sites in tribal territories, along with their traditional knowledge of the environment, to innovate further. Hill also values being able to make his own marketing decisions and to deal directly with the Japanese buyers, whom he takes out to his operations for a first-hand look. Though having a J licence has led a number of BC First Nations to operate their own custom spawn-on-kelp–processing plants employing village band members, Hill prefers to have his harvest processed by an established fish-packing firm in Prince Rupert. This is a hard-nosed business decision. He believes that his customers will have more confidence in the final product.

There is more than a comfortable living at stake for Native holders of J licences. The nature of spawn-on-kelp harvesting – especially that using open ponds – allows them to move easily on a single voyage among the J-licence fishery, 'food fishing' for herring spawn and other marine products for themselves and their villages, and other activities. In addition to harvesting spawn on trees, crew members of the *Western Spirit*, when I was there, also jigged for herring to bait halibut long-lines for household consumption. They harvested cod and set a few crab traps for food. They made side trips by skiff to comb the beaches for relics and suitable logs for building the floating frames for the spawning ponds. They had time to make cultural visits to ancestral grave sites and a pocket-size Indian reserve, at the head of an inlet, traditionally used by the Kitkatla as a fishing village but now reclaimed by the northern rain forest. In one place a new squatter's cabin noticed in the previous spawn-on-kelp season appeared now to be abandoned, but – as in ancient times – the boundaries between 'occupied' and 'abandoned' structures blur in this highly contingent, cyclical, intertidal environment.

For many such cultural, economic, and social reasons, several BC First Nations hoped to obtain licences for herring spawn-on-kelp. But Ottawa last issued new J licences in 1978, by which time only twenty-eight were available and renewable each year. Of these, eighteen had gone to Indians (five issued in 1978 went to five different Indian bands, and the other thirteen went to individuals) and ten to non-Indians. Though the federal Pearse Commission of 1981 (Pearse 1982) recommended that Ottawa expand this industrial fishery and protect Native involvement in it, and though DFO assured the commission that Indian bands would have priority for any new J licences issued, it refused to issue additional licences. Many Indian bands requested new or additional ones for community economic development. The Nishga (Nisga'a) demanded three spawn-on-kelp permits as part of their modern treaty settlement, a requirement that DFO saw as suggesting an alarming trend.[2] By 1988, DFO had received a reported 260 requests from individuals, including Indians, and thirty from Indian bands (Parsons 1993: 211).

Government research indicated that spawn-on-kelp production could expand, with up to seven additional licences, for an annual total of thirty-five, without affecting the price. Everyone connected with this industry knows that whatever threat the commercial spawn-on-kelp fishery posed to the herring or kelp resources, it was nothing in comparison to the magnitude of that posed by the frenzied roe-herring fishery, which DFO seemed to favour. Unlike the spawn-on-kelp harvests, the roe fishery destroys all of the herring – female and male – caught. In spite of such compelling arguments for awarding additional spawn-on-kelp licences, the government held out until 1991.

The major problem was, and has always been, political, involving allocation of additional licences to Indians. Despite many technical improvements, which have helped to conserve both herring and kelp supplies in this fishery, and the excellent markets for spawn-on-kelp, federal fisheries managers remain nervous. Though they have made every effort to place the harvest on a scientific basis and under a rigid set of controls, it nevertheless remains 'on the wild side.' Fisheries managers are especially anxious about the widespread use of open ponds, despite the low mortality rates for herring. They saw, and still see, open ponding as a Native method: mobile and difficult to monitor, with poor spatial control and high risks of illegal harvests. They say that the illegal spawn-on-kelp fishery has been worth as much as $3 million annually – or roughly the value of the annual harvest with eight new licences (Parsons 1993: 211).

Is conservation a serious consideration for the spawn-on-kelp fishery? The issue came to a head in *R. v. Gladstone*, in which two brothers, William and Donald Gladstone, both members of the Heiltsuk Indian Band (successor to the Bella Bella Indian Band) were charged – under section 61(1) of Canada's Fisheries Act (1970) – with offering a large quantity of spawn-on-kelp for sale in 1988 and of attempting to sell herring spawn-on-kelp not caught under a J licence, contrary to section 20(3) of the Pacific Herring Fishery Regulations. In responding to the Gladstones' appeal of their conviction of 3 October 1990, in which they claimed an Aboriginal right to sell spawn-on-kelp, the crown argued that there may not have been a proven threat to the conservation of herring, but there was a potential threat, so the government's regulatory scheme was justified (*Regina v. Gladstone* 1993).[3]

That Ottawa eventually had a change of heart about issuing additional licences was the result largely of the campaign of the Heiltsuk Indian Band. In 1989, the band council filed its suit against DFO for five additional commercial licences to harvest, in an area near Bella Bella, herring 'roe on kelp' for which it claimed an Aboriginal right pursuant to section 35(1) of the Constitution Act, 1982. This band had received one of the early, limited permits in this fishery. Every year from 1980 to 1988, it applied for additional licences to harvest 'roe on kelp' in its traditional harvesting areas – but Ottawa would not budge (Newell 1993: 201). Once the Heiltsuk filed the suit, however, the federal fisheries minister announced that ten new J licences would be awarded and all would go to Indian bands. The band refused to drop the suit, and Ottawa delayed awarding new licences, pending the outcome of that suit.

The case (*Heiltsuk Indian Band et al. v. Canada* 1993), eventually dismissed by the Federal Court in November 1990 (but not officially concluded until 28 February 1993), because the Heiltsuk people had 'not established that commercial harvesting of roe on kelp, was ever, at the material times, an aboriginal right,' seems to have brought to a head the entire issue of federal policy on Indian licensing in the Pacific coast commercial fisheries (see Newell 1993: 200–3; Harris 1996). The fisheries minister immediately made good on his promise to issue ten additional J licences for the 1991 season, all to Indian bands, but none to the Heiltsuk (or any other band already holding a J licence). However, beginning in 1993, one new licence was issued under the Aboriginal Fishing Strategy initiative to the band, which brought the total to thirty-nine in the fishery.

Conclusion

Throughout federal involvement in the spawn-on-kelp fishery, there has been continuing recognition that this fishery would be for Indians. Even so, neither the historical importance of this fishery to specific Native communities, nor the fishery itself, has received serious attention, in or out of court. Government regulations have fashioned a tightly restricted commercial fishery that essentially bans the use of existing Native technology and know-how and puts markets and bureaucratic surveillance ahead of the long-term interests of both the First Nations fisheries communities and the environment.

It must be stressed that Natives initiated, and were granted priority in, this fishery during its developmental, non-profitable stage in the early 1970s: the fishery helped Indians economically by permitting 'a cottage-type industry favouring residents of remote coastal communities,' who were the ones most devastated by the closure of the herring fishery and the introduction of licence limitation in the salmon fishery (Dickson 1977: 28; Newell 1993). Once the fishery became highly profitable in the late 1970s, however, Ottawa was reluctant to permit expansion, citing protection of herring stocks and the market price of spawn-on-kelp.

How long Canada's commercial spawn-on-kelp fishery will survive is not certain. For one thing, the Alaskan product is improving. For another, conservation of herring stocks is once again an issue: the fisheries department closed the east coast of the Queen Charlottes to spawn-on-kelp harvesting and roe-herring fishing in 1995 and 1996, and the Kitkatla district to roe-herring fishing in 1996, citing inadequate supplies of herring in those districts. The days of strong demand from Japanese customers for Canadian spawn-on-kelp may be numbered, because younger generations of Japanese appear to be less interested than their elders in buying this traditional food, and it is unlikely that other commercial markets for the product will be found. In the meantime, herring spawn remains a valuable traditional food in coastal Native societies, and the industrial herring spawn-on-kelp fishery helps Native licence-holders to maintain dynamic traditions within home territories and earn a livelihood, and to achieve cultural and economic goals within the broader political community of which they are a part.

NOTES

1 I am grateful to Cecil Hill, his herring spawn-on-kelp crew for 1992, Karen Hill, and Ingrid Hill for their assistance during fieldwork; Daniel Marshall, for research assistance; Douglas Harris and Arthur Ray, for comments on earlier drafts of this chapter; and the South Coast Division and the Pacific Biological Station, Department of Fisheries and Oceans (DFO), Nanaimo, BC, for providing reports, maps, and standard information – their WAVES-CD ROM (DFO database) and ASFA (Aquatic Sciences and Fisheries Abstracts) were invaluable research tools. The interpretations and opinions expressed here, however, are mine alone. Contemporary First Nations spellings for the historic names of language groups appear in brackets at first mention.

2 The feasibility of a Nishga-owned spawn-on-kelp commercial impoundment was accepted as part of the 'Nisga'a Treaty Negotiations' (1996: 41).

3 The Supreme Court of Canada was asked in *R. v. Gladstone* (1996) whether the men had an Aboriginal right and, if so, whether infringement of that right was justified. It found a commercial right, but six of the nine justices agreed that the evidence and testimony presented in the case were insufficient to allow them to decide on the justification-of-infringement issue and ordered a new trial. Before a new trial could be arranged, the crown dropped the charges, bringing an end to the case.

WORKS CITED

Blackman, Margaret B. 1982. *During My Time: Florence Edenshaw Davidson, A Haida Woman*. Vancouver: Douglas & McIntyre and University of Washington Press.
– 1990. 'Haida: Traditional Culture.' In Suttles 1990: 240–60.
Boas, Franz 1921. *Ethnology of the Kwakiutl, Based on Data Collected by George*

Hunt. 35th Annual Report of the Bureau of Ethnology to the Secretary of the Smithsonian Institution, 1913–14, Part 1. Washington, DC: Government Printing Office.

Bouchard, Randy, and Dorothy I.D. Kennedy. 1989. 'The Use of Herring Roe on Kelp by the Indian People of Central British Columbia.' Ms. report for the Department of Fisheries and Oceans, Pacific Region, Vancouver.

British Columbia Fishery Regulations. Statutory Orders and Regulations (SOR)/55-260, sec. 3, 21A; SOR/74-50, sec. 9, 21A (1).

Brown, Anja, and Clarence Martin. 1985. 'Heiltsuk Herring Roe Harvest: A Living History.' Typescript report, Heiltsuk Cultural Education Centre, Waglisla, BC.

Callon, Michel. 1986. 'Some Elements of a Sociology of Translation: Domestication of the Scallops and the Fishermen of St. Brieuc Bay.' In John Law, ed., *Power, Action and Belief*, 196–233. Boston: Routledge & Kegan Paul.

Canada. Fisheries Act. 1970. *Revised Statutes of Canada* (*RSC*), c. F-14, sec. 61(1).

Carr, Emily, 1965. *Klee Wyck*. First pub. 1941. Toronto: Irwin Publishing.

'Category J – Herring Spawn-on-Kelp,' 1994. *Commercial Licensing Handbook*, 42–3. Ottawa: Department of Fisheries and Oceans.

Constitution Act, 1982. Enacted by the Canada Act, 1982 (UK), C.1911, Schedule B.

Dickson, Frances V., 1977. 'British Columbia Herring Spawn-on-kelp Fishery.' In Dennis Blankenbeckler, ed., *Proceedings of the Third Pacific Coast Herring Workshop, June 22–23, 1976*, Fisheries Research Board of Canada, Ms. Report Series No. 1421, 27–31. Nanaimo, BC: Pacific Biological Station.

Dickson, F.V., G.A. Buxton, and B. Allen. 1972. *Propagation and Harvesting of Herring Spawn-on-kelp*. Technical Report No. 13. Vancouver: Department of the Environment, Fisheries Service, Pacific Region.

Ford, Cellan, 1968. *Smoke from Their Fires: The Life of a Kwakiutl Chief*. First pub. 1941. Hamden, Conn.: Archon Books.

Halpin, Marjorie M., and Margaret Seguin. 1990. 'Tshimshian Peoples: Southern Tsimshian, Coast Tsimshian, Nishga, and Gitksan.' In Suttles 1990, 267–84.

Hanson, William, 1992. 'Kazunoko Konbu.' *Pacific Fishing* Aug.: 40–7.

Heiltsuk Indian Band et al. v. Canada (Minister of Fisheries and Oceans) (1993), 59 FTR 308.

Harris, Douglas. 1996. '"J" Licences and the Heiltsuk Right to a Commercial Spawn-on-Kelp Fishery: Is the Infringement Justified?' Unpublished report submitted to Fisheries Law 396, Faculty of Law, University of British Columbia.

Lane, Barbara. 1990. 'Harvest of Herring Spawn and Commerce in Herring Spawn by the Heiltsuk (Bella Bella) Indians of Central British Columbia from Aboriginal Times to the Present.' Unpublished report: expert testimony in *R. v. Gladstone*.

Newell, Dianne. 1993. *Tangled Webs of History: Indians and the Law in Canada's Pacific Coast Fisheries*. Toronto: University of Toronto Press.

'Nisga'a Treaty Negotiations: Agreement in Principle, 1996.' Canada, British Columbia, and Nisga'a Tribal Council, 15 Feb.

'Pacific Region, 1994 Management Plan: Spawn-on-Kelp,' 1994. Vancouver: Department of Fisheries and Oceans, Pacific Region.

Pacific Herring Fisheries Regulations. SOR/8–876, Sec. 8, 17(a) and (b); SOR/84–324, secs. 17(1) (a) and (b), 20(2) and (3).

Parsons, L.S. 1993. 'Management of Marine Fisheries in Canada.' *Canadian Bulletin of Fisheries and Aquatic Sciences* 225.

Pearse, Peter H. 1982. *Turning the Tide: A New Policy for Canada's Pacific Fisheries.* Final Report of the Commission on Pacific Fisheries Policy (Pearse Commission), 1981. Ottawa: Minister of Supply and Services Canada.

Regina v. William Gladsone and Donald Gladstone [*R. v. Gladstone*] (1996), SCJ no. 79 (Supreme Court of Canada), ordering a new trial (1993) 5 WWR 517 (BC Court of Appeal), which upheld (1991) 13 WCB (2nd) 601 (BC Supreme Court).

Said, Edward. 1993. *Culture and Imperialism.* New York: Vintage.

Shields, Thomas. 1984. 'Examination of a Kelp Suspension System for Use in the Spawn-On-Kelp Fishery.' Archipelago Marine Research, Victoria, BC, Report submitted to Cecil Hill, Western Spirit Fishing Ltd., Kitkatla, BC.

Shields, T.L., and G. Kingston. 1982. 'Herring Impoundment and Spawn-on-Kelp Production in British Columbia.' Archipelago Marine Research, Victoria, BC, Report submitted to Department of Fisheries and Oceans, Pacific Region.

Shields, T.L., and J. Watson. 1983. 'Spawn-on-Kelp Production Using Open Ponds in the Queen Charlotte Islands.' Archipelago Marine Research, Victoria, BC, Report submitted to the Department of Fisheries and Oceans, Pacific Region.

Shields, T.L., G.S. Jamieson, and P.E. Sprout. 1985. *Spawn-on-Kelp Fisheries in the Queen Charlotte Islands and Northern British Columbia Coast – 1982 and 1983.* Canadian Technical Report of Fisheries and Aquatic Sciences No. 1372. Nanaimo, BC: Department of Fisheries and Oceans, Pacific Station.

Stewart, Hilary. 1977. *Indian Fishing: Early Methods on the Northwest Coast.* Vancouver: Douglas & McIntyre and University of Washington Press.

Suttles, Wayne. 1987. '"Coping with Abundance": Subsistence on the Northwest Coast.' In Wayne Suttles, *Coast Salish Essays*, 45–63. Seattle, Wash.: Talon Books and University of Washington Press.

Suttles, Wayne, ed. 1990. *Northwest Coast.* Vol. 7 of *Handbook of North American Indians.* Washington, DC: Smithsonian Institution Press.

PART TWO:
STATE MANAGEMENT AND STATES OF KNOWLEDGE

8

Failed Proposals for Fisheries Management and Conservation in Newfoundland, 1855–1880

SEAN T. CADIGAN

The establishment of a moratorium in the eastern Canadian cod fishery in 1992 has stirred an old debate in Newfoundland about how cod (*Gadus morhua*) stocks should be conserved. Some fishing people fear that controlling effort by restricting the total number of people who can fish may overlook many people's right to fish by 'historic attachment'; others believe that they should be free to catch fish with any gear under strict regulations (*Evening Telegram*, St John's, 1996). Should cod stocks then be protected by controlling the type of gear used or the number of people who fish? Underlying this question is the issue of whether a central bureaucratic authority, or fishers, or both together should decide how to restrict effort.

Surprisingly, this is not just a twentieth-century problem. Recent research is proving that governments and people in Newfoundland were debating these points long before 'traditional' cod traps, gillnets, or even trawl lines were in use. The argument that the population of Newfoundland in the nineteenth century, and the economy that sustained it, had reached the limits of 'extensive growth' in the 1880s (Alexander 1980: 25; 1983: 3–31; Cadigan 1995b; Ryan 1986: xv–75) actually hints at possible ecological disequilibrium at that time. Until now, scholars have identified the problem with factors such as merchant conservatism, truck practices, external market difficulties, the cyclical nature of the resource, and/or the lack of a sufficiently diversified resource base able to support local market diversification and alternative employment. Nowhere has the impact of fishing effort on marine resources been identified as a possible contributing factor to the economic problems of the colony: the assumption has been that fish stocks were able to sustain any effort that could have occurred under the technologies of the day.

Now, there is growing evidence that between 1845 and 1880 increased fishing was harming Newfoundland's marine resources. As early as the 1840s significant public demand forced government to regulate use of new fishing gears so as to protect cod stocks. At that time it was held that, by tradition and law, all ocean resources were common, and therefore all rural Newfoundlanders had a right to fish. Such rights of access to the commons, however, began to have to carry with them the responsibility to establish conditions under which access to ocean resources would be regulated. Many Newfoundlanders had become aware of a growing imbalance between people and the primary marine resources they depended on for their livelihoods, one that threatened widespread impoverishment. Throughout the 1850s and 1860s, fears grew that the newer gears were exacerbating that disequilibrium by destroying cod stocks. In 1863, growing demands for conservation through the restriction of gear types forced the government to introduce legislation for the fishery, but the proposed law failed to pass because government wanted a colony-wide consensus about which gears should be restricted, and to what extent. Without such agreement, and given its limited financial resources, the colony could not hope to enforce such a law.

Regulation of the fishery from above was not the only option available. Throughout the late 1860s the government toyed with the idea of allowing communities to decide for themselves how best to limit access to the fisheries. By then, however, the use of more intensive fishing gear, combined with larger vessels, had become firmly entrenched. Fishing people who used the new technologies had become more mobile and begun to intrude on each other's fishing grounds. The government of the day, unsympathetic to economic regulation, feared stirring up greater social tensions with mandated community self-management of marine resources. Moreover, the new gear types were buttressing faltering fish exports, so essential to colonial economic well-being. Indeed, long before the 1880s, catch failures in the inshore fishery had become common, though adoption of more intensive harvesting gear appeared to alleviate the situation in the short run.

With hindsight and late-twentieth-century awareness, we can now understand that frequent fishery failures and a necessary shift to more intensive technologies, when set beside rapid population increase and large fluctuations in Newfoundland salt cod and seal exports, combine to point to a likely ecological problem. Figure 8.1 reinforces the picture: though overall total exports of salt cod tended to increase to 1884, there were dramatic drops between 1815 and the 1840s, in the 1860s, and after 1874. Exports of seal oil and seal skins also tended to fall. To be sure, prices for salt cod dropped in the post-1815 international depression, but fluctuations in total exports of salt cod do not appear to bear much relation to those in prices, as can be seen in Figure 8.2.

Not surprisingly, Newfoundland fishers increased the volume of salt cod pro-

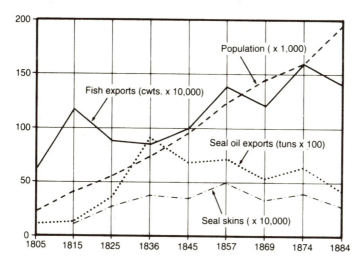

Figure 8.1   Newfoundland's saltfish exports in cwt. (1 = 10,000), seal skins
(1 = 10,000), seal oil in tuns (1 = 100), and human population (1 = 1,000), 1805–84.
Sources: Ryan 1986: 258–60; 1994: 446–9

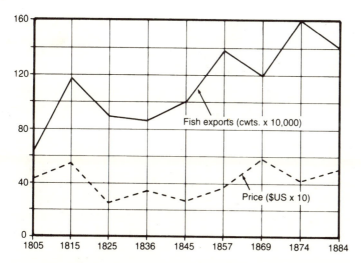

Figure 8.2   Newfoundland's saltfish exports in cwt. (1 = 10,000), fish prices
($U.S. × 10), 1805–84. *Sources*: Ryan 1986: 258–60; Vickers 1996: 92–104

duction to make up for declining returns throughout the first half of the nineteenth century (Cadigan 1995c). They also compensated for declining incomes from the cod fishery after 1815 by relying more on the seal fishery. Unfortunately there are no good price series to permit a secure association between exports of seal products and prices, but known popular dependence on the seal hunt for earnings suggests that reduction in effort would not be the best explanation of declining exports of seal products to 1884: it is more likely that the seal herds were also under pressure from excessive exploitation.

The growing public debate throughout the second half of the nineteenth century about the need for legislation to conserve seal herds tends to confirm the explanation of declining seal-product exports by excessive exploitation (Ryan 1994: 98–117), while constant (human) population growth meant even heavier effort in the fisheries as exports fell. No other option was available: other than fishing, sealing, and supplementary farming, Newfoundlanders had few alternative work opportunities from 1815 to 1884, and fishing remained the primary occupation of over 80 per cent of the working population, as shown by census returns (Newfoundland, various years). The economy was thus plagued by a continuing and serious decline in per-capita fish exports (Ryan 1986: 258–60).

Early Evidence for Stock Depletion

There is some evidence that fishers may have been exhausting discrete bay or substocks of cod in inshore waters in the first half of the century (Hutchings and Myers 1995: 37–93). The diaries of the fish-merchant firm Slade and Kelson (Slade-Kelson Diaries) show that its employees and clients noticed lessening availability of cod in Trinity Bay in these years. William Kelson, the firm's managing agent at Trinity, made almost daily observations about weather, trade, and fishing conditions (among many other things) from 1815 to 1852. In almost every year, the number of days on which Kelson observed fishing to be very bad exceeded by far the number in which fishing seemed good, as Figure 8.3 shows.

According to reports by the firm's clients and employees, the primary reason for Trinity Bay fishers' poor catches was that there were simply no fish to be found in the bay, as Figure 8.4 shows. Even when fishing people knew that cod were on the fishing grounds, there was no caplin (*Mallotus villossus*) or other bait fish available to enable them to catch it. Fewer recorded observations in Kelson's diaries in the 1840s and 1850s may suggest that by then fishing conditions had begun to improve; it is equally likely, however, that Kelson had given up commenting on persistently poor fishing. In 1847, for example, there are only four remarks about fishing, terminating in June, when he reported a 'good

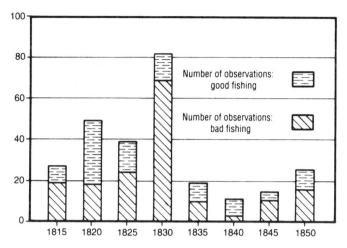

Figure 8.3   Observations on fishing conditions by five-year Intervals. *Source*: Slade-Kelson Diaries

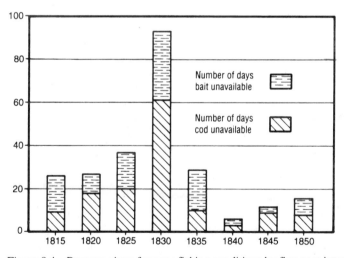

Figure 8.4   Reasons given for poor fishing conditions by five-year intervals. *Source*: Slade-Kelson Diaries

prospect of fish,' but his first comment the next year was that the fishery of 1847 had proved so bad that fishing people could not get credit to begin fishing in 1848 (Cadigan 1995c).

Newspaper accounts support the view that the inshore fishery was failing also in Conception, Trinity, and Bonavista bays regularly in the first half of the nineteenth century. Reports from newspapers for the period 1856–80 suggest that failure continued to plague fishers who exploited the longest-used inshore fishing grounds along the coast.[1] These reports deemed thirteen of those twenty-five years general failures (Cadigan 1996a: Appendix 7). It is thus likely that the conditions of scarcity that Kelson observed were more general and a major reason for the fluctuations in Newfoundland's saltfish exports in these years.

The cyclical recoveries in Newfoundland's total fish exports pictured above in Figure 8.1 might seem to contradict this analysis, were it not for the fact that we must consider some masking factors. First, increasing effort in the Labrador fishery probably disguised inshore failures on the island, as island residents often did not differentiate those returns from local catches. The successful fishing season of 1858, for example, resulted from a productive Labrador fishery that counterbalanced failures throughout the northeast coast, parts of Placentia Bay, and the southern shore of the Avalon Peninsula (*Newfoundlander* [*N*] 16 Aug. 1858). In 1860, when it became apparent that the inshore fishery along the northeast coast was failing early in August, an editor reported 'that several of those having large boats have gone off to the Labrador, whence we are glad to learn the accounts are favourable' (*Newfoundland Express* [*NE*] 4 Aug. 1860). The next year, the Conception Bay fleet encountered new vessels from Green Bay and Fogo, fishing in Labrador to compensate for failure in the inshore fishery, but their combined efforts were insufficient to ensure a good year (*N* 16 Dec. 1861).

Inshore fishing failures were also forestalled by the introduction of more intensive harvesting gear. In the 1840s, the use of cod seines, trawl lines (called 'bultows' at the time), and gillnets had become more widespread than the older hook-and-line method of fishing, and they were essential to the initial recovery and subsequent short-term maintenance of exports. For example, the 1859 inshore fishery on the southern shore of the Avalon Peninsula proved disappointing, with reports indicating that fishers using baited handlines were most frustrated, while the use of cod seines by others prevented complete disaster (*N* 18 July 1859). Similarly, though the fishery at Bonavista proved poor in 1861, one fisher using gillnets had a good year (*Courier* [*C*] 28 Aug. 1861).

In time, of course, even the use of more intensive gears could not guarantee good catches on a regular basis. The fishery of 1862 was very poor inshore because fishers throughout the northeast coast, Placentia Bay, Fortune Bay, and Cape St Mary's found that there were few fish for either cod seines or baited handlines (*N* 10 July, 11, 21 Aug. 1862; *NE* 22, 29 July, 2, 12 Aug. 1862). However, in 1864, when reports indicated that the fishery at Bonavista was failing, newspaper editors wrote that 'little or nothing [was] doing with the hook and line, but a fair catch with seines' (*Standard* [*S*] 13 July 1864). The following year, fishers blanketed the fishing grounds off Cape St Mary's with bultows to improve their catches, with some apparent success (*N* 19 June 1865).

Significantly, recovery in the 1870s coincided with the introduction of cod traps and expansion into newer fishing grounds further offshore, especially in the bank fishery (Hutchings and Myers 1995: 37–93). Many inshore fishers began to respond to the failure by 'fitting out large craft to prosecute it at a distance' (*S* 15 July 1863). Some intended to go to the Labrador fishery if the inshore fishery proved poor in a particular season, but others wanted to expand their mobility in inshore and nearby headland waters to search for newer, more productive fishing grounds (*S* 31 May 1865; *Public Ledger* [*PL*] 27 July 1866).

Such failures of the inshore fishery impoverished rural Newfoundlanders and led to public demands for relief. Episodic improvements in the fisheries between 1856 and 1880 did not alter the inshore failures or even hide the resultant socioeconomic problems from observers. In a final public address on leaving Newfoundland in 1864, for example, Governor Alexander Bannerman pointed out that the long-used inshore fishing grounds could no longer support the growing population. Without other resources to exploit, 'there appears but one remedy for an increasing evil ... namely, an extension of the fishing grounds' (*PL* 13 Sept. 1864).

By 1868 larger boats were doing well in headland fisheries off Lamaline, Cape St Mary's, Bay de Verde, and Grates Cove and from Cape St Francis to Cape Race. Most of the inshore fishing people who used small boats on fishing grounds within Newfoundland's bays did not share in the success of headland fishers. Newspapers judged the 1871 inshore fishery a success for those using cod seines, gillnets, and bultows alone. Through the 1870s even these gears yielded poorer catches inshore, and the general fishery improved over the next two years on the strength of cod seine catches at Labrador. Poor fisheries through the late 1870s led newspaper editors to exhort people to invest more in larger boats and better equipment. Some fishers were using new cod traps by 1876, even as reports of a fish scarcity surfaced around the island, and, though cod traps did not immediately contribute to good inshore fisheries, by 1879 and 1880 newspapers were crediting the gear for good catches at Labrador (*Express* [*E*] 8 Sept. 1868; *C* 29 July 1876; *PL* 30 June, 7 July 1876; *N* 1 Aug. 1871, 23 Aug. 1872, 2 Sept. 1873, 14 July 1876, 29 July 1879, 10 Aug. 1880).

Social Protest and Government Response

The gears increasingly used to boost catch rates were much more expensive than the commonly used handlines. This led to debates about equitable access to the resource and concerns about what we would today speak of as its 'sustainability.' As early as the 1840s poorer fishing people had begun to protest the use of cod seines, and, while some public commentators attributed such protests to envy, there were more complex reasons behind the complaints.

William Kelson (mercantile agent at Trinity), writing under the pseudonym of the famous angler 'Izaak Walton,' began a newspaper correspondence campaign that articulated the conflict in terms of ecology and equity. Kelson believed that cod jiggers, seines, and bultows caused cod scarcity on old inshore fishing grounds by frightening off or diverting habitual fish-migration patterns. He suggested that the greater efficiency of seines and bultows meant that only those few fishers who could afford such gear might be able to catch the increasingly scarce resource, leaving the mass of fishing society to poverty and possible starvation.

Kelson began to lead mass, open-air demonstrations for the prohibition of all gears except baited handlines and in 1849 circulated public petitions that asked the House of Assembly to bring forward such legislation. The colonial government refused to take any measures against cod seines, bultows, and jiggers, denying the need for conservation and accepting the argument that introduction of new technology was the prerogative of private capital. It blamed inshore scarcities on French and American overfishing offshore. Frustrated, some fishing people began to haul up, or obstruct the use of, seines and trawls. Those who persisted in using the new gears sometimes received anonymous beatings (Cadigan 1995c).

Demands for better management of fisheries quickly greeted the new system of responsible government in 1855. Kelson continued his pseudonymous letter-writing campaign for protection of cod stocks, but now he alerted newspaper readers to impending destruction of seal herds. Besides still opposing jiggers and cod seines as harmful, the old Trinity merchant warned that the tendency for sealers to kill younger seals each spring at the ice was threatening the ability of seal herds to reproduce. Kelson called on the legislature for an independent commission of inquiry into the impact of all types of gear on cod stocks and, almost to the end of his life, continued to warn Newfoundlanders that the only method of fishing that would not harm cod stocks was the baited hook and line. He was sure that fishing people would witness a continuously declining fishery as long as the use of jiggers and cod seines increased (*PL* 4 April, 4 July, 22 Aug., 22 Dec. 1856, 20 Jan. 1857, 31 Aug., 21 Sept. 1858, 8 May 1860).

By 1856, some opposition Conservatives in the legislature had begun to take up the cause of fisheries reform. That year, opposition MHA W.H. Ellis engaged in a bitter debate with the Liberal colonial secretary, in which he demanded that the government recognize that the inshore fishery was failing and take measures to deal with the problem. Though the Liberals denied that the fisheries were in trouble, Conservative leader H.W. Hoyles applied more pressure. Hoyles argued that Newfoundland must stop the French from taking mother cod (large, female spawners) in the bank fishery by cutting their bait supplies. The colony must further prohibit use of caplin as manure. Pointing to the almost moribund fishery within Conception Bay, where local people took caplin mostly for manure, Hoyles suggested that the diminution of caplin there had discouraged cod from migrating inshore annually in search of food. Liberal MHAs such as Thomas Talbot would not agree with the Conservatives, especially on the issue of taking caplin for manure. The Liberals denied the need for conservation in a utilitarian manner: 'It is intended by nature that caplin and all other productions of the sea and land should be used for the extension of human food' (*N* 28 April 1856).

Politics and Fisheries Management

The growing importance of the south-coast herring fishery led to more demands for conservation. Some fishing people in Fortune Bay had begun to use large, small-meshed nets that caught many undersized fish, which they discarded. One St John's newspaper editor reported a popular fear that such practices would scare off Atlantic herring (*Clupea harengus*) from their usual migratory routes by the presence of so much dead fish in the water. Concerns about the need for fisheries management fuelled demands for an appropriate colonial government department. St John's newspapers commented on how odd it was that Newfoundland had a Board of Surveyors to manage land resources, with its head, the surveyor general, a minister of government, while, even though fishing was by far its most important economic activity, Newfoundland had no equivalent bureaucracy for the marine resources. Such a deficiency would, the press advised, be best ended by the immediate appointment of a Board of Fisheries (*PL* 15 Sept. 1857).

Governor Bannerman supported the demand for better fisheries management. He noted that fisheries policies were limited to protests about fishing rights extended to the French and Americans by imperial treaties. Puzzled that no colonial law prevented 'the indiscriminate capture of all sorts of fish, nor does any attention seem to be paid to their breeding season,' Bannerman called for fisheries regulations to conserve marine resources. As imperial treaties did not allow the French and Americans fishing rights that would violate colonial law, such legislation would also serve to limit those fisheries (*PL* 29 Jan. 1858). The legislature responded during the spring session of 1858 by passing a law to regulate the size and manner in which seines could be used in the herring fishery. Bannerman applauded the effort but noted that the colony should establish a commission to look into a more general fisheries code such as those of Canada, Nova Scotia, and New Brunswick (*PL* 11 May 1858).

Though the Liberal government demonstrated limited enthusiasm for such an inquiry, at least some fishing people kept up the pressure for action. In 1860, for example, the inhabitants of Bay Bulls petitioned the legislature to establish a four-inch minimum mesh size for cod seines. In presenting the petition to the House, Thomas Glen, the receiver general, pointed out that local people never minded the limited use of small seines in the cod fishery. Now huge seines were being used to sweep fishing grounds, destroying juvenile fish and scaring away others. Hook-and-line fishing people consequently found less fish to catch so that they might earn a living.

Despite much disagreement about how to respond to the petition, the attorney general, G.J. Hogsett, did feel that the Assembly should appoint a select committee to investigate the problem further. The main difficulty with taking any action on seines was that cod seines had become the most important gear in the inshore fishery's annual cod catch; moreover, Hogsett felt that the legislature should not interfere with fishing people's own decisions about how to catch fish (*NE* 21, 25 Feb. 1860). The government launched an investigation of the seal fishery, where declining resources were most acute, and established small fines for those who killed undersized young seals (*S* 4 April 1860; *NE* 23, 30 Oct. 1860; *N* 7, 11, 21, 25 Feb. 1861).

The early impulse for fisheries conservation faced many obstacles. The colony lacked resources to fund fisheries management, but a greater problem was the nascent rivalry between users of different gear types. Small-boat handliners argued that almost all the other types of gear harvested too intensively and were threatening the viability of the inshore fishery. However, not everyone was convinced that fishing was reducing stocks inshore. Many commentators continued to believe that fish appeared or not by the will of God, not by the continued use of certain fishing gears (*NE* 24 Jan. 1862). The Liberals appeared convinced by this argument and left the fisheries largely to their own fate. They concluded that their efforts were better spent in developing landward resources, over which colonial jurisdiction was clear (Cadigan 1995a, 1996b).

The Conservatives won the 1861 general election and accepted that they must investigate possible French overfishing by bultows on the Grand Banks and the related bait trade by south-coast Newfoundlanders. For the first time, conservation rather than limiting foreign competition became the primary object of fisheries management (*NE* 25 Jan. 1862). A new act for the herring fishery in 1862 prohibited the seining of herring during the spawning season, established a minimum mesh size of 2.5 in. (6.4 cm), and forbade the barring of herring in coves or the use of seines within a mile of settlements on the south coast. At the close of the legislative session in March 1862, Bannerman congratulated the House of Assembly on its new interest in fisheries legislation and looked forward to additional greater measures (*PL* 28 March 1862; *N* 24, 31 March 1862).

Newspaper correspondents pushed the government to take greater action, arguing that the legislature must provide the funds for effective enforcement as well and promoting 'pisciculture,' especially in the restocking of salmon rivers (*N* 15 May 1862). One editor believed that the government should appoint a full-time scientist to investigate local fisheries, make recommendations to the legislature for the enhancement of cod stocks, and protect the fisheries from unwise domestic fishing practices. Scientific management was necessary – the time had come for a comprehensive inquiry into the effect of various methods of fishing on the actual fish stocks of the colony (*PL* 6 June, 8 July 1862; *NE* 15 July 1862).

Early Explorations of Science-Based Management

Late in 1862 the Hoyles government bowed to pressure for action by calling for a select committee of inquiry into the state of the Newfoundland fisheries in the next session of the legislature (*NE* 24 Dec. 1862, 8 Jan. 1863). The old view that fish species were divinely inexhaustible, based on their apparent fecundity, was being questioned in the British press, and Newfoundland's papers reprinted debates on the issue for their readers.

Local editors began to realize that any particular fish was merely part of a food chain; most of its eggs were consumed by other marine creatures, and those that hatched were often preyed on before maturity. Taking a dim view of fishers' own knowledge of the habits of cod, the British writers suggested the need for scientific observation on the life expectancy and reproductive habits of cod and turbot in places such as Scotland, the Great Dogger and Rockal Banks, and the Faroe Islands, because 'exhausted shoals and inferior fish tell us but too plainly that there is reason for alarm, and that we have in all probability broken upon our capital stock.' Such study might lead to laws that prevented destruction of spawning fish and to the revitalization of fish stocks by aquaculture and artificial propagation (*NE* 15, 18 Jan. 1863, quote from 10 Jan. 1863).

The Newfoundland inquiry coincided with the formation of a mutual-benefit Society of Fishermen at St John's, but there appears to have been little connection between them. Governor Bannerman addressed the society on the need of fishers to improve the quality of their cure but did not invite the society to assist in the investigation of the reasons for failing inshore fisheries (*PL* 27, 30 Jan. 1863). Some people, moreover, still thought it beyond the ability of government to deal with crises in the fisheries, arguing that only fishers and merchants could immediately respond to shortfalls in catches by lowering costs and bettering cures. Such measures would ensure higher incomes, which would lessen demands for public relief (*N* 2 Feb. 1863). Others argued that legislation could ban the use of bultows if, as many believed, they caught mother fish that would otherwise spawn, while cod seines could also be banned if they did indeed capture too many juvenile fish.

Furthermore, though the government had no way to force the French to observe conservation regulations in their treaty areas, passing such laws would encourage public pressure on the French government to ban the use of harmful gears in Newfoundland waters (*NE* 3 Feb. 1863). While the colony could not ban its competitors, it could prohibit specific gear types that they used to harvest too intensively. However, some observers warned that any laws that the committee did propose would only be evaded by fishers who disagreed with them, because the government could not pay for enforcement (*PL* 27 Feb. 1863; *N* 2 March 1863).

There was public support for, as well as opposition to, greater state intervention in the fishery. As word of the fisheries inquiry spread, the legislature began to receive petitions from around Newfoundland asking for the restraint of specific types of gear. Tensions in one south-coast area grew so bad that the Supreme Court investigated the matter and reported to the government that about 160 fishing people had decided not to use bultows, believing that they 'destroyed the mother fish,' and opposed cod seines as being harmful to the fish. When others continued to use the gear, the opponents took the law into their own hands 'to prevent what might be deemed by some an improper mode of fishing.' The court quieted the situation by advising the fishing people to petition the legislature for a change in the law to back their views (*Journal of the House of Assembly* 1863, 35–6, 42, 45–6, 53, 71, 86, 99, 107; appendix, 1197–8).

Throughout the winter of 1862–3 the Select Committee received evidence from 104 witnesses – merchants, fishers, planters, magistrates, and government officials testified directly before the committee, provided written submissions answering circulars by the committee, or petitioned the assembly to put forward their concerns. Many addressed concerns about specific types of gear and/or practices without commenting on others. By far the most felt that bultows, jiggers, gillnets, and taking caplin for manure were harmful to the inshore cod fishery.

Among a wealth of evidence, too detailed to catalogue here, were comments on such matters as using seines to catch cod, but ending up catching and discarding vast quantities of juvenile fish, thereby hurting cod stocks. Jiggers wounded more fish than they caught and scared other cod off the fishing grounds into deeper water. Fishing people tried to follow the cod to such waters by using deep-water trawl lines. But these bultows tended to catch the largest fish, or spawners. These 'mother fish' were often covered with roe, and no fishery could survive if it harvested most of the breeding stock (*Journal* 1863, Apendix, 441–3). Many fears echoed those we know too well today: 'The great oceanic supply of Codfish is rapidly declining; and that caplin are much less abundant than formerly, we have ample proof' (Appendix, 521). Very few people wished to defend any of these gears or practices, except in the case of cod seines. More witnesses (36.5 per cent) argued that seines did not injure the fishery than felt that they did (30.0 per cent), and by the 1860s cod seines appear to have gained greater respectability among inshore fishing people.

The minority of witnesses, who defended more intensive fishing gear, argued that their opponents were motivated by envy of 'these somewhat expensive but remunerating means of fishing.' Some argued that the use of baited handlines was always preferable, but when they were unsuccessful people still had to provide for their families. It was not the business of the legislature to interfere with their right to catch fish any way they could (1863, Appendix, 445–8, 471–3). Those who used bultows in Fortune Bay avowed in a petition 'that without the use of the bultow they would not be able to obtain a living for their wives and families with the hook-and-line' (Appendix, 514–15).

Sadly, the committee's deliberations were ultimately ineffectual. The chair, MHA John Rorke, introduced an omnibus bill in the assembly in March for a first reading, so that it might be circulated for public discussion at the close of the session. The bill would have banned use of caplin as manure, restricted the time and place for jiggers, abolished use of bultows, restricted herring seines to the period from 20 October to 20 April, prohibited cod seines larger than 100 by 70 fathoms after the next two sessions of the legislature, banned cod seines from fishing grounds used by hook and liners, and not allowed gillnets to interfere with cod seines or hook-and-line fishing. Rorke also proposed legislating the improvement of fish cures and providing premiums for the establishment of smoke-houses (*NE* 21 March 1863).

Rorke never seriously considered that the government might pass his bill. The committee's evidence had not provided him with enough political security to act more boldly. Though more witnesses were in favour of the government's legislating against cod seines, bultows, and gillnets than against, the percentages favouring new laws were much lower than those believing that the gears were harmful to the inshore fishery. Witnesses provided relatively strong support only for proposed limits on the use of cod seines on fishing grounds (called 'ledges') used by handliners and on the size of the seines. Even in the case of seines, opposition to regulation was higher than in the case of bultows and gill-nets. Few proposals were as pointed as the Fishermen's Society of St John's, which explicitly proposed banning gillnets and the hauling of caplin for manure and limiting bultows, cod seines, and jiggers (*Journal* 1863, Appendix, 495–7). The Conservatives were quite aware of the variety of fishing practices used throughout Newfoundland. Rather than risk imposing unpopular legislation from above, they preferred to seek consensus for future legislation in public debate that might be stimulated by Rorke's proposal (*PL* 24 March 1863).

The government none the less continued to receive advice on the negative impact of new fishing technologies on fish stocks. For example, Abraham Gesner (the Nova Scotian geologist, inventor, and naturalist) informed New-foundland's colonial secretary, Robert Carter, that it was 'a startling fact that the fisheries are rapidly falling off along the entire North American coast ... and this decline has evidently resulted, in a great degree, from the modern destruc-tive modes of fishing, and the practices of foreigners' (Provincial Archives of Newfoundland and Labrador [PANL], GN2\2, 1863–64, Gesner to Carter, Hal-ifax, 29 May 1863). Opposition to the bill made it clear that the fishing industry had come to rely on the increased use of newer, more intensive fishing gears to stimulate short-term recoveries in fish catches. Such dependence divided fish-ing communities between those who could afford the new gear and those who could not. Such circumstances made it extremely unlikely that the government could use new laws to manage the industry and conserve fish stocks with unan-imous public support.

Failure and Its Aftermath

Opposition to the act demonstrated to government the unlikelihood that it could expect consensus of all fishing people on even the most direct legislation (*N* 17 March 1864). The government consequently abandoned the fisheries bill.[2] On the close of the legislature in April, Governor Bannerman addressed the body on the proposed bill to the effect that 'there has been so much differ-ence of opinion ... – with difficulties as well as prejudices to overcome, – that I understand, the honourable gentleman who introduced the measure has (I think very wisely) thought it inexpedient at present to ask for legislation on so impor-tant a question' (*N* 14 April 1864).

The difficulties of fisheries management would not go away because the gov-ernment feared to act. Further equivocation only worsened the problem. Fishing people at Renews, led by William Kelligrew, asked the government again to ban bultows in February 1865, to no effect (*Journal* 1865, 31). 'Piscator' wrote in 1865 that cod in Placentia Bay seemed fated to disappear. He or she believed that handliners' and cod seiners' recent persistent poor catches proved that cod were disappearing, since such gear had had no problems years before. Most fishers who relied on baited handlining opposed the use of bultows and gillnets because they feared that these newer types of gear would take the last fish. But in the absence of legal prohibition, 'Piscator' argued, many of the same fishers adopted the new gear against their own better judgment and only because others did so: if they did not, handliners' productivity would fall behind that of users of new gear. Such deficiencies would make the handliners far less likely to secure the merchant credit they needed to begin the fishing season. 'We are not agreed as to the best plan of improving the fisheries,' the correspondent wrote, 'every one has a plan of his own, and those who succeed with cod-net and bultow only think and act for the present and appear to be quite careless as to the future of the fishery.' The longer it was legally possible to use any new gear, the worse the problem of unrestrained intensification of effort by such a strategy would become (*N* 13 July 1865).

The government doubted that a central authority had the knowledge to design laws that would suit the multivariate circumstances of the fisheries. Any laws that it might make would be sure to hurt some participant and lead to widespread violations of the law. Not only would they become unpopular, but such laws would dissipate an already strained treasury on relatively ineffective enforcement. Despite such views, the greatest proportion of petitions to the legislature on the subject of gear regulation supported action. Nine out of eleven petitions from fishing people at Toad's Cove, St Mary's Bay, Shoe Cove, Oderin, Burin, Placentia Bay, Cod Roy, and Torbay asked for government to either ban completely or restrict use of bultows and/or cod seines (*Journal* 1866, 54, 55, 73, 113–14, 121, 125, 165–6, 108, and 114–15), respectively.

The government persisted in its refusal to regulate fishing gear. On 28 February 1866 a new cabinet proposed that it should attempt no further legislation but rather allow individual communities to decide how they might exploit local fishing grounds. 'According to our law,' it argued, 'the sea was common, and every man had a right to fish in it.' The legislature could, however, back community restrictions on the amount and type of gear that each fisher used on local fishing grounds. The attorney general suggested that the government could recognize any regulation of local fishing grounds agreed to by a three-quarters majority of a community. Enforcement would lie in the hands of community members. Anyone who felt misused by such enforcement could bring their case before the local justice of the peace for arbitration. The assembly dropped this novel proposal when one member lambasted the government for even considering fisheries conservation again – something that should be left to the individual decision of fishing people (*N* 22, 26 March 1866).

The tide was beginning to turn against conservation in press and government circles by 1866. A recent report of a British royal commission on sea fisheries denied that there was a problem with declining fish stocks in the North Atlantic because of overfishing. Charging that ordinary fishing people did not have the scientific training to ascertain whether stocks were healthy or not, the commission argued that fish were actually increasing. The proof lay in the manner in which larger vessels now caught more fish than anyone could dream of years ago. All conservation laws must prove ineffectual because no amount of effort could diminish ocean species that fluctuated normally with the availability of prey. Fishing people who complained about newer types of gear, thought the commission, were too lazy to adopt them.

This British report came exactly when some Newfoundland fishers were beginning to expand by using more gear and larger boats in headland fishing. The editor of the *Public Ledger* argued that if there was no need for conservation in the British fisheries, then it was not possible for a few thousand Newfoundland fishers to hurt their own resources. Faltering exports were not the result of failing cod stocks, but were rather the consequence of a lack of local energy and initiative. Newfoundlanders should expand the range of their inshore fishing craft to go 'after the fish that is abundant elsewhere': 'We spend our time debating about "bultows" and "cod-seines" – laying down laws how not to catch fish, with our population starving. Are there no men of energy or enterprise, with clear heads and some little capital, who will initiate a more excellent way?' (*PL* 7 Aug. 1866). Others suggested that the government should turn away from conservation towards to stock enhancement, by means of scientific study and artificial propagation (*N* 25 Oct. 1866).

Fishing people who could not afford larger boats or more equipment were not, however, about to let the issue of conservation disappear. Many continued to believe that their increasing inability to catch fish was the result of more intensive fishing practices. Such fishers further demonstrated their willingness to regulate access to the fishery even without legal support and recognition. In the summer of 1868 fishing people in a number of outports destroyed gillnets when the hook-and-line fishery failed while those using the nets did well. A correspondent from Twillingate wrote that the hook-and-line fishery there improved when the people forced the gillnets out of the water. At Greenspond, Bonavista Bay, fishers committed 'some unwarrantable proceedings against those using gill nets.' The *Public Ledger* called on the government to enforce the laws protecting private property (*PL* 6, 15, 24 Aug. 1868). In Placentia Bay fishing people destroyed bultows and cod nets. The Southern Circuit Court arrested nine men in Fox Harbour and convicted them of such offences. Another fisherman was arrested at Gallows Harbour, Placentia, for attacking a bultow fisher, attempting to sink the latter's boat with him in it and cutting the bultow lines 'at a time when they were most profitable.' The Gallows Harbour man was also convicted (*PL* 11 Sept. 1868). While the press applauded the court's swift punishment of those who destroyed fishing gear, there was little acknowledgment of the oddity apparent in the state's ability to defend property so readily when it had long pleaded its inability to enforce conservation legislation – hardly a more expensive undertaking.

The opposition took up the fight for legal recognition of communities' right to self-regulate access to local fishing grounds. In April 1869 the Carter government asked the legislature to give justices of the peace summary powers to punish on the spot people such as the Gallows Harbour man who destroyed bultows. The opposition protested, claiming that there 'were certain places where they had local customs by which the use of bultows was interdicted, and it would certainly be a great hardship if, under this law, strangers should be at liberty to come and violate those regulations by which the people of the place restrained themselves.' The government did not succeed in gaining additional powers for the justices of the peace. But neither did the opposition gain any greater recognition of community self-management, as political developments pushed the fisheries from centre stage (*E* 27 April 1869).

In the general election of 1869 the battle over Confederation displaced fisheries management and conservation. The Carter government initially proposed that union with Canada would provide Newfoundland with the resources to effect fisheries legislation. The argument that Newfoundland was not capable of managing its own resources alienated nationalist sentiment and helped the anti-confederates, led by C.F. Bennett. Bennett's personal investments in mining made him a target for confederate attacks on his self-interest in wanting to make the colony's supposed interior resources the private preserve of a St John's elite. By arguing that Newfoundland had interior resources that Canadian expertise and capital could help develop, the Carter-Shea alliance played into Bennett's hands. The anti-confederate leader contended that the prospect of internal diversification held out by the confederates was a good reason for not joining, as Newfoundlanders should enjoy the benefits alone. Both sides in the Confederation debate obscured the larger issue of the need for fisheries management by focusing so much on the question of landward diversification (Cadigan 1996b).

The protection of cod had begun to slip from the legislative agenda just as new information suggested that it was possible to threaten them with overfishing. In 1870 Moses Harvey began to write publicly that cod stocks inhabited discrete locations inshore, and these fishing grounds could be exhausted. People had an erroneous belief that the fish were limitless because of the number of eggs laid by spawners, but most eggs never reached maturity because of predation (*E* 16 June 1870). The government chose rather to focus action on the salmon (*Salmo salar*) fishery, which was visibly in decline because of the practice of barring rivers. As barring continued in the herring fishery, the government began to review legislation there as well (*E* 25 Jan. 1871). The new Bennett cabinet also faced fears that the seal fishery was about to be annihilated by the overexploitation of herds in the spring. By 1871, none the less, there was little strong press support for conservation.

Conservation by preventing the depletion of fish stocks through gear regulation was not an easy choice for government, partially because of the importance of cod seines, gillnets, and bultows to the inshore catch. The poor season of 1872 would have been worse, claimed one newspaper, but for these types of gear; the hook-and-line fishery around Newfoundland had caught almost nothing. 'It is well that the fish are taken,' its editorial stated. 'We do not see how legislation can improve matters. If a man cannot catch fish in one way, there is neither right nor reason in preventing him from catching it in another, if he is able.' Some fishers had begun to combine in community societies to adopt local regulations, but the development of more far-ranging fishing craft meant that people 'might come from other bays and harbours, who were not bound by these regulations, and thus render any local law of no effect.' The only solution was for those fishers who could not afford new gear and larger boats to leave the fishery by finding other occupations (*PL* 20 Aug. 1872).

The government did not completcly abandon efforts to regulate the fishery. In 1875 the press greeted favourably new legislation on coastal fisheries and pickled fish designed to protect salmon stocks and improve the quality of cured herring. One paper hoped that the colony might go further by developing a salmon restocking plan for its rivers (*C* 1 May 1875). The editor of this paper, the *Courier*, went further at the end of an extremely poor inshore fishing season in 1875. He recalled William Kelson's early protests about the decline of the inshore fishery and the use of new types of gear. Now that Americans had greater rights to fish in Newfoundland waters by the Treaty of Washington, there seemed little hope that fish could bear up under the greater pressure. It was time for Newfoundlanders to aid the fisheries by 'legislation, experience, science, and every other means at their command.' Assistance meant artificial propagation and restocking as the Atlantic Canadians were doing in their salmon fisheries. Newfoundland should further diversify its economy landward to draw surplus labour from the fisheries (*C* 11 Dec. 1875, 22 April 1876).

Others preferred that legislation focus on the protection of bait fish so that the cod fishery could go on unchecked (*PL* 14 Jan. 1876).

Controversy about types of gear such as bultows did not go away; government simply no longer responded to it. The poor fishery on the south coast from Cape St Mary's to Burin in 1877, during which even the seines failed, led many people there to protest that bultows had taken too many spawning mother fish (*PL* 24 July 1877). Worries about the need for conservation still surfaced around the seal hunt, as stocks there continued to decline even as merchants introduced larger steamers to search out the remaining animals. Throughout the 1870s and 1880s the government made half-hearted efforts to pass conservation legislation for the sealing industry. Constant demands for employment and the amount of investment by merchants in steamers brought constant public pressure to relax such laws. The net effect was that 'the legislative efforts to cope with the decline in the seal stocks were feeble and enjoyed very little success' (Ryan 1994: 117).

Conclusion: Later Years

In 1879, the Society of United Fishermen asked the Newfoundland government to create a department of fisheries or a fisheries science commission to take charge of any further investigation of, and recommendations for, the fisheries. There was by that time, however, little interest in fisheries management, as the press and government turned to landward diversification though a comprehensive plan to open up mining and forest resources by a trans-island railway (*N* 28 March 1879; *C* 10, 24 Feb., 10 March, 4 Aug., 6 Oct., 29 Dec. 1877, 12 Jan., 2, 9 March, 4 May, 1 June 1878; *PL* 30 April, 4 May 1878). Though the Whiteway cabinet paid lip service to the need for a commission of inquiry on the fisheries, it took no action until 1888, when it established a Fisheries Commission (Hewitt 1993: 58–80).

Government fisheries policy otherwise concentrated on industrial expansion by encouragement of the bank fishery, the discovery of new fishing grounds off the northern coast of Labrador as older ones showed signs of exhaustion, and use of telegraph lines to allow fishers to shift effort rapidly to those areas reporting good catches. The government continued to promote construction of larger inshore fishing craft that could roam further out into headland areas looking for newer fishing grounds, the offshore bank fishery, and diversification into the herring and lobster fisheries (*N* 26 Aug. 1870, 22 Dec. 1876, 13 June, 1 July 1879, 13 Feb. 1880).

The Fisheries Commission of 1888, supervised by Adolph Nielsen, demonstrated greater interest in cod-stock enhancement by the artificial propagation of cod than in conservation by restraint of their depletion in the first place (Hewitt 1993). Though support for marine-resources conservation was very strong in Newfoundland earlier than in the United States or the rest of British North America, by the late 1880s the colony had abandoned it and fallen a decade or so behind in other areas of fisheries management. Canada had put in place rudimentary regulations in the 1870s, as had the Americans with the U.S. Fish Commission. While both Canadian and American fisheries management focused more on licensing and stock enhancement through artificial propagation and exotic transplants than on conservation, Newfoundland was much slower to reach even this point, partially because of the government's preoccupation with landward diversification (Cadigan 1996b; Gough 1991: 7–11; McEvoy 1986: 101–19).

The loss meant even more than Newfoundland's simply lagging behind Canada and the United States. Newfoundland depended far more than they on the exploitation of marine animals. But by waiting too long to explore community self-management as an alternative approach, which would have allowed communities to control access according to local needs, the colony ensured that, in the long term, such management would be just as difficult as directives from St John's. In little more than a decade not only had gillnets, trawl lines, and cod seines become fixtures of the inshore fishery, but cod traps had also appeared, as well as larger, more wide-ranging vessels. Access to the inshore fishery had become more open while government procrastinated, making community jurisdiction over local fishing grounds less clear.

By 1880 many of the types of fishing gear today considered 'traditional' had become established without any alternative commitment emerging to control entry to a common-property resource. There still were abundant warnings against the wastefulness of dumping immature cod taken by seines or the folly of using bultows to take large breeding fish covered in roe. Claiming that three-quarters of the island's fishers would obey banning of such gear, such voices prophesied that failure to act would one day mean that there might be no more fish to catch by any means (N 20 May 1879).

Government did not act and claimed, as justification, that many fishers were willing to use more intensive gear. Such willingness suggests that nineteenth-century Newfoundlanders conformed to the Gordon-Hardin thesis about fishers' tendency to fish without regard for conservation because of the risk of rent dissipation in an industry characterized by a common-property, supposedly open-access resource – the tragedy of the commons (Gordon 1954; Hardin 1977). The real tragedy, however, has been the general acceptance of this 'bio-economic' model of human behaviour, with its impoverished historical perspective, which has dominated fisheries management in Canada for too long (McCay 1994–5). The constant public voicing of concerns about possible marine-resource depletion, popular protests against more intensive harvesting technologies, and demands for either legislative action or recognition of community rights to self management add further weight to recent work that undercuts the notion that fishers are Hobbesian competitors who act reasonably only when compelled to by the state (Matthews 1993).

The failure of the Newfoundland state to respond to popular demands for restrained access is a more likely explanation of the tragedy of the commons. State management of the cod fishery during the nineteenth century pursued extensive economic growth fuelled in large part by more intensive harvesting of marine resources. Newfoundland fishers had always been part of highly commodified staple production (Cadigan 1995b). They had to keep fishing no matter what the resource and/or market conditions, because they had to have access to merchant credit to purchase necessities in a coastal ecology that could not support much import substitution. This compulsion to fish limited people's ability to remain committed for long to the prohibition of new types of gear to conserve fish stocks. Once that moment of commitment passed because of government inaction, the greater productivity of new gear used by a few forced other fishers to adopt it in order to survive in an increasingly competitive credit environment (Hiller 1990). The state's ineffectuality consequently contributed to the growing economic culture of the open-access resource in the cod fishery. That DFO and fishers in 1996 renewed such an old debate about how best to conserve cod stocks testifies to the opportunity lost 130 years earlier to establish the basis of an ethic of community responsibility and nurturing of marine resources.

NOTES

1 I read every surviving issue of four St John's newspapers: the *Public Leger* (*PL*), the *Newfoundlander* (*N*), the *Newfoundland Express* (*NE*), and the *Courier* (*C*), recorded any comments that they made about general fishing conditions, locations of failed and successful catches, types of gear used, and explanations offered. I supplemented this data by reading the only Newfoundland paper from outside St John's to have been published in the period for which copies exist: the *Standard* (*S*), Harbour Grace. I recorded and examined the data using a simple database program: Paradox 5.0.

2 H.A. Innis mistakenly thought that the 1863 bill became law (Innis 1954: 396–8).

WORKS CITED

Alexander, D. 1980. "Newfoundland's Traditional Economy and Development to 1934." In James Hiller and Peter Neary, eds., *Newfoundland in the Nineteenth and Twentieth Centuries: Essays in Interpretation*, 17–39. Toronto: University of Toronto Press.

– 1983. 'Development and Dependence in Newfoundland, 1880–1970.' In David G.
   Alexander, compiled by Eric W. Sager, Lewis R. Fischer, and Stuart O. Pierson,
   Atlantic Canada and Confederation: Essays in Canadian Political Economy, 3–31.
   Toronto: University of Toronto Press.
Cadigan, S. 1995a. 'A "Chilling Neglect." The British Empire and Colonial Policy on
   the Newfoundland Bank Fishery, 1815–1855.' Paper presented at the Annual Meeting
   of the Canadian Historical Association, Montreal.
– 1995b. *Hope and Deception in Conception Bay: Merchant–Settler Relations in Newfoundland, 1785–1855*. Toronto: University of Toronto Press.
– 1995c. "Marine Resource Exploitation and Development: Historical Antecedents in
   the Debate over Technology and Ecology in the Newfoundland Fishery, 1815–1855."
   Paper presented at the Conference on Marine Resources and Human Societies in the
   North Atlantic since 1500, St John's, Memorial University, Oct.
– 1996a. "The Sea Was Common, and Every Man Had a Right to Fish in It." Occasional
   paper. St John's: Eco-Research Project, Memorial University.
– 1996b. 'A Shift in Economic Culture: The Impact of Enclave Industrialization on
   Newfoundland, 1855–1880.' Paper presented at the Atlantic Canada Studies Confer-
   ence, Moncton.
– *Courier (C)*. 1844–78. St John's.
Gordon, H. Scott. 1954. 'The Economic Theory of a Common-Property Resource: The
   Fishery.' *Journal of Political Economy* 62: 124–42.
Gough, J. 1991. *Fisheries Management in Canada 1880–1910*. Halifax: Department of
   Fisheries and Oceans.
Hardin, Garrett. (1977). 'The Tragedy of the Commons.' In John Baden and Garrett J.
   Hardin, eds., *Managing the Commons*, 16–30. San Francisco: W.H. Freeman and Co.
Hewitt, K.W. 1993. 'The Newfoundland Fishery and State Intervention in the Nine-
   teenth Century: The Fisheries Commission, 1888–1893.' *Newfoundland Studies* 9:
   58–80.
Hiller, James K. 1990. 'The Newfoundland Credit System: An Interpretation.' In Rose-
   mary E. Ommer, ed., *Merchant Credit and Labour Strategies in Historical Perspec-
   tive*, 86–101. Fredericton: Acadiensis Press.
Hutchings, J.A., and R.A. Myers. 1995. 'The Biological Collapse of Atlantic Cod off
   Newfoundland and Labrador: An Exploration of Historical Changes in Exploitation,
   Harvesting Technology and Management.' In R. Arnason and L. Felt, eds., *The North
   Atlantic Fisheries: Successes, Failures and Challenges*, 37–93. Charlottetown: Insti-
   tute of Island Studies.
Innis, H.A. 1954. *The Cod Fisheries: The History of an International Economy*. First
   pub. 1940. 2nd ed. Toronto: University of Toronto Press.
McCay, Bonnie J. 1994–5. 'The Oceans Commons and Community.' *Dalhousie Review*
   74, 3: 310–39.

McEvoy, A.F. 1986. *The Fisherman's Problem: Ecology and Law in the California Fisheries, 1850–1980*. Cambridge: Cambridge University Press.

Maritime History Archive, Memorial University. Slade Collection, box 1A-2, Slade-Kelson diaries, 1815–50.

Matthews, David Ralph. 1993. *Controlling Common Property: Regulating Canada's East Coast Fishery*. Toronto: University of Toronto Press.

Newfoundland. Various years. *Census of Newfoundland*. 1845, 1857, 1869, 1874, 1884.

– 1863. *Journal of the House of Assembly of Newfoundland*. 8, III.

– 1865. *Journal of the House of Assembly of Newfoundland*. 8, IV.

– 1866. *Journal of the House of Assembly of Newfoundland*. 9, I.

Newfoundlander (N). 1827–84. St John's.

Newfoundland Express (NE, later *E)*. 1851–76. St John's.

Provincial Archives of Newfoundland and Labrador (PANL). GN2.2, Colonial Secretary's Incoming Correspondence, 1863–4.

Public Ledger (PL). 1827–82. St John's.

Ryan, S. 1986. *Fish Out of Water: The Newfoundland Saltfish Trade, 1814–1914*. St John's: Breakwater.

– 1994. *The Ice Hunters: A History of Newfoundland Sealing to 1914*. St John's: Breakwater.

Slade-Kelson Diaries. Maritime History Archive, Memorial University, Slade Collection, box 1A-2, Slade-Kelson diaries, 1815–50.

Standard (S). 1859–80. Harbour Grace.

Vickers, D. 1996. 'The Price of Fish: A Price Index for Cod, 1505–1892.' *Acadiensis*, 25, no. 2 (spring): 92–104.

9

An Ojibwa Community, American Sportsmen, and the Ontario Government in the Early Management of the Nipigon River Fishery

J. MICHAEL THOMS

Current fishery laws throughout North America and in many other settler societies are antithetical to systems and technologies of Aboriginal fisheries, especially spearing, netting, and harvesting times. Recent court decisions about Aboriginal fisheries have, however, forced all levels of government in Canada to acknowledge Aboriginal harvesting prerogatives, rights, and needs, and many Ontario First Nations are preparing to assert these rights. Tensions mount as government bodies and non-Native commercial and sport fishers predict that Aboriginal control will destroy the fisheries.

In 1991, the Hunting and Fishing Committee of the Opwaaganising First Nation of the Nipigon River area, officially the Red Rock Indian Band, began discussing development of an agreement on co-management of natural resources with the Nipigon District Branch of the Ontario Ministry of Natural Resources.[1] Band members have been repeatedly prosecuted for violations of Ontario's game and fish acts. The committee hopes that an agreement will facilitate its people's control over acceptable forms and rules of use of natural resources in its region and over traditional lands (Thoms 1996).

The history of the introduction of exogenous practices, laws, and regulations for the Nipigon River demonstrates the problems and provides background needed to formulate future decisions and directions about use and management of a historically important, ancient freshwater fishery. The reconstruction of the Nipigon River's non-Native management history is possible because the river was a world-famous trout-fishing location for half a century (1873–1920). Indeed, the river was both a jewel in provincial fishery management and the subject of inaugural and ambitious conservation plans during the formative stages of Canadian ideas about resource management.

The Traditional (pre-1850) Use and Management of Fisheries

The Opwaaganising First Nation's traditional hunting and fishing grounds

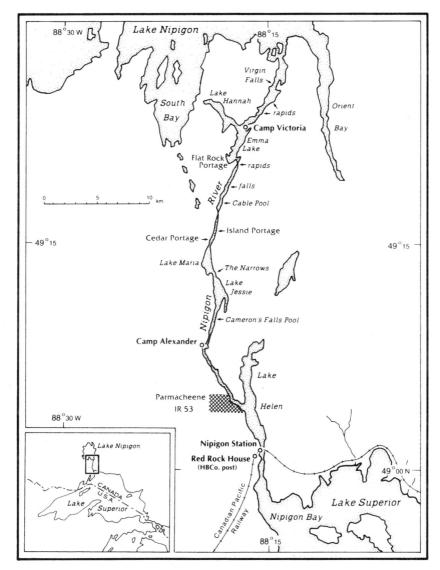

Figure 9.1   Map of Nipigon River, depicting the river before damming in 1921. The author has labelled the location of the original reserve ('Parmacheene'). *Source*: Adapted from *Rod and Gun in Canada* 7, no. 2 (July 1905): 177

surround and include the Nipigon River, near the head of the upper Great Lakes (see Figure 9.1 and Ray, Figure 5.1, this volume). Lake Nipigon, at the top of the system, is the largest inland lake in Ontario (tertiary watershed of 38,129 km^2) and drains into Lake Superior exclusively through the Nipigon River. The river is fifty-one kilometres long, falls seventy-five metres over its course, and is interspersed by five lakes. Prior to the 1920s, the river contained sixteen kilometres of swift-flowing white water and several small waterfalls (MacCallum 1989: 4). The river and lake sit in a northern boreal wilderness and form a distinct biological region known as the Nipigon Basin (Waters 1987).

The name Opwaaganisining, meaning 'pipestone,' refers to the fact that these people lived along the Nipigon River, where the soft red rock, which they used to make pipes, could be removed from the tall cliffs that line the river's outlet into Lake Superior. The Opwaaganisining Ojibwa spent winters on family hunting areas distributed across the group's traditional land base. In spring, the families would gather along the shores of the Nipigon to fish the large spawning runs of sturgeon (*Acipenser fulvescens*) (Umfreville 1929: 13). The availability of a variety of other fish throughout summer offered them an opportunity to socialize and engage in games. By autumn, they would disperse to camp at traditional family fishing places located along the length of the river, where they netted lake trout (*Salvelinus namaycush*) and lake whitefish (*Coregonus clupeaformis*) migrating to spawning grounds up river. Families smoked and dried their catches to take to their hunting grounds in the interior as winter provisions for themselves and their dog teams. Collectively, all the family hunting areas combined to form the people's traditional land base; the heart and hub of this traditional hunting area was the Nipigon River.

The importance of the fisheries to these people may be understated from the scholarly literature. Game was crucial, but so too was fish in this 'fish-rich' pocket of the central subarctic. Oral interviews and a closer reading of the historical data here and at Rainy River, where sturgeon fisheries were likewise a major resource base within the nineteenth-century regional economy, reveal that by the early nineteenth century, some northern Ojibwa groups subsisted primarily on fish, along with small game, birds, and natural rice, as the seasons allowed (Holzkamm, Lytwyn, and Waisberg 1988).

The earliest historical records for the region note the importance of the fishery to the northern Lake Superior Ojibwa, though most visitors came through during the main fishing seasons, beginning with a Jesuit priest in the late seventeenth century and continuing throughout the second half of the nineteenth century. In 1859, a Canadian fishery overseer reported that 'a great portion of the Indians, and the half-breeds depend upon fish, from September till sugarmaking' (Canada 1860: 85). In 1865, a sportsman wrote that lake trout were annually captured in fall by Aboriginal harvesters and, 'when smoked, furnished the principal food supply of the Indians' (Roosevelt 1865: 62). As far as one Lake Superior travel guide in 1880 was concerned, 'without the fish, the native tribes would soon perish' (Wyatt 1880: 58). In terms of the Nipigon River itself, Norval Morriseau's Ojibwa legends stress that lake trout were the 'staff of life among the Ojibway living there' (Morriseau 1965: 64).

The fish composition of the Nipigon River was complex (MacCallum 1989). Sturgeon, whitefish, lake trout, speckled (or brook) trout (*Savelinus fontinalis*), northern pike (or Jackfish, *Esox lucius*), pickerel (or walleye, *Stizostedion vitreum*), lake herring (*Coregonus artedii*), suckers (*Catostomidae*), slimy sculpin (*Cottus cognatus*), and arctic grayling (*Thymallus arcticus*) are known to have been present. Waters (1987: 137) points out that the fertile estuaries of rivers flowing into Lake Superior often led early writers to think mistakenly that the lake contained an extraordinary abundance and complexity of fish throughout its breadth. In truth, Lake Superior cannot sustain a fishery as complex as that of the rivers feeding into it.

These Ojibwa organized their fisheries around the spring and fall fish spawns, when fish were available and easily captured in their ascent of the Nipigon River. As throughout most of the central subarctic, sturgeon, pike, and pickerel were caught in spring, and lake trout and whitefish (particularly valuable for their high oil content), during their fall runs. Throughout summer and winter, suckers and other fish were exploited as needed. Contact with non-Natives began when European fur traders entered the area in the late seventeenth century, but long before that the Opwaaganisining Ojibwa had developed a sophisticated set of local fishing technologies. The earliest historic records for Nipigon show the well-established use of gillnets, spears, weirs, and baited gear. The women wove gillnets from strands of willow bark (MacCallum 1989: 16). Spearing, a largely misunderstood practice, is a precise technology for harvesting fish on shoals and rivers where the gillnets could not be set. One elder talked to me of spear fishing, 'hunting' fish, when he was young, in the 1930s: 'My people would spear fish in the rocks near the railway bridge. I also remember my family going out around the shores of lakes, spear fishing, and building a fire out of little birch logs in the bottom of the canoe. It was our way of life. We used to hunt pike.' They also baited rabbit leg bones to catch lake trout in winter through holes in the ice. And they used weirs constructed from natural materials for spring fishing in the streams tributary to the Nipigon.

Their descendants today explain that each family had a particular spot along the river, which the others recognized to be theirs for the purpose of fishing. In spring and fall, families would camp at their traditional locations on the banks of the river and set nets in the adjacent water to capture the fish running at that time. These locations were passed down from generation to generation. Snatches of late-nineteenth-century tourist accounts record the continuing pres-

Figure 9.2   Aboriginal fishing camp, Nipigon River, 1867, by William Armstrong.
Source: Royal Ontario Museum, Toronto

ence of fixed Aboriginal family fishing camps. In 1887, one tourist remarked of the camps on Isle St Ignace and smaller islands in the Nipigon Bay that lake trout 'is taken only in this locality, and only in the fall of the year as a rule. The Indians come from Nipigon expressly to fish for it' (Thompson 1887: 2). Tourists described Native families camped in small groups along the shores farther up the Nipigon and at its major lake, Lake Helen. A wigwam or two and racks with fish and nets laid out to dry marked each family site (see Figure 9.2).

Orderly exploitation of fish was possible only because families located themselves at historically determined and designated spots. As the spawning runs passed through, each family was obliged to allow some fish through so that families situated up river would not be deprived of their fair share of the resource. This system guaranteed fish to all families and assured the escape of sufficient supplies to the spawning grounds and hence future recruitment to the resource base. Each family's use was based on and directed by need and a specific and perennial knowledge of the fishery. This practice of harvest, as researchers have shown for Aboriginal fisheries elsewhere, effectively ensured equitable allocation of the resource to all members of the community (Newell 1993: 40; Schmalz 1992: 6; Van West, pers. comm., Feb. 1995). Indeed, as Dianne Newell (1993) notes for Aboriginal fisheries in British Columbia, this was not a system of 'open access' or of private property in the European tradition, but rather one based on the equitable sharing of the resource, directed by traditional ecological knowledge and community consensus. A family's traditional fishing area was recognized to be theirs, and thus restricted, only during times of harvests. At other times of the year, these areas would have appeared unoccupied or indistinguishable from the rest of the environs. So only non-Natives visiting the Nipigon in the late fall, as the trout began to move up the river to their spawning grounds (which was 'out of season' for angling), could have observed and recorded the band's distinctive organization, execution, and allocation of this major 'late' fishery for the Ojibwa. Newell (1993) notes the same pattern of 'invisibility' for BC Aboriginal fishing systems during non-harvesting times.

As was the custom for Aboriginal fisheries throughout Canada, a system of community arrangements or rules existed to maintain this allocation system. The rules prohibited use of another family's fishing location during periods of fish spawns, complete obstruction of the river, and use of nets in streams tributary to the river. 'In these creeks, you couldn't use a net in spawning season,' explained one of the elders whom I interviewed. 'We'd use a weir. It would pull away when you get a few fish. In a net you would block the creek and get too many.' In summer, between spawning times, however, it was acceptable to fish in any part of the river. There were also important rules for the sharing of catches and strong ethical values about preventing wastage of captured fish. Family elders were responsible for enforcing the rules and educating children about how to use the fishery resource.

During my 1993 interviews about the band's traditional harvesting values, stern disapproval for the wasting of fish and wildlife resources was cited as fundamental: deterring waste and 'taking only what is needed' were basic to contemporary and past hunting and fishing education within the community. Most subarctic 'small-scale' societies had reciprocal economies, in which sharing was the norm. Opwaaganisining First Nation elders identify a reciprocal process of sharing game and fish harvests within the community, called *nagoonsai,* as a central community value and practice that prevented greed among individual harvesters and avoided waste of resources. In *nagoonsai,* harvesters would share their fish and other resources with families that were short of food. 'Later, the favour would be returned.'

Stories and legends were also integral to education and the reinforcement of rules and values for use of resources. One elder recalls that when he was young: 'At night, in the wintertime, we were in a shack with no electricity. We had no light at night, and we would go to bed very early. Grandpa would beat his drum and tell us stories. It was a ritual. The stories were about life – explaining that killing was necessary, but respect it, and don't waste. He would tell us these things in stories. He would tell us these animals were created for us to live off, don't feel badly, because they were given to you. But, there were limits. Grandpa would tell us not to hunt any females in the spring and not to waste.'

Another explains: 'There is a moral to every story. Stories will tell you what to do when you kill an animal. White people have to be taught these things, but the Indian grows up with these things. Indian legends were how we learned. How to live involved how not to waste. It is different from you people. Our stories tell us how we live.'

Some of Norval Morriseau's published stories represent the oral tradition of the ecological knowledge of the local Ojibwa. One popular story recounts how the composition of the Nipigon River sturgeon population changed with the invasion of a foreign and poisonous type of sturgeon. The Ojibwa maintained this knowledge about the poisonous sturgeon in a story as 'a lesson to future generations' (Morriseau 1965: 37). The foreign sturgeon is reported to have disappeared from Lake Hannah about 1910.

For the Ojibwa community the Nipigon River is a sacred place; it is home to a populous underwater spirit world. These spirits were known to punish the people for transgressions of traditional beliefs, misuse of the fishery, and failure to perform rituals. 'Offering rocks' existed at the openings of lakes along the river, where travelling Ojibwa would place tobacco to thank the spirits for their safe passage and pray for successful fishing.[2] The Maymaygwashi, a mischievous underwater spirit people, were particularly populous in the river, and an effigy of one is depicted in a series of glyphs (prehistoric rock paintings) at the mouth of the river (Dewdney and Kidd 1967: 13–14). Aboriginal fish-based societies throughout Canada believed that aquatic spirits controlled the fishery and that these spirits, if not shown proper respect, could choose to withhold fish from the nets. Just prior to the fall fishery, says Morriseau (1965: 27), the band would form a circle with its canoes around a particularly sacred offering rock in the Nipigon and make extensive offerings to the spirits. The offerings included dogs and European trade goods: guns, traps, alcohol, and large amounts of tobacco.

That fishing was an essential factor in the economy of this band is obvious from the nineteenth-century historical record, which suggests that Aboriginal use of the Nipigon fishery was intensive, though it is possible that outsiders may have exaggerated accounts of Ojibwa 'overfishing' to serve other purposes. Nevertheless, the traditional Aboriginal harvest must have been substantial, as it was designed to provide local Ojibwa families and their dogs with a supply of food to last throughout the winter.

What enabled the Ojibwa to harvest impressive quantities of fish was their techniques of fish preservation. Because most species of fish were abundant and intensively caught only during short periods in spring and fall each year, it was necessary to preserve the catch for later use. In effect, to live by nature's cycles required a technology to manage this timing, for the most abundant species were available in large quantities only seasonally. Most of the catch was dried and smoked over a fire, then stored in birch-bark containers for winter use. Some dried trout were pounded into a powder and eaten with berries as fish pemmican (Morriseau 1965: 65). Fish caught in winter were frozen in the snow or ice, where they would keep until spring.

An ability to preserve fish was also essential for trade. Elders maintain that their ancestors belonged to a fish-trading network, and geographers and historians are now citing the work of anthropologists to show that this Ojibwa trade system existed prior to contact (Lytwyn 1990: 30, Schmalz 1992). As the fur trade expanded into northern Ontario on a regular basis and the European post network became established in the first half of the eighteenth century, there is ample evidence of a fish trade among Aboriginal groups and between them and fur traders (see Ray, Tough, this volume). By 1821, well after the subarctic economy had been transformed, Nipigon Aboriginal people provided the Hudson Bay Company (HBC) with essential provisions of 'country' foods, including fish. Without question, these people had organized their fisheries in an expert and sustainable manner long before non-Natives developed their first conservation laws for the region after 1850. Indeed, after centuries of intensive harvesting and use, in the 1870s the Opwaaganisining Ojibwa would bequeath to the new discoverers of the Nipigon fisheries a rich and well-managed resource.

Even before Canada and the province of Ontario assumed jurisdiction over the entire Great Lakes region in 1867, the Nipigon Basin had come under the Robinson-Superior Treaty, signed in 1850 (see Ray, Figure 5.1, this volume). Ojibwa people had fought for this agreement. Ojibwa chiefs reminded the colonial government that the Ojibwa possessed sovereignty over this region, and they travelled to Montreal to demand redress for damages to their land by mining prospectors. The Robinson-Superior Treaty was quickly negotiated and signed. Central to the negotiations, in what would become a common pattern for all later treaties, were issues related to the disposition and continued use of natural resources. The Ojibwa chiefs protected their right to resources in the treaty statement, 'to allow the said chiefs and their tribes the full and free privilege to hunt over the territory now ceded by them, and to fish in the waters thereof as they have been in the habit of doing, saving and excepting only such portions of the said territory as may from time to time be sold or leased to individuals, or companies of individuals, and occupied by them with the consent of the Provincial Government' (*The Robinson-Superior Treaty* 1964 [1850]: 7).

The treaty explicitly states that Aboriginal people retained the right to hunt and fish 'as they have been in the habit of doing' (and see Ray, this volume). The Opwaaganisining Ojibwa were not a signatory to the treaty and thus never ceded their traditional lands. However, in 1885 the band acquired a reserve land base on the Nipigon near its traditional fishery and at the centre of its traditional hunting grounds. Its selection of this site was consistent with land selection patterns expressed by other Ojibwa groups who began to protect their traditional fisheries as a critical food resource for the future in the treaty period. As non-Natives (other than fur traders, who entered much earlier) arrived, they too found the fisheries to be reliable and abundant relative to other resources in the area. Conflict over the fisheries was imminent.

Development of Non-Native Management

American sport fishers – anglers – caused most of the ensuing conflict. These men belonged to a distinct culture, whose origins were to be found in Isaac Walton's 1653 text, *The Compleat Angler; or Contemplative Man's Recreation: Being a Discourse on Rivers, Fish-Ponds, Fish, and Fishing.* Walton was among the first to record in writing the English rules and directions for taking fish with a hook and line. The book itself was said to be a mixture of 'enchanting pastoral poetry' and 'fine morality,' written to promote a love of the outdoors. It helped foster the growth of a leisured, sporting class (Walton 1653: Introduction by Sir John Hawkins: 17). The book identified trout as the finest of sportfish and claimed that the values and laws governing its pursuit were among the noblest achievements of the leisure class.

Charles Hallock (1873) announced that the Nipigon River possessed a speckled trout fishery unrivalled in North America. The 'discovery' of the Nipigon formed a major part of the new information in his highly anticipated book, *The Fishing Tourist: Angler's Guide and Reference Book*, and garnered international attention for the fishery of this Ojibwa territory (Hallock 1873: 203). Hallock subsequently launched *Forest and Stream*, which was the first significant, long-running American magazine aimed at identifying angling and shooting locations in North America, lobbying for protection of game, and serving as a forum for 'inculcation in men and women of a healthy interest in out-door recreation and study.' In its inaugural edition (Aug. 1873), it featured a story on sport fishing on the Nipigon. The romance with this river would last for four decades, during which time the magazine ran countless articles and government fishery reports about the river, generated debate on acceptable uses of the river, and lobbied the Ontario government on behalf of American sport fishers. Thanks to Hallock and a handful of other writers, the Nipigon quickly became widely known as a 'world-famous' trout stream.

The Canadian Pacific Railway (CPR) facilitated the deluge of tourists – mainly American – who arrived. Where the line crossed the river near its outlet into Lake Superior, the company built a station. A fierce promoter of tourism, the CPR developed a series of hunting and fishing pamphlets that featured the Nipigon and even commissioned one of Ontario's leading figures in fisheries management, W.F. Whitcher, a former dominion commissioner of marine and fisheries, to prepare a book about trout fishing on the Nipigon (Whitcher 1888).

The Nipigon River HBC post, Red Rock House, also joined in promoting tourism, initially by building a steamship dock at Nipigon and also by organizing tourist trips. From the outset, the dominion fisheries department required tourists to purchase a licence and register their names, origins, catches, comments, and concerns in the HBC post's log book. While this record has long since disappeared, *Forest and Stream* frequently reprinted the annual statistics, so the information remains available. The data are useful for constructing a profile of early visitors and their catches. From 1887 on, the data also formed part of the Ontario fishery reports and licensing records, which are still accessible. According to these data, the river's popularity began abruptly and steadily increased until it peaked in 1888, with 155 visitors, levelling off at eighty-five anglers per year until after the turn of the century. The visitors were mostly wealthy Americans – initially from port cities on the Great Lakes and later from all over the United States – and there were Canadians and European visitors too.

Credit for the river's popularity among American sportsmen must go to *Forest and Stream*, which promoted the Canadian area as one of the few surviving wilderness areas available to Americans (Archives of Ontario [AO] 1887; Roland 1887:b; Whitcher 1888: 32). In the words of one American writer visiting the Nipigon, 'Canada is now the goal for American sportsmen, as with cashiers' because the United States was no longer wild, the Adirondacks were 'populous with inns,' Pennsylvania had been logged over, and the Appalachians were 'prisoned in tame preserves' (Macdonough 1889: 271). American anglers were invading the Nipigon as though it were a U.S. fishing hinterland.

The anglers' visits showed a consistent pattern. Trips were generally prearranged through correspondence with the HBC post and consisted of a four-to-ten-day guided canoe trip, featuring a slow ascent of the river over three major portages to Virgin Falls and finishing with a quick descent of the river, in which the rapids were often 'shot.' Tourists would fish along the river in the various pools promoted in the travel literature. Guides were necessary, and the Opwaaganisining Ojibwa filled this role. Aboriginal guides prepared the fish, cooked meals, made camp, transported anglers, portaged gear, and helped tourists to fish. As is seen below, these guides were exceptionally innovative when it came to getting tourists to spend additional money.

Though only a partial record of the anglers' trout catch survives, we know that the amount taken was large. For the first fourteen years, the dominion government did not restrict catches, and wholesale exploitation was the norm. 'During my ten days sojourn upon the Nepigon, I took perhaps one hundred speckled trout,' reported Havelock in *Forest and Stream*, and he added, 'I might as well have taken a thousand, scarcely one of which weighed less than three pounds' (Havelock 1873: 4–5). A writer who had fished the Nipigon in 1867 reported filling half-barrels with trout totalling eight hundred pounds (363 kilograms). Another noted a decade later: 'I am afraid to say how many trout we had at the finish, but I know that we packed in ice more than three hundred pounds weight to take home with us, and gave away more' (Thompson 1877: 9). Later still, two New Yorkers sought 'to prove what could be done' and in one hour of fishing caught twenty-nine trout weighing a total of 50.8 kilograms (Anonymous 1882: 51). The belief was that the large size of Lake Nipigon guaranteed a perpetual supply of speckled trout that would for ever descend into the Nipigon River to replenish the large harvests.

Though the CPR would claim in 1888 that 'North of Lake Superior, the Nipigon, Steel, Jackfish and other streams are almost inexhaustible, and many of them have been scarcely touched yet' (CPR 1888: 17), the actual exhaustibility of the famous Nipigon trout fishery was already obvious. In 1886, a veteran Nipigon angler and *Forest and Stream* writer announced that he no longer visited the Nipigon, 'as fishing is wearisome there from its success' ('H' 1886: 190). In 1888, W.F. Whitcher took Lady Macdonald, the prime minister's wife, to task for her comment that the fishery was inexhaustible; Whitcher warned that the river had in fact been the subject of 'abusive' fishing practices and that the speckled trout were now 'scarce' (Whitcher 1888: 34).

Then A.R. Macdonough, a professional angler and *Scribners Magazine* writer, argued in a twelve-page feature on the Nipigon that overfishing was causing a rapid decline and called for strict conservation and severe restrictions on access to the river. Macdonough charted and compared catch records for a single rod for two hours a day, from 1886 to 1888 (Macdonough 1889: 282–3, 276), to illustrate his contention that the threat to the fishery came from non-Aboriginal anglers, particularly settlers, to whom he refers as 'pot-hunters' (literally meaning a person who hunts for food, or for profit, without regard for the rules of the sport), and low-class anglers, whom he calls 'pseudo-sportsmen.' He blamed the CPR for opening the area to such undesirables. Elite sport fishers such as Macdonough believed that conservation lay in protecting the river for their own kind, and they meant business: on several occasions they would attempt to purchase the banks of the Nipigon, privatize the property, and then restrict access to the select members of a sportsmen's club.

The practice of sports clubs' purchasing river banks to control use and access to trout streams was already a popular strategy in the United States. By the 1880s U.S. courts had ruled that riverbeds and fish remained public property but established that the fish, once removed from the water, were the property of the shore owners. Thus privatization of shores by sport clubs could be used to establish exclusivity over the capture of the fish within the river. By 1885, the Ontario Fisheries Act accepted that 'streams or lakes leased to individuals or clubs cannot be fished by the public' (CPR 1899: 163; see *Ontario Statutes* 48 Vic., c. 9, s. 3[5], and 55 Vic., c. 10, s. 19). Ontario allowed privatization for some rivers and lakes, but it sought to keep the Nipigon open because of large revenues generated by the river's great popularity, its importance to the railway and steamboat companies, and its ability to lure into northern Ontario wealthy easterners and Americans, all of whom might be tempted to invest in the future development and exploitation of the region's rich resource base.

Tourism led staff of the dominion department of fisheries to monitor fishing using the system of the fishery log book and licensing administered by the HBC postmaster at Nipigon. In the late 1880s, when concerns about overfishing of the Nipigon were first voiced, the commissioner of crown lands for Ontario appointed the succeeding HBC postmaster to be an official provincial fishery overseer to enforce fishery laws, administer licences, and collect fees for the Nipigon River, Lake Nipigon, and adjacent waters (Ontario, Crown Lands, 1887: 21).

In the 1870s, all traditional technologies for fishing on the Nipigon were considered legal: netting, spearing, weir and trap fishing, or fishing with hook and line (angling), of which there were two forms, fly-fishing and fishing with bait or spoons. Because only the elite sportsmen fished with the fly (a cultural trait), one way to exclude all others from trout fishing was to ban all methods other than fly-fishing. To this end sports fishers, whether domestic or foreign, had to lobby Ontario through the Nipigon fishery overseer (Morris 1900: 512). Hallock himself launched the campaign in *Field and Stream*, arguing: 'True sportsmen will scorn to use' bait, spoons, and imitation minnows. The fly-fishers would also defend themselves against impersonations and infiltration of their culture by arguing that fly-fishing was an 'art.'

The definition of acceptable modes of capturing fish remains at the heart of objections to Aboriginal spear fishing and netting today. This nineteenth-century elite sportsmen's strategy to restrict access to the river along class lines, based on the fishing technology used by each class, is where it all started for the Ojibwa. In 1888, when many still believed that Lake Nipigon was the source of the Nipigon River's almost 'limitless' stock of speckled trout, the local fishery overseer reported 'the very earnest and frequently expressed desire' of anglers that the government ban all netting and commercial use of Lake Nipigon, and he could report, 'Fly fishing and angling with hook and line have been the only means practiced here, no nets of any kind or other ways of taking fish have been resorted to' (Ontario, Crown Lands, 1889: Appendix 13). By 1889, netting was clearly defined as 'an illegal means of procuring fish' (Ontario, Crown Lands, 1890: 21), but there is no clear statement that the ban on net fishing was meant to include Aboriginal netting.

With netting outlawed, it remained only to ban fishing with spoons and worms and to legitimize fly-fishing. This goal seemed simple enough, as earlier overseers' reports indicated that the river was almost 'entirely' fished by artificial flies (Ontario, Crown Lands, 1887: 21, 1889: 21). A decade or so later, the overseer would give tacit acceptance to the fly-fishers' arguments against fishing with spoons and worms, reporting that it was the fly-fishermen's contention that 'the various artificial baits are barbarous, and not fit to use for taking the lordly Trout ... [while] on the other hand, the bait-fishermen claim that the larger fish do not rise to the fly, in fact, if they had to depend on what they caught with the fly, their supply of fish would be very small.' This concern about maximum efficiency in fishing technology had always been the bait-fishers' issue, prompting fly-fishers to accuse the bait-fishermen of only 'wanting something easy' (Ontario, Fisheries Branch, 1903: 36). The same year, at a sportsmen's convention in Ottawa, the overseer claimed that outside of 'his duty' to enforce provincial fishery laws, he was also to facilitate 'in any way possible, by information and courtesy, the pleasure of the anglers' (63).

What effect did these new regulations outlawing so many traditional fishing technologies, including netting, have on the fisheries of Aboriginal people, specifically the Opwaaganisining Ojibwa? Not much, at that time, since Aboriginal people openly continued to net spawning fish in spring and fall all through the period (Anonymous 1913; Hewitt 1918; Macdonough 1889; Ontario, Fisheries Branch 1906). During the first two decades of tourism, from 1873 to 1893, a degree of co-existence developed between the two resource-user groups on the river. Large non-Native settlement had failed to occur in this region, and those who were there were restricted from fishing without a licence. Band members used the fishery in their traditional manner, as Havelock's second editorial for *Forest and Stream* observed: 'Opposite [our] camp is a low, flat island. There are two wigwams on the point. These are occupied by the families of two of the Indians who are to take us up stream. Their canoes are just visible protruding from the alders that skirt the shore. Up the bank, nets and clothes are drying' (1873, 7 no. 2: 19). This passage clearly represents the design of Aboriginal harvesting at the beginning of the sport-fishing boom. It was August, and Aboriginal families were strung out along the river at their traditional camp locations. They were using nets.

From this time on, the record situates Aboriginal fishing camps below Lake Helen, around the lake, and along the Nipigon River above the lake. There are no descriptions of a collection of Aboriginal people living on the 500-acre (202-hectare) Indian reserve surveyed on the river in 1885, but rather repeated mentions of their dispersal along the river in small family groupings at traditional family harvesting locations. In 1887, one angler observed Aboriginal fishing camps on the islands in the Nipigon Bay (Thompson 1887: 2). After the turn of the century, family fishing camps were described on the islands at the head of the Nipigon River, above Virgin Falls, in Lake Nipigon (Ontario, Fisheries Branch, 1905: 23–6). This same pattern was visible in 1913 and 1924 (Anonymous 1913: 1139, Longstreth 1924: 60–70). Families were concurrently practising agriculture and constructing more permanent dwellings at these locations, as Macdonough's trout-fishing essay (1889: 276) documents: 'better shanties, now and then built on some cleared half-acre yielding a handful of potatoes or hay.'

Many accounts exist of Aboriginal people continuing their tradition of netting from small boats under sail in Lake Helen and in the waters off their camps on the Nipigon River (Longstreth 1924; Macdonough 1889; Millard 1917). Aboriginal spearing was also observed throughout the period (Millard 1917). In 1891 Hewitt noticed one local Native using a spoon on a line to lure fish to his net. The fisher was 'securing his winter supply of smoked fish' and had 'taken a large number, cleaned them, and cut off the heads, and was preparing to smoke them when I happened to come along' (Hewitt 1948: 26). So, while sport tourism and government regulation of fishing were proceeding in the Nipigon River drainage basin, the Aboriginal people continued subsistence fishing for lake trout, whitefish, sturgeon, and pickerel using traditional techniques.

It appears that here, as in Aboriginal homelands elsewhere in Canada, when government fisheries regulations first came in, the laws were intended for the newcomers, not Aboriginal people; the fishery overseer wrote as much in his report for 1887 (Ontario, Crown Lands, 1887: 21). Aboriginal people were apparently not seen as a threat to the resource in this early, sport fishing period, from 1873 to at least the 1890s. Not until late in the 1890s would an Aboriginal threat to the resource be invented, and then the laws about technology would begin to be specifically amended to prevent Aboriginal fishing.

The Opwaaganisining Ojibwa not only maintained their fisheries but were quick to enter the tourist trade as fishing guides, providing essential services and knowledge to tourists. Band members were also part of the overall attraction of the river, because then, as now, tourists sought a 'wilderness experience' complete with 'exotic' Natives. Band members responded by fabricating a repertoire of Aboriginal lore to satisfy the tourists' need for a cross-cultural thrill.

Some of the elders remember that guiding was not a good job, as they were expected to paddle, haul gear, make camp, cook, and cater to all of the tourist's needs. As one of them recalls of his grandfather: 'You were working, you were fishing, you were hunting, you did everything for them. It was everything. First thing in the morning you get up and make coffee. That's before they wake up, so that when they get up, they get a cup of coffee. You were on call almost 24 hours. If in the middle of the night, one of those bastards wanted a drink of water, you are going to get it.' Another suggests that guiding was akin to being a butler; in fact, a tourist once described the Nipigon guide to be 'as complete a valet of the woods as could be desired' (Macdonough 1889: 276).

Most non-Natives saw things differently; they generally described Aboriginal skills in romantic language (see Longstreth 1924: 158). Tourists seldom acknowledged the extent to which they depended on the skills of these Aboriginal guides, or how much the tourist industry relied on indigenous knowledge, or the entrepreneurship of band participants in the tourist industry.[3] Examples of crucial indigenous knowledge, besides canoeing, included Aboriginal medicines for warding off insects (especially blackflies), 'cockadoosh' (slimy sculpin, a small, bottom-dwelling fish) for bait, and techniques for preserving fish.

At portages, particularly the arduous haul of 2.5 miles (four kilometres) at Camp Alexandra, one band member with a horse and cart operated a shuttle service for an additional charge not included in the guide's fee. This service operated for over sixty years, and it was not cheap. Strict negotiators, the Ojibwa guides were uncompromising on the issue of wages (Longstreth 1924: 66; Alexander 1911: 67–9). Writers complained that the guides' charges were 'exorbitant,' and the literature recommended fixed wages in order to undermine the guides' abilities to find tourists' upper spending limits. Band members also bolstered the local Aboriginal economy by conducting tourists to band farms along the river for purchases of potatoes and eggs for their camping meals (AO 1887; Longstreth 1924; Millard 1917). Band elders and records of the Department of Indian Affairs indicate that this band was once known as one of the most self-sufficient in the region (NA, RG 10, 1937).

Application of Game and Fish Laws to the Band

In 1890, the sportsmen of Ontario managed to have the province set up the Ontario Game and Fish Commission to look into the state of game and fish. The commission's hearings, with over one thousand sportsmen interviewed, signalled the latter's triumph in having fish and animal conservation in the province reflect their own values and needs (Tough 1992). The findings of the commission showed that sportsmen were beginning to think that the game and fish laws of the province applied to Aboriginal people (Ontario, Game and Fish Commission, 1892). Nevertheless, regardless of what was going on in the minds of sportsmen, the commissioners made it clear in the revision of the laws that 'nothing' in their laws 'shall prejudicially affect any rights specially reserved to or conferred upon Indians by any treaty or regulation in that behalf made by the Government of Canada' (*Ontario Statutes*: 55 Vic., c. 10, 14). In effect, Aboriginal communities possessing fishing rights protected by treaty could not be forced to abide by the laws being developed to protect sportsmen's interests in the resources.

As sportsmen increasingly sought to control more of the province's resources, including those harvested and managed by Aboriginal people, they successfully campaigned to vilify Natives' use of natural resources, arguing that their right was to 'subsistence,' not to commercial use, and that conservation laws and external control over their use of resources was in their best interest (Tough 1992; Newell 1993 and this volume). The alleged Aboriginal destruction of the Lake Nipigon fishery was hypothesized in the early 1890s by one Alex Starbuck, a writer for *Forest and Stream* who had initiated a semi-regular column on Lake Superior fishing. Starbuck, using the racist language of the day, repeatedly stressed that the lake's fishery was being 'decimated' and 'depopulated by the tawny savages' (Starbuck 1891: 286–7; 1892: 226; 1894: 289, 339, 536). His arguments focused on the '*modus operandi*' of Aboriginal harvesting. The technology used by Aboriginal people and sport fishers differed along cultural lines, and Starbuck, like his angling friends, sought to impose a new fishing regime by defining traditional Aboriginal technologies of using gillnets and setting lines baited with cockadooshes as 'destructive,' 'treacherous,' 'deadly,' and a 'terrible pot-hunting tactic' – the ultimate insult from an angler (Starbuck 1892: 226). He challenged readers of *Forest and Stream* in 1894 to fight the exemption of Aboriginal people and commercial fishers from fishery laws: 'I do not believe that any man of leisure could find a more benevolent cause then in that of a crusade against the use of small-meshed nets in Lake Superior' (Starbuck 1894: 289). 'Small-meshed nets' was code for the technology used by Aboriginal people and commercial fishers. Some sportsmen now began to interfere personally in Aboriginal fish harvesting on the Nipigon.

When in 1898 Ontario forced the dominion government to relinquish control over inland rivers and lakes in Ontario, the province formed a new fisheries branch, which fully embraced sport fishers. The agency quickly targeted Lake Nipigon and the Nipigon River for special protection through a set of unique fishery regulations that created a two-week licence with a fee (the Nipigon region was the only place in the province to have a fee), made anglers more accountable to the local fishery overseer, and created more rigid sanitation laws.

The truly significant change lay in a new clause for Aboriginal fishing: 'These regulations shall apply to Indians who may act as guides, boatmen, canoemen, camp assistants or helpers of any fishing party or persons or persons who may hold a fishing license or permit during the time they are engaged with such party, person or persons, but not otherwise to Indians' (CPR 1899: 65).

The intention was to force Aboriginal compliance with Ontario fishery laws by linking employment as guides to fishery regulations. More directly aimed at limiting the Aboriginal right to fish by traditional means on the Nipigon River, the second part reads: 'But no Indian shall fish with net or trap or night line or otherwise than by angling in the said River Nepigon or any other of the creeks or streams tributary thereto' (CPR 1899: 65). The province apparently had accepted the sportsmen's lobby to restrict all forms of harvesting, by all cultures on the Nipigon, to angling. Fortunately for the band, the second part of the Aboriginal clause does not appear in the final version of the Ontario Fisheries Act (*Ontario Statutes*: 63 Vic., c. 50, 52).

In the absence of a clear-cut legal means to prohibit the Aboriginal treaty right to net, spear, and use other traditional technologies on the Nipigon, a full-scale campaign against traditional fishing methods ensued in which Aboriginal people were blamed for reckless destruction of fish stocks. Trout fed to dogs attracted special criticism. In 1905, the Nipigon fishery overseer reported that Aboriginal people were netting thousands of pounds of speckled trout for dog food, 'one Indian alone having 2000 brook trout, weighing from two to seven pounds [0.9 to 3.2 kg] each, in his possession for dog feed' (Ontario, Fisheries Branch, 1905: 24–5). He argued that the best measure to make Aboriginal people subject to Ontario's fisheries laws and regulations was to license Aboriginal guides, whereby an Aboriginal person could hold the licence only through personal adherence to provincial fishery laws.

Elders believe that it was through the development and imposition of various licence requirements (especially for guiding) that non-Aboriginals attempted to control the band's unlimited right to fish. Lise Hansen's research (1991) confirms that Ontario's laws on Aboriginal fishing fall silent at this time, as field interpretations of Aboriginal rights and prerogatives begin to take precedence. Also, in 1905, the Nipigon overseer, though recognizing the enormous difficulties of enforcing fishery legislation on Aboriginal harvesters, recommended extensive enforcement during fall months, when Aboriginal people were harvesting, and suggested that Nipigon Forest Reserve rangers be given the powers of fishery guardians to assist in restraining the Aboriginal fishery.

Band members today argue that the band was managing the Nipigon River through a culturally distinct paradigm. When I asked one elder why the Aboriginal people cited in the 1905 report did not feed suckers or catfish, which many of us assume are of low value, to their dogs, instead of speckled trout, he explained: 'Now that is a question only a whiteman can make. The whiteman classifies fish – classifies them as sport fish, coarse fish, and so on. The Indian does not do this. The Indians were living on time [*sic*]. The speckled trout runs in the fall and the speckled trout is fat in the fall. Fat is important in winter food because when it is burnt in bodies, it generates heat. Fat is for heat. Dogs were very important for getting to the trap lines. So, because the speckled trout is most available in the fall and is fatty, it was ideal for netting in the fall and feeding to our dogs over the winter. In Ojibwa, you can translate the word for speckled trout to mean dog food.'

Needless to say, sport fishers valued speckled trout as a high-status 'trophy' fish, whereas Ojibwa prized it as essential dog food. In 1905, fishery legislation attempted to ban its use for dog food in order to preserve the financial revenues from the river.

Aboriginal Responses

This band and other Ojibwa bands in the area quickly responded to the sportsmen's rather transparent efforts to criminalize their use of traditional fishing and hunting resources. They demanded protection under the Robinson-Superior Treaty and asked Ottawa to protect their fisheries when they came under threat from resource development projects that Ontario had sanctioned (AO 1987). When, in the late 1890s, a pulpwood company began to drive logs down the Nipigon, Band Chief Deschamps advised the dominion fisheries department that the environmental impact would destroy the trout fishery, but the department considered the Aboriginal concern unfounded (NA, RG 23, 1897). Chief Deschamps persisted: 'We are prepared to prove that we have seen and picked up hundreds of fish that were killed by dynamite,' and 'should this pulp business, and dynamite using be allowed without let or hindrance on these waters, it will destroy our fishing and stop tourists from coming here, thereby taking away, to a great extent, our means of making a living' (NA, RG 23, 1898). Aboriginal people spoke, but their voices were seldom heard, and, because they lacked the support of the Department of Indian Affairs in this matter, they did not have the political clout with which to gain attention.

Though Aboriginal harvesting technologies were perceived as illegal, and the right to hunt, fish, and sell furs outside seasons was also defined as illegal (see Ray, Newell, this volume), elders explain that from the 1910s onward (the period within living memory), their people were still living on their family hunting grounds in winter and returning to the Nipigon River and Lake Helen for fishing and socializing in summer. Band members thus found themselves in the stressful position of facing prosecution each time they engaged in these essential traditional enterprises. My interviews with elders in 1993 produced a litany of accounts of raids on Aboriginal camps, prosecutions, confiscations, fines, harassment, imprisonment, fear, and family efforts to conceal their harvesting times, practices, and economies.

With tourism on the decline in the Nipigon by 1900, Ontario's Fisheries Branch determined that the increased presence in the river of pike, pickerel, and suckers, which crowded out and devoured the game fish, was to blame (Ontario, Fisheries Branch, 1901: 50) and so decided to eliminate these species from the Nipigon. A year later, the local overseer could report: 'The war against the pike has been a decided success' (1902: 36). The wholesale destruction of the unwanted species continued for several years and often made headlines in *Forest and Stream*, which, not surprisingly, endorsed the removals and boasted that the river was being recovered for fishers of speckled trout. The overseer continued to call for 'still more radical measures' and suggested that 'it will be found necessary' to fish out all the pike, pickerel, suckers, and whitefish from Lake Nipigon, whence the fish descend (1905: 26)! In effect, to promote tourism, Ontario tried to manicure the Nipigon watershed into a trout-only ecosystem: 'Then indeed we would have a trout stream to surpass the fondest dreams of [Izaac] Walton' (1903: 35). Ontario simultaneously attempted to reproduce the Nipigon trout fishery elsewhere. In 1904, the CPR shipped five hundred live Nipigon trout to the Bow River in Alberta and then another five hundred to the Kicking Horse River in British Columbia (1905: 24).

In 1921, despite the Nipigon's 'world-famous' reputation and its choice for Ontario's earliest and most ambitious conservation efforts, the river was dammed by the Hydro Electric Commission of Ontario, which tamed the formerly fast-flowing river to create a series of still-water reservoirs. The dams prevented the free passage of fish throughout the length of the river, thus destroying the band's traditional fishery and, in fact, any type of fishery. The end of the fishery meant the collapse of the tourism on which band members also relied for income. Hydro further damaged the band's chances for economic recovery by discriminating against Aboriginal people in its hiring practices for the next seventy years. The band's diversified economy and self-sufficiency, which had survived well beyond the period of the fur trade and endured the sport fishers' lobby to monopolize the fishery, could no longer be maintained.

Conclusion

Elite sportsmen had sought, with some success, to limit access to the Nipigon fishery by lobbying the fishery overseer to legitimize fly-fishing, wipe out angling with bait and spoons, and criminalize all other technologies, including especially those of the Ojibwa who occupied the territory. The Ojibwa were never able to convince governments to reinstate the legitimacy of their fishing traditions, especially spearing and netting. Despite the early and ambitious plans of governments to protect the Nipigon River fishery, they mismanaged the river. Many fish assemblages, including speckled trout, have collapsed. Ontario devalued and tried to criminalize an indigenous management system that offered alternative values and approaches to understanding and harvesting the river's biology and violated its obligations to the band.

When I interviewed the Nipigon district manager for the natural resources ministry in 1993, he seemed unaware of the band's long fight for acknowledgment of its rights to the region's natural resources. He complained that every prosecution of an Aboriginal person for a violation of the Game and Fish Act had become a political issue: 'In my mind, the problem is that Natives are too legal oriented – not like old pacifist, common good Ojibwa. Everything gets politicized, even a small charge.' However, the band had been resisting interference with its harvesting methods for a very long time, adjusting its tactics to suit the times. From a non-Aboriginal perspective, a new arrangement of laws and regulations, based on an innovative meshing of Aboriginal and non-Aboriginal values and ecological knowledge, might be a first effort to creating a uniquely Canadian management system.

NOTES

1 Special thanks to the Opwaaganising First Nation's Hunting and Fishing Committee and its current chair, Terry Bouchard; to the band chief and council; to elders Mona Cormier, Andrew Hardie, Willie John, Lawrence Martin, Dave Quackageesick, Abe Quackageezick, Ron Sault, Frank Skinaway, Ed Thompson, and Agnes Wawia; to Nora Bothwell, former chief, Alderville First Nation; and to the staff of the Nipigon District, Ontario Ministry of Natural Resources. The views and interpretations are, however, mine alone.

2 One offering rock was so important to the band that tourists often asked the guides to explain their veneration for the spot. Comprehending the place to be the home of an evil fish spirit, non-Natives dubbed the place 'Devil's Rapid.'

3 In 1915, the reputation of the Nipigon River as the 'world's greatest' for speckled trout was guaranteed when the world's largest speckled trout was caught there. It was immortalized in a highway monument and modern travel literature and is an enduring bit of Nipigon angling trivia. The official credit for the catch went to a non-Native doctor, but it was a guide from the Red Rock Indian Band, Andrew Alexie, who caught the fish. The doctor was asleep at the time. The elders are unanimous on this point.

WORKS CITED

Alexander, Kirkland B., 1911. *The Log of the North Shore Club: Paddle and Portage on the Hundred Trout Rivers of Lake Superior*. New York: Knickerbocker Press.
Anonymous. 1875. *Picturesque Canada, The Northern Lakes Guide to Lakes Simcoe and Couchiching: The Lakes of Muskoka and Lake Superior, via The Northern Railway of Canada, Giving a Description of the Lakes and River Scenery with the Best Spots for Waterside Summer Resorts, Hotels, Camping, Fishing, and Shooting Distances and Costs of Travel*. Toronto: Rose & Co.
– 1882. 'Fishing in the Nepigon.' *Forest and Stream* 19, no. 3: 51.
– 1886. 'The Nipigon.' *Forest and Stream* 27, no. 13: 247.
– 1913. 'The Giant Trout of Nipigon.' *New York Herald*, n.d., reprinted in *Rod and Gun in Canada* 14, no. 11: 1142.
Archives of Ontario (AO):
– 1887. 'Diary of Hiram Worcester Slack.' Pamphlet file no. 54.
– 1897. RG 8, no. 3257-1892, William McKirdy to G.M. Gibson, Provincial Secretary, 2 July 1892.
Canada. 1860. Crown Lands. *Report of the Commissioner*.
Canadian Pacific Railway (CPR). 1888. *Canadian Pacific Primers: Summer Tours*. 2nd ed. Montreal: CPR, Passenger Department.
– 1893. *Fishing and Shooting on the Canadian Pacific Railway*. Montreal: CPR, Passenger Department.
– 1899. *Fishing and Shooting along the Lines of the Canadian Pacific Railway in the Provinces of Ontario, Quebec, British Columbia, the Prairies and Mountains of Western Canada, and the Maritime Provinces, the State of Maine and in Newfoundland*. 12th ed. Montreal: CPR, Passenger Department.
Dewdney, Selwyn, and Kidd, Kenneth E. 1967. *Indian Rock Paintings of the Great Lakes*. 2nd ed. Toronto: University of Toronto Press.
'H.' 1886. *Forest and Stream* 27, no. 10: 190.
Hallock, Charles. 1873. *The Fishing Tourist: Angler's Guide and Reference Book*. New York: Harper and Brothers.
Hansen, Lise C. 1991. 'Treaty Fishing Rights and the Development of Fisheries Legislation in Ontario: A Primer.' *Native Studies Review* 7, no. 1: 2–21.
Havelock, 1873. 'Trout Tails from the Nepigon.' *Forest and Stream* 1, no. 1: 4–5; 1, no. 2: 19; 1, no. 4: 49–50.
Hewitt, E.R. 1948. *A Trout and Salmon Fisherman for Seventy-five Years*. New York: Charles Scribner's.
Holzkamm, Tim, Victor Lytwyn, and Leo Waisberg. 1988. 'Rainy River Sturgeon: An Ojibwa Resource in the Fur Trade Economy.' *Canadian Geographer* 33, no. 3: 194–205.

Hunter, Martin. 1909. 'Nipigon.' *Rod and Gun in Canada* 11, no. 6: 533.

King, Aldolph. 1971. 'History of Commercial Fishing on Lake Nipigon.' Unpublished report, Ontario Ministry of Natural Resources, Nipigon District files.

Longstreth, T. Morris. 1924. *The Lake Superior Country.* New York: Century.

Lytwyn, Victor. 1990. 'Ojibway and Ottawa Fisheries around Manitoulin Island: Historical and Geographical Perspectives on Aboriginal and Treaty Fishing Rights.' *Native Studies Review* 6, no. 1: 1–30.

MacCallum, Mary Ellen. 1989. 'The Nipigon River: A Retrospective Summary of Information about the Fish Community.' Unpublished report to the Ontario Ministry of Natural Resources (Ontario), Division of Fish and Wildlife. Technical Report Series and North Shore of Lake Superior Remedial Action Plans Series. Toronto.

Macdonough, A.R. 1889. 'Nepigon River Fishing.' *Scribner's Magazine* 6: 271–96.

Millard, E.E. 1917. *Days on the Nipigon.* New York: Foster & Reynolds.

Morriseau, Norval. 1965. *Legends of My People the Great Ojibway.* Ed. Selwyn Dewdney. Toronto: Ryerson.

Morris, Robert. 1900. 'Flies versus Other Lures for Nepigon Trout.' *Forest and Stream* 55, no. 26: 512.

National Archives of Canada (NA):

– RG 10, Department of Indian Affairs (DIA).

– 1937. 8040/492/30-8-53, pt 1, Burke to Superior, 25 Feb.

– RG 23, Department of Fisheries, 320/2672.

– 1897. 320/2672. 'A.N.W.,' Assistant Secretary, DIA, to F. Gourdeau, Deputy Minister of Marine and Fisheries, 28 Sept.; F. Gourdeau to Secretary, DIA, 15 Nov.

– 1898. 320/2672. Chief Deschamps to Indian Agent J.F. Hodder, 6 Feb.

Newell, Dianne. 1993. *Tangled Webs of History: Indians and the Law in Canada's Pacific Coast Fisheries.* Toronto: University of Toronto Press.

O'Brien, L.R. 1882. *Picturesque Canada: The Country As It Was and Is.* Toronto: Belden Bros.

Ontario, Crown Lands. 1887–90. *Report of the Commissioner*, 'Report of the Nipigon Fishery Overseer.'

– 1899–1906. Fisheries Branch. *Annual Report.*

– 1892. Game and Fish Commission. *Commissioner's Report.*

– 1892. Fisheries Act.

– 1898. Forest Reserve Act.

The Robinson-Superior Treaty. 1964. Signed 1850. Ottawa: Queen's Printer.

Roland, Walpole. 1887. *Algoma West: Its Mines, Scenery and Industrial Resources.* Toronto: Warwick.

Roosevelt, Robert B. 1865. *Superior Fishing: Or the Striped Bass, Trout, and Black Bass of the Northern States, Embracing Full Directions for Dressing Artificial Flies with the Feathers of American Birds: An Account of a Sporting Visit to Lake Superior, ETC.* New York: Carleton.

Schmalz, Peter S. 1992. 'The European Challenge to the First Nations' Great Lakes Fisheries.' Paper presented at the Annual Meeting of the Canadian Historical Association, Charlottetown, PEI.

Starbuck, Alex. 1891. *Forest and Stream* 37, no. 15: 286–7.

– 1892. 'Nepigon Trout.' *Forest and Stream* 39, no. 11: 226.

– 1894. *Forest and Stream* 42, no. 13: 26–7; no. 14: 288–9; no. 16: 339; no. 25: 536.

Thompson, W. 1887. 'A Trout Trip to St. Ignace Island, 1887.' Mimeograph, Nipigon Museum, Nipigon, Ont.

Thoms, J. Michael. 1996. 'Illegal Conservation: Two Case Studies of Conflict between Indigenous and State Natural Resource Management Paradigms.' MA thesis, Trent University.

Tough, Frank. 1992. 'The Criminalization of Indian Hunting in Ontario, ca. 1892–1930.' Paper presented at the Commonwealth Geographical Bureau Land Rights Workshop, Christchurch, New Zealand.

Umfreville, Edward. 1929. *A Canoe Voyage through Western Ontario, with Extracts from the Writing of Early Travellers through the Region.* Ottawa: R. Douglas.

Walton, Izaac. 1653. *The Compleat Angler: Or Contemplative Man's Recreation: Being a Discourse on Rivers, Fish-Ponds, Fish, and Fishing.* Ed. Sir John Hawkins. London: Samuel Bogster, 1808.

Waters, T.F. 1987. *The Superior North Shore.* Minneapolis: University of Minnesota Press.

Whitcher, W.F. 1888. *Nepigon Trout, an Ottawa Canoeist's Experience on the Northern Shore of Lake Superior.* Montreal: CPR, Passenger Department.

Wilson, Leona. 1991 'Historical Literature Review of the Nipigon Area with Emphasis on Fisheries from 1654 to 1990.' Typescript, for Ontario Ministry of Natural Resources and the Department of Fisheries and Oceans, Canada. Technical Report Series and North Shore of Lake Ontario Remedial Action Plans Series. Toronto.

Wyatt, , G.H. 1880. *The Traveller's and Sportsman's Guide to the Principal Cities, Towns, and Villages near the Hunting and Fishing Grounds of the Great Northern Lakes in Canada and Manitoba, Their Best Localities for Game and Fish.* Toronto: Liverpool Printing and Stationery.

10

Estimating Historical Sturgeon Harvests on the Nelson River, Manitoba

PETER J. USHER AND FRANK J. TOUGH

Since the late nineteenth century, Indian fisheries on Canada's northern lakes and rivers have frequently been disrupted by habitat destruction, depletion, pollution, and expropriation. Indians were unable to prevent these activities and for most of this period had no effective remedies for either the loss of harvest or the disruptive effects on their way of life, their communities, and their culture.

One obstacle has been the lack of legal recognition of Aboriginal and treaty rights to fish, and in particular the unwillingness of the courts to characterize harvesting rights as a form of property right that provides for defences against, and remedies for, nuisance, trespass, or expropriation (Usher and Bankes 1986). Another has been the difficulty of proof of loss, even where the right is recognized (Usher 1997:72–6). To obtain compensation for damages, a claimant must prove that the loss resulted from the defendant's actions and must be able to specify and quantify the loss.

This chapter[1] outlines certain methods for quantifying the loss of a subsistence (domestic) fishery, where, as in this case, the product is directly consumed rather than sold on the market and there are no records of prior catch. It also compares the use of social scientific and historical evidence relied on by the Indian claimants with the evidence from fisheries biology used by the Manitoba Department of Natural Resources.

The Case

Cross Lake is an Indian reserve community (population 3,106 in 1994), located on the Nelson River system in northern Manitoba. Until the mid-1970s, Cross Lake was unconnected to the provincial highway system, and its economy relied heavily on fishing, trapping, and hunting. Fishing was·a particularly important activity, as it supplied both the major source of food for the community and cash from commercial production. Lake sturgeon (*Acipenser fulvescens*) was especially valued for both purposes, and sturgeon fishers were highly skilled and respected members of the community. Cross Lake fishers customarily harvested the entire Nelson system from where it enters Cross and Pipestone lakes through to Sipiwesk Lake (see Figure 10.1). The main sturgeon fishing areas were located at the various rapids on the Nelson itself, where sturgeon gather to spawn.

In the early 1970s, Manitoba Hydro began a major program of hydro-electric development involving impoundment, diversion, and regulation of the major northern rivers, called Lake Winnipeg Regulation and Churchill River Diversion. As part of this program, Manitoba Hydro erected the Jenpeg control and hydro dam on the Nelson, just above the Cross Lake Indian Band's traditional fishing area, to regulate flow so as to maximize power production on the river. Its downstream effects were to reverse the natural seasonal flow and increase the flow and level range, and hence substantially reduce the level and volume of Cross Lake in summer. This adversely affected the domestic and commercial fishery at Cross Lake and virtually eliminated the sturgeon fishery by disrupting spawning sites (a similar blow to the Nipigon River trout fisheries is touched on in Thoms, this volume).

After much protest, Cross Lake (along with five other northern Indian communities) was able to negotiate the Northern Flood Agreement (NFA) of 1977, with Manitoba, Manitoba Hydro, and Canada (Waldram 1988). The NFA was intended to provide remedies for adverse effects, through mitigation, compensation, and priority access to fish and wildlife.[2] In the outcome, these provisions went largely unfulfilled, and, as negotiations with the other parties faltered, the Northern Flood communities began filing claims under the NFA's arbitration provisions (Larcombe 1997). One of these was a claim by the Cross Lake Indian Band for the loss of its domestic sturgeon fishery.

Cross Lake's sturgeon claim alleged that the sturgeon fishery had been disrupted by Manitoba Hydro's activities and sought compensation. The band retained Symbion Consultants of Winnipeg to quantify the loss by comparing pre-disruption amounts of sturgeon consumption (i.e., normal levels prior to Jenpeg) with post-disruption levels, based on a recall survey (self-reporting of past activity) of community residents (Symbion Consultants 1990). The band subsequently retained Peter Usher to assess the validity of Symbion's estimates of weights and values (Usher 1991). In November 1991, the arbitrator heard evidence in Cross Lake from former sturgeon fishers about the effects of disruption and the level of their fishing effort and catch prior to Jenpeg (Northern Flood Agreement [NFA] 1991).

Manitoba had never obtained data on subsistence harvest for sturgeon or any

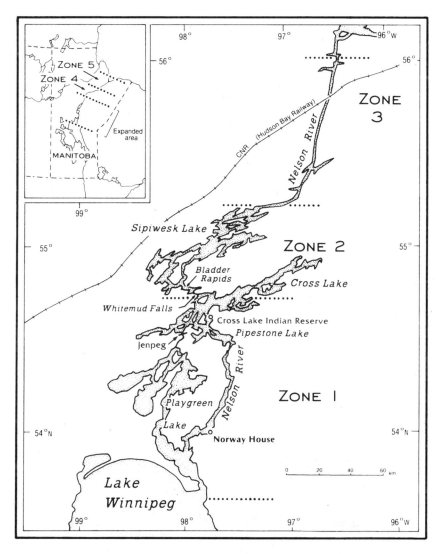

Figure 10.1   Location map, including management zones established by Manitoba Department of Natural Resources for Nelson River commercial sturgeon fishery. *Source*: Patalas 1988: 5

other species of fish.[3] Impact studies for the hydro project conducted in the early 1970s, which consisted of either speculative estimates or consumption surveys, also failed to estimate sturgeon harvests specifically at Cross Lake (for a review of these estimates, see Usher and Weinstein 1991). Consequently the only data available for pre-disruption harvests were from Symbion's consumption survey and the fishers' own testimony, both given in the context of a compensation claim. Manitoba Hydro responded to the band's claim by alleging that because the estimate made by the provincial Department of Natural Resources of the maximum sustainable yield (MSY) of the Nelson River sturgeon fishery was lower than Symbion's consumption estimate, that estimate, and hence the band's claim for harvest loss, were excessive.

Following the Cross Lake hearings, the band retained Usher to provide an opinion on the relative merits of each of the estimates before the arbitrator, and Usher in turn retained Tough to assess the conflicting historical data on the commercial sturgeon fishery, in order to develop a comprehensive analysis of all catch and consumption data. What follows below is a summary of that analysis.

Though numerous claims relating to harvest disruption and loss had been heard by the NFA arbitrator, none had turned on the technical problems of quantifying these losses. The case thus constituted a practical test of the problems of quantifying and attributing fisheries losses where there are no prior recorded catch data. The challenge was to reconstruct historic harvest levels, using the best available evidence, and to assess the validity of various sources, including the biological management data. (This discussion does not address technical questions of the value of the harvest, proof of cause of disruption, and the issue of property rights and legal cause of action). Technical evidence on historic harvests was heard before the arbitrator 13–15 April 1992 (NFA 1991).

Three estimates of, or surrogates for, pre-project sturgeon harvests at Cross Lake were provided to the arbitrator: Usher's harvest estimate, based on a small sample of fishers (Usher 1992, based on the testimony of the Cross Lake sturgeon fishers given at arbitration hearings, 4–7 November 1991); Symbion's consumption estimate, based on a recall survey (Symbion Consultants 1990); and Manitoba's estimate of productivity, which the province alleged constituted an upper bound on the long-term average harvest.

We evaluated each estimate/surrogate on the basis of data integrity and soundness of estimation procedure and then compared them with other available estimates from similar communities. None of these data was ideal. The objective was, however, to ascertain which of these estimates, on the balance of probabilities, should be accepted as the best available indicator of historic catch levels. We first discuss the two direct estimates – Usher's and Symbion's – based on social survey principles and then compare these with the line of evidence based on a conventional resource-management approach focusing on biological data.

Evidence for Pre-Disruption Sturgeon Catch from Social Survey Data

Methodological Issues

In the absence of contemporary recorded data, as was the case for the Cross Lake fishery, there are two methods of reconstructing historical subsistence catch levels. Each involves recall surveys, because rarely are there adequate recorded observations on which to base independent catch estimates. One method is to reconstruct the harvest itself (how many fish were caught), and the other is to reconstruct levels of fish consumption (how much fish was eaten).[4]

Harvest surveys measure the amount (in number or weight) of fish caught and retrieved. The methods for obtaining subsistence harvest data, and for estimating total subsistence harvest volumes, have been described by Berkes (1983) and Usher and Wenzel (1987). They have been applied recently in connection with Indian freshwater fisheries by the James Bay and Northern Quebec Native Harvesting Research Committee (1982), Michalenko, Marcogliese, and Muskrat Dam Band (1989), Hopper and Power (1991), Berkes et al. (1994), and Tobias and Kay (1994). However, estimating subsistence fish catch from harvest surveys poses special difficulties. Subsistence fishing is typically a frequent and even regular activity, in which large numbers of fish are taken. Because fishers rarely count their fish or commit these numbers to memory, surveys must incorporate certain aids to recall such as recording calendars, bulk container equivalents, and short recall periods. If no survey data are available for a given historical period, it is still possible to reconstruct historical harvests, and minimize bias, through use of carefully designed recall surveys (viz. Usher et al. 1979).

Consumption or diet surveys measure the amount of fish eaten and employ methods similar to those of harvest surveys (Gibson 1990, Wein 1995). However, consumption quantities are always less than harvest quantities, for reasons that include spoilage, cooking (which may account for weight loss of up to 25 per cent), and 'plate waste,' or table scraps (Wein and Freeman 1995). Some earlier consumption surveys relied on incorrect assumptions with respect to units and portion size of consumption (Usher and Weinstein 1991). Harvest and consumption data must obviously be standardized for direct comparison.

Both harvest and consumption surveys are subject to certain sources of error or bias, which can be identified and minimized, but not always accurately estimated. This is of particular concern in compensation claims such as the one advanced by the Cross Lake Indian band. A supplementary method of validation is by comparison with other data. For example, if the actual or theoretical productivity of the fish stock in question can be estimated, it follows that har-

Table 10.1 Pre-Jenpeg subsistence sturgeon harvest (kg edible weight)

1	2	3	4	5	6
Witness number	Trips per season	Sturgeon per trip*	Average weight per sturgeon	Total number of sturgeon (2×3)	Total weight of sturgeon (4×5)
1	3–4	6–8	7.3 +	18–32	131–234
2		10–20			
3		20	5.5–14.6		
4	6	7.5–15	10.9–14.6	45–90	491–1314
5	3–4	7.5–10	18.2–21.8	22.5–40	410–872
Mean	4.25†	12.20†	12.53†	41.25†	517
Range	3–6	6–20	5.5–21.8	18–90	131–1314
Sample size	n = 3	n = 5	n = 4	n = 3	n = 3

Source: Northern Flood Agreement 1991.
Data given for spring fishery (spawning season, at rapids).
*Per person (assumption: party = 2).
†Calculated as mean of individual means.

vests cannot be consistently larger (i.e., exceed sustainable yield) over any length of time. Harvest data can also be compared for consistency with survey results from similar fisheries at other locations.

Harvest Estimate

The closest approximation on record to a retrospective harvest survey for Cross Lake is the sworn testimony of five sturgeon fishers at the arbitration hearing (NFA, 4–6 Nov. 1991). Usefulness of this sample is limited by its small size and non-random nature. Quality of the data was limited by the context in which they were obtained: recall was elicited by legal examination rather than by formal harvest-survey interview. The hearing transcript indicates that questions relating to pre-disruption harvest were neither asked according to a consistent protocol nor probed to clarify ambiguities or non-response. None the less, the line of questioning was sufficient to form a basis for reconstructing harvests. Most witnesses were asked, and responded to, a series of questions relating to gear use, number of trips per season, number of sturgeon taken per trip or per unit of effort, and average weight of sturgeon obtained (Usher 1988; Usher and Wenzel 1987). Table 10.1 summarizes the 'survey results' extracted from the transcript.

If the witnesses called were representative of the core group of sturgeon fishers, then the mean annual domestic catch per fisher was 517 kg during the peak fishing season. While fishers indicated that some sturgeon fishing took place in late summer and under the ice in winter, it is reasonable to assume that the spring fishery accounted for 90 per cent of the total subsistence sturgeon catch. The total annual catch per fisher would thus have been 570 kg.

The number of sturgeon fishers prior to disruption can be estimated based on the following information. Witnesses who fished at Bladder Rapids, the location most favoured by Cross Lake people, spoke of up to twenty parties (a party consisting of two fishers in one boat) there at a time. In the last few years before Jenpeg came into operation, a pattern seems to have emerged of weekend fishing, as a result of increased wage employment that kept many people from fishing during the week. Assuming that effort was concentrated on the weekends, then the witnesses were probably seeing a high proportion of those who actually fished. However, there were several other locations at which people fished, and it seems unlikely that much more than half of fishing effort was concentrated at Bladder Rapids.

We therefore estimate that there was a core group of fishers numbering perhaps thirty to fifty. Some probably went as partners, others probably took younger or more casual fishers. Household specialization in harvesting is widely known in northern communities (viz. Wolfe and Walker 1987), especially with respect to those types of harvesting calling for specialized skills and gear, as is the case for sturgeon. In such instances, older, more experienced persons, who have accumulated the required gear and knowledge, constitute the core group of fishers. A casual group of fishers – individuals who go once or twice a year and may not be as successful – also participates. But the core group also teaches other younger, interested men, who come along as partners. Taking all these factors into account, we suggest that a conservative estimate of total fishing effort in the early 1970s was equivalent to about thirty fishers typical of the witnesses. Total annual domestic harvest would therefore have been 17,100 kg. Obviously we cannot assign confidence limits to this estimate.

This is, unfortunately, the only estimate of pre-project subsistence catch on record. It suffers from uncertainty as to the representativeness of the sample and insufficient precision in interview procedure. However, unlike many other estimates of subsistence harvest (especially those from Manitoba – see Usher and Weinstein 1991), it has the advantage of known method, sufficient detail to provide a basis for estimation, and a basis in sworn testimony.

Consumption Estimate

Symbion Consultants (1990) developed a recall survey–based estimate of pre-disruption sturgeon consumption at Cross Lake of 6.45 kg/capita/yr raw edible weight, or 12,611 kg total (if this were obtained by thirty fishers, each would have taken 420 kg). Symbion's figure probably underestimated portion size, so, assuming that the number of sturgeon meals eaten was correctly estimated, the figure is conservative (Usher 1991, Usher and Weinstein 1991). If we take into account the normal method-based disparities noted above between harvest and consumption estimates, the estimate of harvest derived from fishers' testimony and Symbion's survey-based estimate of consumption seem reasonably consistent.

Comparative Estimates from Other Locations

The most comparable data available are those obtained by Michalenko, Marcogliese, and Muskrat Dam Indian Band (1989; Michalenko pers. comm.) for the community of Muskrat Dam on the Severn River in northwestern Ontario (see Ray, Figure 5.1, this volume). This sturgeon fishery was undisrupted by river regulation or pollution, though it had previously suffered some depletion from commercial fishing. Two separate estimates produced the following results: catch per fisher, 89 and 118 kg/yr; available for consumption, 10.01 and 12.54 kg/capita/yr; and sturgeon as a proportion of all fish, 19.6 and 16.7 per cent.

Over 10 per cent of the population at Muskrat Dam was engaged in sturgeon fishing, while most of the harvest at Cross Lake was accounted for by about 2 per cent of the population. This difference would account for the lower catch per fisher at Muskrat Dam, even though per capita consumption there was apparently higher. (No information is available on possible differences in lake productivity that might affect the interpretation of these data.)

Unfortunately, other recent fisheries surveys have covered communities where sturgeon are rare or absent. While no direct comparison can be drawn, they provide some indication of common levels of fish consumption in Aboriginal communities. A recent analysis of ninety-six subsistence fishery estimates for Aboriginal communities across Canada (Berkes 1990) found average consumption of 42 kg/capita/yr (edible weight, all species combined), for communities with relatively undamaged fisheries. More comprehensive and reliable estimates by the Alaska Department of Fish and Game indicated an average of 86.7 kg/capita/yr (all species) in interior subarctic villages (Wolfe and Walker 1987).

The Cree villages of northern Manitoba are located on waters that are relatively productive by northern Canadian standards, and it is thought that a higher-than-average part of the traditional diet consisted of fish. If we apply the Muskrat Dam ratios of sturgeon to all fish to a hypothetical but conservative pre-project rate of fish consumption of 50 kg/capita/yr, then sturgeon consumption rates would have been 8.35 kg–9.80 kg. These figures are substantially higher than Symbion's consumption estimate at Cross Lake, which may be a further indication of the conservativeness of that estimate.

Evidence for Pre-Disruption Sturgeon Catch from Biological Data

The government of Manitoba presented a very different case: it alleged that local sturgeon stocks had been so depleted by overfishing prior to Jenpeg that MSY at the time could not possibly have sustained the level of consumption documented by Symbion, or the level of fishing reported by the Cree fishers themselves at the arbitration. This 'depletion hypothesis' was crucial to Manitoba Hydro's case.

The provincial Department of Natural Resources estimated that the productivity of the Cross Lake Indian Band's historical harvest area was 0.099 kg/ha/yr (raw edible weight equivalent). Hence the maximum sustainable yield was 10,596 kg/yr, for both subsistence and commercial fishing (Macdonald 1990, 1991). This productivity estimate is substantially lower than that for undepleted riverine stocks in undisturbed waters (see Table 10.3, p. 210).

The department based its estimate on an annual average recorded commercial production (post-1970) of 2,719 kg (headless dressed) on Sipiwesk Lake, multiplied by 1.5 to account for the estimated subsistence harvest (see below). It then applied the resulting level of exploitation of 0.083 kg/ha/yr (headless dressed) to the total presumed area of the Nelson River system traditionally harvested by the Cross Lake Indian Band (about 107,000 ha), for an estimate of 8,830 kg/yr headless dressed weight.

The depletion hypothesis in Manitoba's evidence is asserted rather than demonstrated. Macdonald (1991: 1) and North/South Consultants (n.d. no. 4: 1–2) cite the classic Canadian literature on sturgeon biology, which in fact refers to documented cases of depletion in Lake of the Woods, Lake Nipissing, and Lake Winnipeg, not to the Nelson River. The first biological studies of Nelson River sturgeon by Sunde (1959, 1961) – whose recommendations, according to the province's expert, had 'formed the basis for the successful management of the Nelson River commercial fishery since 1970' (Macdonald 1991: 2) – also cite these sources. Later reports (for example, Patalas 1988; Sopuck 1987) simply reiterate the depletion hypothesis as though the earlier literature 'proved' it.

The data on which the province based its depletion hypothesis were originally prepared by Sunde for the years 1907–59 and published by Sopuck (1987: 3) with additional data for 1971–8. These data are said to show commercial sturgeon production for 'the Nelson River,' without the authors' specifying what is actually included in this body of water, which begins at Lake Winnipeg and ends at Hudson Bay, about 600 km downstream, and includes several large lakes (Figure 10.1). What these data are purported to show is a major collapse from an early peak, followed by two repeated and rapid collapses after the province temporarily reopened the fishery in 1937 and 1953 (Patalas 1988: 3) – a pattern consistent with the classic cases of sturgeon depletion noted above. Only the strict management regime imposed after the 1970 reopening, Manitoba alleged, had since maintained the fishery at a sustainable level.

These historical data were crucial to Manitoba's depletion hypothesis, because no one has ever made a comprehensive assessment of Nelson River sturgeon stocks. The biological assessments of Sunde (1959, 1961), Sopuck (1987), and Patalas (1988), on which Manitoba relied at arbitration, only infer changes in stock status and productivity from commercial catch records and test netting. In the absence of direct estimates of total biomass, the province's case required it to assume that recorded commercial harvests plus its own estimate of domestic harvest equals the sustainable harvest.

**Problems with the Depletion Hypothesis and
Manitoba's Biological Estimate**

There are at least five difficulties with the depletion hypothesis and Manitoba's estimate of productivity and hence with its conclusions about pre-disruption subsistence catch levels: incomplete and inaccurate historical data on Nelson River's commercial sturgeon fishery, including lack of geographic specificity and of effort data; failure to identify the discrete stocks to which these data apply; lack of understanding of the historical context of the fishery; failure to recognize the magnitude and significance of the subsistence fishery; and inadequate consideration of socioeconomic reasons for variation in effort.

Commercial Catch Data

The most critical data, but the least reliable, are those pertaining to the years up to Manitoba's second closure of the Nelson River fishery in 1930. The most commonly cited data set is attributed to Sunde (Sopuck 1987: 3) and is shown in column 1, Table 10.2. However, another data set prepared by Harkness in 1936 for the period 1915–1933 (Harkness 1980: 17; see column 2, Table 10.2 here) indicates that with the exception of the years 1917 and 1918, the sturgeon production assigned to the Nelson in Sunde's data set is actually from the area of The Pas, and no production is reported from the Nelson.

In view of this confusing evidence, Tough reviewed the original data sources

Table 10.2 Historical catch data (lb.) for the Nelson River commercial sturgeon fishery, 1901–46 (lb. headless dressed weight)

Year	Sunde (S)	Harkness (H)	Original sources (O)	Comments, by source
1901			?	O: Nelson River included in Lake Winnipeg
1902			135,000	Reported in Saskatchewan District
1903			220,000	As 1902
1904			180,000	As 1902
1905			120,000	As 1902
1906			?	O: Nelson River included in Lake Winnipeg
1907	7,000	7,000	7,000	
1908	8,000	8,000	8,000	
1909	25,700	25,700	20,000	S, H: Includes Saskatchewan River
1910	9,000	9,000	.	O: No commercial fishing in Nelson River district, reported catch under S and H is from Saskatchewan River
1911–16				Sturgeon fishery closed
1917	150,000	150,000	150,000	
1918	68,000	68,000	78,000	S, H: Does not include Cross Lake
1919			?	O: Nelson River probably included in The Pas district
1920			?	O: As 1919
1921	36,600	*	?	S: The Pas district, includes Saskatchewan and Nelson rivers; O: no separate data for Nelson River
1922	62,300	*	?	As 1921
1923	124,300	*	?	As 1921
1924	146,800	*	105,000	S: The Pas district includes Saskatchewan and Nelson rivers.
1925	97,800	*	86,000	As 1924
1926	76,900	*	45,000	As 1924
1927	48,200	*	32,000	As 1924
1928	500		500	
1929	4,400		4,200	S: Apparent typographical error O†
1930			200	O†
1931–6				Sturgeon fishery closed
1937	15,000		15,000	O†
1938	30,000		30,000	O†
1939	26,000		26,000	
1940	26,400		36,200	S: Does not include Playgreen Lake
1941	25,200		33,200	As 1940
1942	15,400		17,800	As 1940
1943	10,600		13,000	As 1940
1944	9,200		11,400	As 1940
1945	13,400		13,700	As 1940
1946	5,800			S: Northern Lakes–The Pas; no separate data for Nelson River

Sources: Sunde: cited in Sopuck 1987: 3; Harkness: Harkness 1980: 17; 'Original sources': Canada, *Annual Reports for the Department of Fisheries, Sessional Papers*, 1901–16; Canada, Dominion Bureau of Statistics, *Fisheries Statistics*, 1917–46; Canada, *Annual Reports for the Department of Fisheries*, 1924–9
Notes: S: geographical area listed as 'Nelson River'; H: geographical area listed as 'Nelson River and tributaries, Hudson Bay etc.'; O: Nelson River from Playgreen Lake to Hudson Bay inclusive, unless otherwise specified
*Catch attributed to Nelson River by Sunde is attributed to 'The Pas Area' by Harkness.
†Includes only specified parts of Nelson River
All weights are given in lb. as in the original data sets but converted to kg equivalents for Figure 10.2.

to provide an amended data set up to 1946 (column 3, Table 10.2). The geographical confusion is elucidated but cannot be entirely eliminated from the original data sources, because geographical reporting units and administrative districts vary over time and are not always clearly stated in the annual government reports on fisheries (Table 10.2). For example, there was a brief period prior to the northward extension of Manitoba's boundaries in 1905 in which Nelson River catches were reported in the Saskatchewan district of the North-West Territories, which probably explains the absence of pre-1905 data in the data sets of Sunde and Harkness. The chief problem with Sunde's data set is the confusion over both the waters and the stocks to which it applies. The description 'Nelson River' is all-inclusive and, for the early period of the commercial fishery, undoubtedly included at least the current management zones 1 and 2, and possibly even zones 3 and 4 (Figure 10.1). Harkness's data set does not resolve this confusion, but it does separate Nelson and Saskatchewan River catches (one possible source of confusion in early provincial fisheries data is that there were two Cross Lakes, the other one being on the lower Saskatchewan).

Early Nelson River catch reports are seldom broken down by specific lake or reach and sometimes include other, unconnected northern rivers or are subsumed in other districts. Data for Playgreen Lake are sometimes included in Lake Winnipeg reports, and other times (for example, in the two peak years of the 1920s) in the Nelson River report (and see Tough, this volume).

The amended data set does not clearly confirm the pattern that Manitoba asserted. The earlier period of 1902–5 shows very high production (see also Tough, this volume), but apart from those years, production levels were not as high as the province alleged. Only the isolated peak year of 1917 is consistently reported in all three data sets. The prolonged high production levels of the 1920s indicated by Sunde are not supported by the other data sources. Harkness reports no production from the Nelson during the 1920s, while the original data are substantially lower than Sunde's. Thus much of the production attributed by Sunde to the Nelson in that period may have occurred on the lower Saskatchewan.

Production data are difficult to interpret if there are no corresponding data on fishing effort. While such data have been kept since the 1920s,[5] they tend to be unreliable in the relatively unsupervised fisheries of the north (viz. Usher and Weinstein 1991). Like the production data, they are geographically unspecific and, in any event, have not been analysed, except by Sunde for the period 1937–60. Normally the records consist only of number of fishers, not gear type or amount, and therefore cannot be used to reconstruct catch per unit of effort.

No one can be sure what the commercial harvest was on the Nelson River, or exactly where it occurred, in the critical early years when the original depletion is said to have occurred. Even the standard citations used in Manitoba's evidence do not, on inspection, support the depletion hypothesis. Harkness and Dymond, for example, singled out the Nelson *in contrast* to the typical cases of depleted sturgeon fisheries (1961: 76) because, according to the data they used, recent (i.e., late 1950s) catches there were a much higher percentage of early peak catches (13.9 per cent on the Nelson, compared to 0.3 per cent on Lake Erie and 0.0045 per cent on Lake of the Woods). They took the increasing percentage of young fish as a sign of reproductive viability and were encouraged by the rapid recovery of the fishery after closure. (The Nelson River sturgeon fishery was closed on four occasions: 1911–15, 1930–6, 1947–52, and 1961–9.) Finally, they noted that, in contrast to the classic depleted fisheries, original conditions on the Nelson had been little affected by industrial development.[6]

The biological literature on the Nelson sturgeon fishery is by no means consistent in its analysis or interpretation of the available biological data. Harkness and Dymond's observations have been noted above. Neither Sunde's own account, nor the actual record of commercial production, bears out Harkness's early, pessimistic predictions for the fishery (see also Macdonald 1991: 1).

Stock Identification

A critical problem with using aggregated historical data for the Nelson as a whole is that in such a river sturgeon stocks are separated by the major rapids, which they cannot ascend, though some fish probably descend to augment downstream stocks. These semi-isolated stocks can be depleted only separately. They must therefore be managed separately, and reported on separately, but this fact was formally recognized by the province only with the establishment of the present management zones after 1970 (Figure 10.1). There are at least two and perhaps more discrete stocks in the Cross Lake area (those above Bladder Rapids and/or Whitemud Falls and those below them, to somewhere below Sipiwesk Lake). These in turn are separate from Playgreen Lake and Lake Winnipeg stocks.

While there is no doubt that very high levels of sturgeon fishing occurred for a few years in the early twentieth century, the historical data are not reliable enough, or specific enough, to confirm the depletion hypothesis. These data were not collected for management purposes and may not be salvageable for management purposes. If reported catches cannot be assigned to specific stocks, then statements about depletion and its causes are more difficult to make, because different stocks may have been affected at different times.

If the Nelson's early commercial catch has been exaggerated, or wrongly attributed to particular stocks, then the depletion model based on the well-documented cases of Lake of the Woods and Lake Winnipeg may not apply to the Nelson River, particularly not to all of its separate stocks. It seems likely that there were at least four years of high commercial catch (1902–5), though at what locations and from which stocks cannot be confirmed. This may well have had a temporarily deleterious effect on stocks, but there is little if any evidence that intensive commercial sturgeon fishing lasted long enough, or was widespread enough, to have resulted in long-term depletion of sturgeon stocks, especially in the Cross Lake area.

Historical Context

While the historical catch record is ambiguous at best, the historical context provides some basis for estimating where the greatest effort and catch may have occurred.

Macdonald speculates that during the early years of the commercial fishery, pressure on Playgreen Lake and Cross Lake stocks (zone 1) was greater than that on the Sipiwesk Lake stock (zone 2). He provides no evidence in support of this idea, however, and does not specify the dates of occurrence and duration. In all probability, Playgreen Lake and the upper Nelson in fact formed the initial zone of exploitation, because they were most accessible to Lake Winnipeg, but this brief peak appears to have taken place over ninety years ago. However, even if Playgreen Lake stocks were depleted in those years, that did not necessarily entail depletion of downstream stocks. Sturgeon were taken commercially on Sipiwesk Lake and transported by river in the early years of the century (Tough, this volume) but became more accessible with the construction of the Hudson Bay Railway, starting in 1912. The isolated production peak of 1917 may be related to that event as well as to wartime demand and higher prices. However, the stock most heavily relied on by the Cross Lake Indian Band was the least accessible to commercial exploitation in the early years of the twentieth century and therefore the least likely to have suffered depletion from it.

A further indication that the depletion model of the larger lakes may not apply to the Nelson River lies in the actual historical development of its sturgeon fishery (Tough 1996: 239–43). Two factors stand out. First, poundnets, commonly regarded as a more destructive form of gear than gillnets, were almost never used on the Nelson. Second, most commercial fishermen were themselves local Cree (in contrast to earlier fisheries in Ontario and on Lake Winnipeg), which may have moderated the intensity with which the fishery would otherwise have been prosecuted by non-Native labour.

There is good evidence that the closure of the Nelson River sturgeon fishery in 1930 was the result less of depletion than of falling prices. According to departmental correspondence (Provincial Archives of Manitoba, RG 17, B1, box 50, file 44.5.4), Manitoba declined to reopen the fishery in 1935, despite local requests, for two reasons. First, low prices and the lack of assured markets and competitive fishing reduced the economic viability of a licensed operation. Second, sturgeon were suspected of becoming sufficiently plentiful in the Nelson that under high-water conditions sturgeon might actually be able to restock Lake Winnipeg and the Saskatchewan River 'and provide a more widespread and lasting benefit to the settlers than can presently be obtained by even a limited commercial fishery.' In other words, Manitoba officials at that time understood the Nelson to be not permanently depleted but, if managed as a refuge, capable of replenishing depleted stocks elsewhere.

The Subsistence Fishery

For many years, Manitoba managed the commercial fishery as though the subsistence fishery did not exist, indeed as though the Cree were nothing more than historical curiosities. Sunde refers to Indians only once, noting that they associate spawning time with leaf emergence (1961: 53). Nowhere does he indicate that there was a continuing Cree subsistence fishery, let alone estimate actual catch levels. His estimates of productivity are based on the commercial catch alone. Macdonald acknowledges that the size of the subsistence harvest is unknown but claims that 'the effort has not been observed to be anywhere near the effort that the commercial fishery puts in' (1991: 2). No supporting evidence is provided. In view of the history of enforcement related by Cree fishers at the arbitration proceedings (4–7 November 1991), it seems unlikely that the Cree would have done anything to make such observations easy. Manitoba's assumption that the subsistence harvest is 50 per cent of the commercial harvest must be taken as the unreliable 'guesstimate' that resource agencies often have to make and hence should not be given much weight.

Socioeconomic Determinants of Effort

Critical to Manitoba's hypothesis that sturgeon were already depleted before Jenpeg is the assumption that the decline in catch per fisher before that event was a direct consequence of there being fewer adult fish. Sunde asserted that the determinants of effort were more or less constant and discounted the possibility that declines in catch per fisher were the result of changing fishing intensity. He hypothesized that intensity should have increased in response to price increases (Sunde 1961: 71), but in fact the closure in the 1930s was a response to low prices, rather than overfishing caused by high prices.

The actual determinants of effort, especially for Indian fishers, are complex. Commercial fishing was traditionally, for Indians, not a single enterprise but part of a mixed economy. Thus many factors other than price affect effort. Among the most important are alternative resource harvesting opportunities (i.e., scarcity or abundance of other valued resources relative to sturgeon) and, from the 1950s onward, alternative economic opportunities to commercial fishing. Any attempt to relate fishing effort to fish prices must control for these factors, which Sunde's analysis does not.

Neither the determinants of effort nor actual fishing practices are the same for the Indian subsistence fishery and the commercial fishery. The former is highly efficient and takes place during spawning time, when sturgeon are concentrated in and below rapids. The latter occurs after spawning, when they are more spread out. Besides, not all the subsistence fishers were involved in the commercial fishery. Provincial biologists drew their samples only from the commercial fishery or from their own test netting. The degree of cooperation from Indian commercial fishers is not reported, but one Indian fisher testified at the arbitration that he did not cooperate with sampling because it required that his fish be cut open before scheduled delivery, and this he did not want done. Also, Aboriginal harvesters commonly observe that biologists do not always know where to fish, and this observation has often proved correct. The sampling method used by provincial biologists to develop management recommendations was thus not necessarily representative of the subsistence fishery. It follows that changes in harvest levels and in catch per fisher, which are interpreted by Manitoba as evidence of depletion, may be as well or better explained by unrelated natural or socioeconomic factors.

Other Evidence

If the depletion hypothesis were correct, we should expect to find historical accounts of scarcity from the Cree themselves. Just such accounts exist for well-documented cases of sturgeon depletion, such as that at Lake of the Woods (Holzkamm, Lytwyn, and Waisberg 1988; Van West 1990) and Lake Winnipeg (Tough 1996: 239–43). However, published Indian agents' reports for the Cross Lake district for c. 1900–15 (Canada, various years, Annual Report for the Department of Indian Affairs) reveal no such concerns, even though these are precisely the years we should expect them to be reported.

Further, the fishermen who testified at the arbitration related that within living memory sturgeon were never scarce or difficult to obtain prior to the construction of the Jenpeg control structure. It is known that sustainable mixed subsistence-commercial sturgeon fisheries existed on northern rivers long before the rise of modern commercial fishing for distant markets (viz. Holzkamm and McCarthy 1988). Other indicators that are inconsistent with the depletion hypothesis are the accounts of gear size, sturgeon size, and sturgeon condition reported by the fishers who testified before the arbitration. Reported mesh size of nets ranged from 10 to 13 in. (25 to 33 cm), with most at the upper end of this range. This is consistent with reported fish sizes of 42 to 60 in. (about 100–150 cm fork length, equivalent to about 9–25 kg – Sunde 1961: 41). There was thus no indication of declining gear size, typically associated with depletion, prior to Jenpeg.

Given these facts, along with the several lengthy sturgeon closures on the Nelson and the relative absence of environmental degradation on its upper reaches prior to Jenpeg, it is entirely reasonable to suggest that MSY in the early 1970s, prior to construction, was not substantially lower than it had been in the late nineteenth century.

Conclusions and Observations

Two types of estimates of pre-disruption sturgeon catch were submitted at the arbitration. One employs social survey data and is expressed as production (kg/yr). The other uses biological data and is expressed as productivity (kg/ha/yr), based on an estimated pre-disruption surface area used by the Cross Lake Indian Band of 107,000 ha. Table 10.3 compares Cross Lake production estimates (harvest and consumption) with productivity estimates for Cross Lake and two other water bodies. We have added average recent commercial catch (Macdonald 1990: 3) to subsistence catch to show total production for these waters.[7] All data are standardized as raw edible weight. To facilitate comparison between the two types of data, we have inferred productivity from the production estimates and sustainable yield from the productivity estimates.

Symbion's consumption estimate exceeds Manitoba's productivity estimate by about 50 per cent, and our harvest estimate exceeds it by about 100 per cent. For comparison, Table 10.3 shows the frequently cited sturgeon productivity estimate by Folz and Meyers (1985) for Wisconsin and an estimate for the Abitibi River in northeastern Ontario by Payne (1987). These are substantially higher than Manitoba's and fall between the harvest and consumption estimates for Cross Lake.

Manitoba's estimate is lower than all the others in Table 10.3 because it is based on the assumption of prior depletion, shown here to be doubtful. For this and other reasons, Manitoba's estimate is low by an unknown factor (possibly

Table 10.3 Estimates of sturgeon production (kg/yr) and productivity (kg/ha/yr) for Cross Lake traditional harvesting area (all data converted to kg edible weight)

	Production estimates			
	Subsistence	Commercial*	Total	Inferred productivity
Harvest	17,100	3,563	20,663	0.193
Consumption	12,611	3,563	16,174	0.151

	Productivity estimates		
		Inferred sustainable yield	Productivity
Cross Lake		10,596	0.099
(Macdonald 1990)			
Lake Winnebago		19,688	0.184
(Folz and Meyers 1985)			
Abitibi River		17,976	0.168
(Payne 1987)			

*Since 1970 (Macdonald 1990: 3)

half) and cannot be accepted as a theoretical cap on sustainable pre-project harvest levels. If the Nelson River stocks were healthy and relatively undepleted before Jenpeg, as we have shown there is good reason to believe, then both the consumption and harvest estimates for Cross Lake are within the range of reported productivity estimates of other comparable systems.

We think that Symbion's estimate of consumption may be low for methodological reasons, but by an unknown factor. The harvest estimate is based on a very small sample and should be seen as suggestive or indicative but not necessarily accurate. While none of the data can be characterized as robust, on the balance of probabilities the survey-based production data are stronger, and the case for disruption, and for the need to compensate the Cross Lake Indian Band for the loss of its fishery, is made.

This case study also highlights the deficiencies of traditional, top-down management practices of government resource management agencies. Manitoba managed the sturgeon fishery on the Nelson as though the Cree subsistence fishery did not exist and, until 1970, as though there were a single stock on the river instead of several largely discrete ones.

If the Cree subsistence fishery were accounted for, a graph of historical catch would look rather different from the one based solely on the inaccurate tabula-

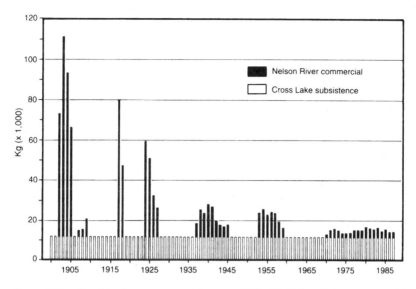

Figure 10.2   Combined sturgeon catch by year, 1901–87: Nelson River commercial (Table 10.2, column 3, converted to kg) and hypothetical constant Cross Lake subsistence (estimated at 12,000 kg, headless dressed equivalent). The base subsistence catch for the Nelson River does not include Norway House or other communities on the river. The Nelson River commercial sturgeon fishery was closed in four periods: 1911–15, 1930–6, 1947–52, and 1961–9.

tions of commercial production so frequently reproduced in the provincial literature (viz. Patalas 1988: 3). Figure 10.2 shows our amended data for commercial catches, along with a hypothetical constant annual Cross Lake subsistence harvest of about 12,000 kg (headless dressed equivalent), which is the approximate mean of the harvest and consumption estimates. The subsistence harvest (as estimated for the early 1970s) had probably been slowly rising with Cross Lake's population growth, so that earlier levels may have been lower. However, Figure 10.2 compares Nelson River's commercial catch only with the Cross Lake subsistence catch, not including that for Norway House or other communities. Figure 10.2 does not clearly illustrate a history of repeated depletion and collapse. The periodic declines may have resulted from commercial fishing too far in excess of the sustainable subsistence catch, which was always the base of the fishery. However, they may also have been the product of fluctuations in prices and alternative opportunities, and hence changing effort.

It would appear that the assumptions on which Manitoba has managed the

Nelson River sturgeon fishery are questionable. This is the result of three factors: the ministry's failure to ascertain the basic biological, economic, and social parameters of the fishery, which are in principle discoverable; its failure to make use of the knowledge of local Cree fishers and of the methods of the social sciences, in combination with fisheries science; and its unwillingness to recognize and acknowledge the obvious deficiencies in its own environmental and social information base and to correct these deficiencies in a cooperative and timely manner.

The methods for constructing acceptably reliable statistics on subsistence harvests (and reconstructing historical statistics) are in fact well established. Some other fish and wildlife management agencies in Canada and Alaska routinely construct and use such statistics. Freshwater subsistence fisheries are one of the least well-documented aspects of Aboriginal subsistence harvesting but need not continue to be. Recent comprehensive Aboriginal claims settlements in northern Canada have established the basis for improved research and monitoring and for cooperative management, in which Aboriginal harvesters and government managers are equal partners (Usher 1997).

NOTES

1 This case study was originally prepared for the Cross Lake Indian Band, Cross Lake, Man., and Savino & Co., Winnipeg (legal counsel to the band). Dr Greg Michalenko provided comparative information on the Muskrat Dam sturgeon fishery, Dr Fikret Berkes commented on a draft, and Wayne Wysocki and Pat Larcombe of Symbion Consultants provided information and comment. Part way through the presentation of the defendant's evidence, the parties agreed to an out-of-court settlement. The agreement included provision for a harvest survey at Cross Lake, creation of a Nelson River sturgeon co-management board, and a cash settlement. The Northern Flood Agreement (NFA) arbitrator thus did not rule on the evidence submitted.

2 Like the James Bay and Northern Quebec Agreement of 1975, which resulted from similar large-scale hydro developments, it can be characterized as an out-of-court settlement for adverse effects. Both agreements were unprecedented (even if in the outcome inadequate) measures to protect the Indian interest in resources and habitat. The NFA thus provided the basis for Cross Lake to lodge a damage claim for fisheries losses.

3 Except for some speculative estimates of total catch, undifferentiated by species, reported by conservation officers in the 1950s (Manitoba 1954–7).

4 To facilitate direct comparison, all source data were standardized to kg/capita/year raw edible weight (Usher and Weinstein 1991, Wein and Freeman 1995), where raw weight = 1.33 cooked weight, and edible weight = 0.80 round weight. The last is a

conservative estimate. Michalenko and Marcogliese (1989) found that only 15 per cent of round weight was discarded in a contemporary subsistence sturgeon fishery in northwestern Ontario. Symbion (1990) found that virtually all parts of the sturgeon are used at Cross Lake.

5 Commercial catch and effort data were maintained by the dominion government until 1930, when control of freshwater fishing was turned over to Manitoba under the Natural Resources Transfer Act of that year. The data are currently retained by the Manitoba Department of Natural Resources in Winnipeg.

6 In the most notable cases of sturgeon depletion in Canadian freshwater systems, initial stock depletion was followed by industrial effects such as altered flow regime and industrial pollution, which permanently inhibited stock recovery (chiefly by despoliation of spawning areas). This was not the case in the Cross Lake area of the Nelson River prior to the construction of Jenpeg in 1976.

7 The total does not, however, include some possible additional subsistence harvesting by residents of nearby non-reserve communities of Wabowden, Thicket Portage, and Pikwitonei.

WORKS CITED

Berkes, F. 1983. 'Quantifying the Harvest of Native Subsistence Fisheries. In R.W. Wein, R.R. Riewe, and L.R. Methven, eds., *Resources and Dynamics of the Boreal Zone*, 346–63. Ottawa: Association of Canadian Universities for Northern Studies.

– 1990. 'Native Subsistence Fisheries: A Synthesis of Harvest Studies in Canada.' *Arctic* 43, no. 4: 35–42.

Berkes, F., P.J. George, R.J. Preston, A. Hughes, J. Turner, and B.D. Cummins. 1994. 'Wildlife Harvesting and Sustainable Regional Native Economy in the Hudson and James Bay Lowland, Ontario.' *Arctic* 47, no. 4: 350–60.

Canada. 1901–16. *Annual Report for the Department of Fisheries. Sessional Papers.*

– 1900–15. *Annual Report for the Department of Indian Affairs. Sessional Papers.*

– 1917–46. Dominion Bureau of Statistics. *Fisheries Statistics.* Ottawa.

Folz, D.J., and L.S. Meyers. 1985. 'Management of the Lake Sturgeon, *Acipenser fulvescens*, Population in the Lake Winnebago System, Wisconsin.' *Environmental Biology of Fishes* 14, no. 6: 135–46.

Gibson, R.S. 1990. *Principles of Nutritional Assessment.* New York: Oxford University Press.

Harkness, W.J.K. 1980. 'Report on the Sturgeon Situation in Manitoba.' Manitoba Department of Natural Resources, Fisheries Branch Ms Rep. no. 80-3.

Harkness, W.J.K., and J.R. Dymond. 1961. *The Lake Sturgeon: The History of Its Fishery and Problems of Conservation.* Toronto: Ontario Department of Lands and Forests, Fisheries and Wildlife Branch.

Holzkamm, T.E., V.P. Lytwyn, and L.G. Waisberg. 1988. 'The Rainy River Sturgeon: An Ojibwa Resource in the Fur Trade Economy.' *Canadian Geographer* 32, no. 3: 194–205.

Holzkamm, T.E., and M. McCarthy. 1988. 'Potential Fishery for Lake Sturgeon (*Acipenser fulvescens*) as Indicated by the Returns of the Hudson's Bay Company Lac la Pluie District. *Canadian Journal of Fisheries and Aquatic Sciences* 45: 921–3.

Hopper, M., and G. Power. 1991. 'The Fisheries of an Ojibwa Community in Northwestern Ontario.' *Arctic* 44, 4: 267–74.

James Bay and Northern Quebec Native Harvesting Research Committee. 1982. *The Wealth of the Land – Wildlife Harvests by the James Bay Cree, 1972–73 to 1978–79.* Quebec City.

Larcombe, P. 1997. *The Northern Flood Agreement: Implementation of Land, Resource, and Environmental Regimes in a Treaty Area.* Report prepared for the Royal Commission on Aboriginal Peoples – Land, Resource, and Environment Regimes Project. For Seven Generations: An Information Legacy of the Royal Commission on Aboriginal Peoples (CD-ROM). Ottawa: Libraxus.

Macdonald, D. 1990. Unpublished memorandum to B. Scaife, Fisheries Branch, 11 March.

– 1991. 'An Overview of the Sturgeon Population in the Cross Lake, Sipiwesk Lake and Nelson River Area.' Unpublished. Fisheries Branch, 31 Oct.

Manitoba. 1954–7. 'Annual Registered Trapline Conference of Conservation Officers.' Manitoba Department of Mines and Natural Resources, Game and Fisheries Branch. The Pas. Unpublished annual reports.

Michalenko, G., L. Marcogliese, and Muskrat Dam Band. 1989. 'The Subsistence Lake Sturgeon (*Acipenser fulvescens*) Fishery of the Indian Village of Muskrat Dam in Northern Ontario, Canada.' Paper presented to the First International Conference on the Sturgeon, Bordeaux, France.

North/South Consultants Inc. n.d. 'Review of the Manitoba Commercial Sturgeon Fishery.' Technical Paper no. 4. Unpublished report.

– n.d. 'Comparison of Reported and Estimated Consumption Levels to Maximum Sustainable Yield.' Technical Paper no. 5. Unpublished report.

– n.d. 'Life History of the Lake Sturgeon (*Acipenser fulvescens*) with Emphasis on the Nelson River.' Technical Paper no. 6. Unpublished report.

Northern Flood Agreement (NFA). 1991. *Proceedings*. Arbitration Re: Claims 110 and 44, Cross Lake, vols. I–III, 4–6 Nov.

– 1992. *Proceedings*. Arbitration Re: Claims 110 and 44, Cross Lake, vols. I–III, 13–15 April.

Patalas, J.W. 1988. 'The Effects of Commercial Fishing on Lake Sturgeon (*Acipenser fulvescens*) Populations in the Sipiwesk Lake Area of the Nelson River, Manitoba,

1987–1988.' Manitoba Department of Natural Resources, Fisheries Branch Ms. Rep. No. 88–14.

Payne, D.A. 1987. 'Biology and Population Dynamics of Lake Sturgeon (*Acipenser fulvescens*) from the Frederick House, Abitibi and Mattagami Rivers, Ontario.' In C.H. Olver, ed., *Proceedings of a Workshop on the Lake Sturgeon (Acipenser fulvescens)*. Ontario Fisheries Technical Report Series no. 23. Toronto: Ontario Ministry of Natural Resources.

Provincial Archives of Manitoba (PAM). RG 17, B1, box 50, file 44.5.4.

Sopuck, R.D. 1987. 'A Study of the Lake Sturgeon (*Acipenser fulvescens*) in the Sipiwesk Lake Area of the Nelson River, Manitoba, 1976–78.' Manitoba Department of Natural Resources, Fisheries Branch Ms. Report no. 87–2.

Sunde, L.A. 1959. 'The Sturgeon Fishery in Manitoba, with Recommendations for Management (Analysis of Nelson River Data 1953–1956).' Report for the Manitoba Department of Natural Resources, Fisheries Branch.

– 1961. 'Growth and Reproduction of the Lake Sturgeon (*Acipenser fulvescens* Rafinesque) of the Nelson River in Manitoba.' MSc thesis, University of British Columbia.

Symbion Consultants. 1990. 'Cross Lake Band of Indians Domestic Sturgeon Fishing Claim, Proposed Basis for Settlement (Final Report).' Winnipeg.

Tobias, T.N. and J.J. Kay. 1994. 'The Bush Harvest in Pinehouse, Saskatchewan, Canada.' *Arctic* 47, no. 3: 207–21.

Tough, F. 1996. *'As Their Natural Resources Fail': Native Peoples and the Economic History of Northern Manitoba, 1870–1930*. Vancouver: University of British Columbia Press.

Usher, P.J. 1988. 'Gathering and Interpreting Native Subsistence Fisheries Statistics.' Paper presented at Subsistence Fisheries Symposium, American Fisheries Society Meetings, Toronto.

– 1991. Northern Flood Agreement – Cross Lake Band Claims Nos. 44 and 110 Respecting Domestic and Commercial Sturgeon Fishing Losses as a Result of the Hydro Project. Unpublished Memorandum to Savino & Co. 21 Aug.

– 1992. Estimating Historical Sturgeon Harvests on the Nelson River, Manitoba. Unpublished memorandum to the Cross Lake Band of Indians and Savino & Co.

– 1997. 'Contemporary Aboriginal Land, Resource, and Environment Regimes: Origins, Problems, and Prospects.' Report Prepared for the Royal Commission on Aboriginal Peoples – Land, Resource, and Environment Regimes Project. For Seven Generations: An Information Legacy of the Royal Commission on Aboriginal Peoples (CD-ROM). Ottawa: Libraxus.

Usher, P.J., P. Anderson, H. Brody, J. Keck, and J. Torrie. 1979. 'The Economic and Social Impact of Mercury Pollution on the Whitedog and Grassy Narrows Indian Reserves, Ontario.' Report prepared for the Whitedog and Grassy Narrows Indian Bands. P.J. Usher Consulting Services, Ottawa.

Usher, P.J., and N.D. Bankes. 1986. *Property, the Basis of Inuit Hunting Rights: A New Approach.* Ottawa: Inuit Committee on National Issues.

Usher, P.J., and M.S. Weinstein. 1991. *Towards Assessing the Effects of Lake Winnipeg Regulation and Churchill River Diversion on Resource Harvesting in Native Communities in Northern Manitoba.* Canadian Technical Report of Fisheries and Aquatic Sciences no. 1794. Winnipeg: Department of Fisheries and Oceans.

Usher, P.J., and G. Wenzel. 1987. 'Native Harvest Surveys and Statistics: A Critique of Their Construction and Use.' *Arctic* 40, no. 2: 145–60.

Van West, J.J. 1990. 'Ojibwa Fisheries, Commercial Fisheries Development and Fisheries Administration, 1873–1915: An Examination of Conflicting Interest and the Collapse of the Sturgeon Fisheries of the Lake of the Woods.' *Native Studies Review* 6, no. 1: 31–65.

Waldram, J. 1988. *As Long as the Rivers Run: Hydro-electric Development and Native Communities in Western Canada.* Winnipeg: University of Manitoba Press.

Wein, E.E. 1995. 'Evaluating Food Use By Canadian Aboriginal Peoples.' *Canadian Journal of Physiological Pharmocology* 73: 759–64.

Wein, E.E., and M.M.R. Freeman. 1995. 'Frequency of Traditional Food Use by Three Yukon First Nations Living in Four Communities.' *Arctic* 48, no. 2: 161–71.

Wolfe, R.J., and R.J. Walker. 1987. 'Subsistence Economies in Alaska: Productivity, Geography, and Development Impacts. *Arctic Anthropology* 24: 56–81.

11

An Interdisciplinary Method for Collecting and Integrating Fishers' Ecological Knowledge into Resource Management

BARBARA NEIS, LAWRENCE F. FELT,
RICHARD L. HAEDRICH, AND DAVID C. SCHNEIDER

Most of the world's major fish stocks are overfished, many to the point of collapse (FAO 1989; Pauly, this volume). This situation has prompted debate concerning whether fisheries science and associated management, in their present forms, can provide a basis for the sustainable use of fishery resources (Holling 1993; Ludwig, Hilborn, and Walters 1993; McCay and Acheson 1987; McGoodwin 1990; Rosenberg et al. 1993; Gallaugher and Vodden, Hutchings, Pinkerton, this volume).

Some argue that finding ways to incorporate fishers' participation would improve the capacity to manage fisheries sustainably. Researchers from many disciplinary backgrounds have argued that users' traditional ecological knowledge (TEK) represents at least a critical supplement to scientific understanding and perhaps an alternative foundation for sustainable resource management (Berkes 1993; Freeman 1992; Gadgil, Berkes, and Folke 1993; Johannes 1981, 1993; Johnson 1992; Kloppenburg 1991; Mailhot 1993; Nakashima 1993; Richards 1985). One of the primary barriers to greater use of TEK lies in the absence of agreed-on methods for gathering information on it. The development of a framework for collecting and using fishers' TEK is the central focus of this chapter. We review the literature on fisheries TEK from a methodological perspective, provide a sample method designed for research among commercial fishers in Newfoundland, and illustrate some potential benefits associated with combining data derived from fishers' TEK with data derived from more traditional sources.

Traditional Ecological Knowledge

The concept TEK appeared in the mid-1980s but draws on two older and somewhat separate research traditions – ethnoscience and cultural ecology (Mailhot

1993). Research on ancient and current ethnoscience, and growing awareness of the history of non-Western science, have contributed to an increased interest in TEK (Berkes 1993; Goonatilake 1984). While there is no universally accepted definition of TEK, Mailhot (1993) offers the most common working definition – 'the sum of the data and ideas acquired by a human group on its environment as a result of the group's use and occupation of a region over many generations' (11). As such, TEK contains empirical and conceptual aspects, is cumulative over generations, and is dynamic, in that it changes in response to socioeconomic, technological, and other changes.

Though no single research project is likely to combine all elements of TEK, five major components would be the categories used by groups to classify components of the environment and the organization of these categories into systems of representation; empirical data on the environment, including spatial distribution of the components, behaviour, relationships between species, and interpretation of natural phenomena; the use that groups make of their environment; the management system covering natural resources, including conservation practices and mechanisms for assessing the state of the resources and management of them; and the worldview of groups (Mailhot 1993: 12).

TEK is 'an integrated system of knowledge, practice and beliefs' that is linked to a wider social context different from the social context of normal fisheries science (Berkes 1993: 4–5; see also Johnson 1992: 7–8). A number of differences between TEK and 'normal' science have been identified (Berkes 1993; Johnson 1992). One lies not so much in the types of observations on which they depend but rather in the 'organization of the observations and the physical recording of them' (Gun, Arlooktoo, and Kaomayok 1988: 25). One of the methodological hurdles associated with combining science and TEK is finding ways to combine the two types of data. It is our view that TEK can be integrated with fisheries science and management and that the potential gains from such integration would be significant.

TEK and Fisheries Research

Mailhot's five elements provide a useful framework for classifying existing research on fisheries TEK. As for systems of classification, we are unaware of any TEK fisheries research that provides an example of a fully developed ethnotaxonomic system for marine species. However, Johannes (1981) includes a glossary of Palauan terms for fishing gear, fish behaviour, parts of the marine environment, and different species of fish (and for Cree and Ojibwa fishers of the Canadian central subarctic, see Ray, Thoms, this volume). Eythorsson (1993) provides an example of a local Sami taxonomy for fiord-spawning cod stocks in Norway (136). A 1922 study based on fishers' knowledge of cod migration patterns in Newfoundland also contained some ethnotaxonomic elements (Munn 1922), as does a recent inventory of northeast coast Newfoundland fishers' knowledge of coastal resources (Graham 1993).

Fishers' knowledge typically includes not only categories of fish but also information on behaviour, annual cycle, winds, tides, and references to time and space that may differ from those customarily recognized by fisheries science (Eythorsson 1993). Thus, while in the 1980s Canada's Department of Fisheries and Oceans (DFO) classified all the cod (*Gadus morhua*) contained in the areas 2J3K3L (from Cape Race to Labrador) as 'northern' cod and managed coastal and offshore fish as one unit, fishers in Trinity Bay distinguished among 'herring' fish, 'caplin' fish, 'sunburnt' or 'blacked' fish, 'deepwater' fish, 'shoalwater' fish, 'mother' fish,' and 'foxy' or 'iron ore' fish (Neis et al. 1996). Stock assessment scientists monitored the changes in the biomass of cod in 2J3K3L, whereas coastal fishers monitored local variations in catch rates, fish size, and season length (Neis 1992).

Because locating fish is central to their economic survival, fishers can be expected to have detailed knowledge concerning the spatial and temporal distribution in the areas in which they fish of at least those components of the marine ecosystem on which they depend. Johannes (1981) drew on Palauan TEK to document the lunar spawning cycles of some Pacific species. Freeman (1989a) documented the Inuit understandings of bowhead whale behaviour that allowed the Inuit to challenge successfully the population estimates of the International Whaling Commission.

Understanding the interaction of a species with its environment can also increase fishing success. Eythorsson (1993) emphasizes the importance of knowledge of tidal currents to successful gillnet fishing among the Sami. Recent research among fishers from Placentia Bay, Newfoundland, found that they have detailed knowledge of winds and tides and the influence of both on fish movements. They also associate the back and forth movement of fish around headlands and between subregions of the bay with particular winds and differentiate between 'dirty' water, which is generated by offshore winds and is poor for fishing, and 'clear' water, produced by onshore winds and associated with fish moving out of shoal water areas.[1] Gillnets are strategically combined in fleets of varying length and set according to knowledge of tidal current, depth, and fish movement patterns (Hutchings, Neis, and Ripley 1997; Paul Ripley, pers. comm.; see also Newell, this volume, for BC Native herring spawn-on-kelp fisheries).

Fishers measure ecological change on the basis of deviations from known, previously observed patterns, and their measurements may differ from those made in fisheries science because of different spatial and temporal scales (Eythorsson 1993). If the intention of fisheries management is to maintain the population structure and ecological basis for stocks, having precise information on where and when fish populations congregate can help in identifying these and monitoring the effectiveness of management regimes (Nakashima 1993; Wilson et al. 1994).

A growing body of fisheries literature has documented the existence of indigenous or traditional management systems based on TEK within many small-scale fisheries. This literature highlights the mechanisms that have regulated access to resources and often points to the long-term 'success' (as measured by stability and persistence) of such management systems until they are confronted by externally imposed management models or the effects of uncontrolled fishing by others outside the managed area (Berkes 1989; Freeman 1989a; Johannes 1978; McCay and Acheson 1987; Pinkerton 1988, and this volume; Thoms, this volume).

McGoodwin (1990) has identified a range of controls associated with indigenous management systems, including limiting access, reducing effort on stocks perceived as under pressure, limiting catches, releasing fry, discouraging overzealousness by fellow fishers, setting up closed seasons, and enhancing the marine ecosytem (131; see also Johannes 1978). The ecological understandings that informed the introduction of these management systems and their precise ecological effects need to be explored among all groups of fishers.

Methods for Collecting TEK

If TEK is to be used to develop fisheries science and resource management, frameworks need to be developed for gathering that knowledge, fusing it with formal science, and using it to inform policy concerns. A number of recent overviews of TEK have addressed some of the methodological questions associated with its collection and use. Mailhot (1993) points out that the field of TEK is enormous and no single method or study can expect to capture a population's entire TEK. She also argues that TEK must be collected in the language of the population and linked to an understanding of the way in which that population 'codifies and organizes its knowledge' (38).

The methods used in existing TEK research include discourse analysis, 'formal eliciting,' the 'triad test,' and tape-recorded semi-directed interviews. Discourse analysis includes recording of conversations between informants and discussions with informants concerning the terms and expressions they have used. Formal eliciting is employed primarily in ethnoscience and refers to 'the ethnographer learning how to formulate – in the informants' language – questions which make sense in the informants' own culture' (Mailhot 1993: 37). Used in the development of ethnotaxonomies, the triad system involves giving groups of three terms to informants and asking them to pick the one that is 'most distinct from the others, and to explain why' (Maihot 1993: 37).

Mailhot agrees with other reviews of the literature in arguing that the semi-directed taped interview based on open, broad questions is the most effective instrument for gathering TEK. She points out, however, that much of the existing research has depended on a very small number of informants, thus raising questions about individual variation and generalizability (1993: 37). Johannes (1981) found that fishers differ in their devotion to observation in fishing, with some learning to seek fisheries knowledge 'for its own sake,' as well as a means to improve their fishing. Searching out such fishers would be wise, but of course research that relies on small samples has limitations.

Johnson points out that 'the quantity and quality of traditional environmental knowledge varies among community members, depending upon gender, age, social status, intellectual capability, and profession' (1992: 4). Fishers' TEK is probably also influenced by region, technological environment, crew structures, and fishing practices. It seems to reflect the influence of the gear that they use (Kent Martin 1973), local topography (Butler 1983), local differences in natural and fishing-induced changes in fisheries, and competitive relations among fishers.

Commercial fishers derive part of their understanding of fisheries ecology from broad social networks, the media, and other general sources. However, as competitive fishers, they trust the knowledge derived from direct observation, from the members of their crew, and from friends and family more than that drawn from other sources of knowledge (see Newell, this volume). Contrary to common assumptions in management models, there may be significant variation in fishing patterns within, as well as among, fleet sectors (Hanna and Smith 1993; Kearney 1993). Fishers with formal training, those who do not come from fishing families, and those in fisheries where there has been a lot of technological change, particularly that resulting from applications of externally derived techniques, are likely to have different understandings of what is happening in the marine ecosystem from those who learned through apprenticeship, who came from fishing families, who fish in a milieu where there have been only limited changes in fisheries technologies and fishing patterns and where any changes have been locally developed (Hanna and Smith 1993). Fisheries labour forces also tend to be divided by generation and gender (DeWees and Hawkes 1988; Nadel-Klein and Davis 1988), which also leads to pattern variation in TEK. Older fishers, for example, who fished stocks prior to the effects of overfishing, may have very different perceptions of what constitutes a

'healthy' stock than do younger counterparts who never experienced the stocks during a period of higher abundance (Cabot Martin 1992). Extensive intergenerational communication would, however, mitigate this difference in experience.

One important consequence of the differentiation of TEK among groups in a fishery is that there exists an ever-changing range of views about the contemporary state of the resource. The TEK of individual fishers or of communities of fishers will reflect their own fishery history and might also be influenced by the wider context of conflict and debate. In Newfoundland, for example, considerable variation in fishers' views of the state of the fishery resource emerged in presentations to both the British royal commission of 1863 on the state of the fishing industry (Cushing 1988) and to a recent independent inquiry into the state of Newfoundland's northern cod stocks (Harris 1990). Gillnets, which can be moved to follow fish concentrations, provide fishers with a different sense of local populations than do immobile types of gear, such as cod and capelin traps (Rose 1992; Rose and Leggett 1989, 1991). Catches from trawls and seines are insensitive to changes in the abundance of schooling species and so provide a different sense of the status of stocks than do less mobile gear. Another consequence of differentiation is that TEK is neither monolithic nor static (Felt 1994).

The complex range of factors that probably influences fishers' TEK means that reliance on a small sample could result in limited and perhaps biased data. A larger, stratified sample would be better. Felt's (1994) approach was to start at a community or regional level and keep interviewing until he reached some coherent patterns in terms of views. Nakashima (1993) used a sample of fifty Inuit informants from three villages. His study examined a wide range of topics, including 'the hunting, use, classification, anatomy and ecology of elders' (100). Sampling strategies might vary with the emphasis of the research. For example, if the focus relates to use of the fishery resource and the kind of intimate understandings that are derived from processing, shore workers and those who prepare fish for home consumption may have relevant ecological information.

Interviewing within subsamples should probably start with those fishers whom several people identify as 'knowledgeable' and continue until a broad consensus within subsamples has been identified. Questions should probe for clues regarding how someone has come to know something: 'Observations are rarely wrong even if the interpretations may be' (Gunn, Artooktoo, and Kaomayok 1988: 24). Where possible, follow-up interviews help capture insights triggered during the period after the first interview and those lost because of fatigue associated with the interview process. Combining an element of fieldwork with recurring contact with some fishers can greatly strengthen the research.

Researchers studying TEK need also be concerned about the truthfulness of responses. TEK research is often initiated because of conflict regarding management and scientific understanding, and this 'politicized' context may influence responses. Johannes (1981) tested the validity of responses by asking two types of questions: those to which he knew the answer and those that sounded reasonable, but which he knew these fishers could not answer. Interviews of this kind assume considerable knowledge of the fishery on the part of interviewers. At the same time, eliciting information on knowledge systems and categories requires input from somewhat different disciplinary backgrounds. Knowledge of fisheries science can help ensure that questions are interesting to fishers, but such knowledge might also act to block receptivity to explanations that do not fit with conventional understandings. The use of interdisciplinary teams and taping interviews will reduce this risk.

In politicized contexts, it is also important to consider the organizational identification of the researchers (i.e., are they associated with an organization considered by many to be 'opposed' to fishers' interests?) and the risks of conducting interviews in a group context. Fishers may be more honest if they are interviewed alone, and observations of group discussions can provide an impression of the impact that public conflict may be having on 'stated views.' Comparing fishers' maps of the physical environment with those of fish locations where there is conceivably less risk of information being withheld is also useful. Moreover, the level and extent of fishers' organizational involvement may influence their expression of ecological knowledge, and so this information should also be collected.

Researchers on indigenous peoples' TEK sometimes map land use and occupancy by constructing map biographies of individuals and asking informants to map biogeographical information (Hrenchuk 1993; Nakashima 1993). In most cases, these maps are constructed using the toponyms of indigenous users. In the case of inshore fishers, their knowledge may involve a local and contingent mapping of the environment, which expresses how they use it to make a living. Attempting to elicit their cognitive maps of what to fish, where, and when, using charts and overlays, may be the most helpful way to isolate key elements in their ecological knowledge. Our research in Newfoundland has found that most fishers were comfortable with charts and often looked for very fine details, sometimes describing features not on the chart that were important as marks or locations in their fishery. They also made extensive use of the bathymetric lines to find particular grounds and ranges.

Fishers transmit ecological knowledge orally; it may have some largely unconscious elements; and it is commonly transmitted in the course of fishing itself. TEK research would be substantially strengthened by conducting interviews on the fishing grounds and combining them with observation and follow-up interviews. Observation and discussion on the grounds help researchers to identify issues otherwise overlooked and can provide another check on the accuracy of their understanding of responses. Some fishers have maintained detailed log-books for years that can provide information on catches, fish size, weather, and so on. Some fishing households have retained receipts for fish sales and boat and gear purchases going back years, sometimes decades. These records should be used, when available, in interviews or in conjunction with interview data as a way of jogging memories and querying fishers' responses based on memory. Since it is often fishers' wives who manage the books and receipts (see Neis, this volume), it might be useful to involve these women in portions of the interview related to the history of gear changes and landings. Where TEK refers to historical change it should ideally be combined with insights contained in archival and other historical or perhaps scientific documents written during the period under discussion (see, for example, Hutchings, Neis, and Ripley 1996; Windsor 1994; Usher and Tough, and Cadigan, this volume).

Finally, a central element of TEK research should be the identification of the types of information most appropriate to be elicited from fishers as opposed to from other sources. While the precise type of information will vary somewhat by the type of fishery and the technology on which it is based, scientists have found TEK useful in answering a broad range of questions related to migration, predator–prey relationship, spawning, and the status of the resource – including perceptions of long-term trends.

A Sample Research Framework

This section describes a research framework developed on the northeast coast of Newfoundland for collecting fishers' knowledge and combining it with science. The fishery in the province of Newfoundland and Labrador is in a sorry state. The northern cod stocks have all but vanished, to the point of being declared commercially extinct (Bishop et al. 1993; Steele, Andersen, and Green 1992; Hutchings, Villagarcia et al., this volume), and moratoria have closed many other ground fisheries. A number of possible causes, including oceanographic change, change in predator–prey relations, overfishing, and various combinations of the above have been suggested for this dramatic demise (Hutchings and Meyers 1994; Lear and Parsons 1993; Hutchings, this volume). Important socioeconomic factors and differences in the timing of perceptions of resource decline, as well as its causes, have attracted social scientific scrutiny of science and management (Cabot Martin 1992; Finlayson 1994; Neis 1992; Steele, Andersen, and Green 1992).

Enormous uncertainty exists about the extent of the long-term damage done to the stocks, their future resilience, and hence the ideal scale, technologies, and management structures for the fisheries of the future. Research prompted by the stock crisis has often returned to basic ecological questions such as the dynamics of spawning and its location, factors influencing recruitment, and migration patterns (Ellerston et al. 1989; Hutchings, Myers, and Lilly 1993; Myers et al. 1993; Rose 1993; Valerio et al. 1992). This research finds gaps and errors in existing science of the northern cod stocks and has begun to address directly questions related to their collapse off the northeast coast of Newfoundland and Labrador (Haedrich 1995; Hutchings and Myers 1993; Hutchings, Myers, and Lilly 1993; Myers, Mertz, and Barrowman 1995; Oceans Ltd. 1990; Rose 1993; Villagarcia et al., Hutchings, this volume). All these matters underscore the need to identify new data sources so as to broaden understanding of stock dynamics and stock history and of the relationship of both to fishing activities. A constructive response to these challenges has been greater formal interest from scientists and managers in input from fishers and their knowledge (FRCC 1993; Pinkerton, Gallaugher and Vodden, Newell, this volume). Our research on fishery workers' TEK is designed to facilitate this process.

Our methodological framework is based on the assumption that TEK is differentiated by a range of factors, including location of fishing effort, fishing technologies and strategies, education, gender, age, and commitment to the industry. We selected a region of Newfoundland – the Bonavista headland and the west side of Trinity Bay (Figure 11.1) – to try to capture ecological differences between headlands and bays and a range of different types of fisheries and fishing technologies. The primary research instrument consisted of in-depth, semi-structured taped interviews, with a mapping component. Interviews were conducted, whenever possible, by a team consisting of a natural scientist and a social scientist. The composition of these teams changed, depending in part on the availability of individual team members. Ideally, however, turnover of this kind should be avoided, because it can lead to substantial divergence in questions asked.

One of our goals was to embed fishers' TEK within the context of their own cognitive understandings of their universe (on embeddedness, see McCay, Sinclair, Squires, and Downton, this volume). No research had in a detailed fashion identified the categories used by Newfoundland and Labrador fishers to classify components of their environment. However, the *Dictionary of Newfoundland*

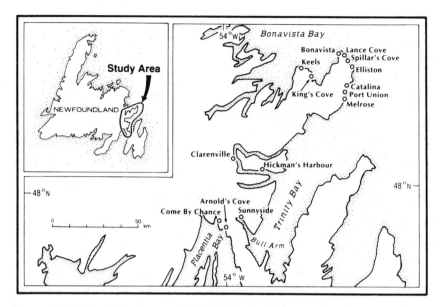

Figure 11.1   Map of study area: Bonavista Peninsula of Newfoundland

English (Story, Kirwin, and Widdowson 1990) included many categories for fish, weather, tides, and so on, from historical documents and interviews. Scott and Scott (1988) provide a detailed description of Atlantic fishes, including scientific and local names, behaviour, distribution, and ecology. Some classifications of Newfoundland birds providing scientific and local names also exist (Gaston 1984), and existing nautical charts contain some of the toponyms used by fishers to locate themselves within their local marine environment.

These sources provided a point of departure from which to develop a taxonomy/toponymy interview schedule. This schedule combined a brief overview of the fishery history of the interviewee with fish and bird species identification, which asked the interviewee to identify species from black-and-white drawings in Scott and Scott (1988) and Gaston (1984) and some colour drawings from Pethon (1985) and with chart work designed to identify toponymies, spawning locations, and so on, for the different species. For cod, the historically dominant species, we also sought to identify the system of classification used by fishers to classify within this species.

We conducted ten of these taxonomy/toponymy interviews with a sample of fishers. We selected older fishers from communities spread throughout the range of the study area and with involvement in both the inshore trap and longline and nearshore longline/gillnet fisheries. These interviews helped us identify the range of species encountered in the different fisheries and regions as well as local names for species and grounds and some data about habitat and behaviour. We then used this information in a larger set of general interviews to prompt respondents concerning local species fished and location of fish.

The main body of the interviews (sixty-one, as of August 1996) was carried out with inshore, nearshore, and offshore fishers, with the interview schedule for inshore and nearshore fishers organized around phases in their fishing careers. We modified the second schedule, for interviews with offshore 'draggermen,' in order to accommodate the different technological and social dynamics of their larger-scale fishery. The sampling strategy was based on an attempt to cover the full region, all gear sectors, and different age groups. We sampled more intensively from communities, such as Bonavista, with larger numbers of fishers. In this area we drew our sample from a list of fishers provided by the union. Selection was guided by names recommended by the local union leadership and by individuals' availability and willingness to participate in the research. In other areas we used a snowball sample (Babbie 1989: 268–9).

A key element in these interviews involved the construction of local maps of fishing grounds, including information on depth, gear type, species fished, seasonal shifts, and changes in annual cycles over the course of individual fishing careers. Fishers deal on a regular basis with three-dimensional mental maps. For both inshore, fixed-gear fishers and those who fish offshore routinely in the same places, its features are well-known to them and form a part of their vernacular world. The land under their keels is composed of banks and gullies, deeps and knolls, ledges and grounds. These have names or toponyms, some of them very descriptive: Hole in the Wall, Northeast Peak, Stony Ground, Old Harry. In among these places are located the fishing berths. In the inshore, these are often found using landmarks, which also have names (Butler 1983; Kent Martin 1973; Ommer, this volume). These marks and corresponding locations provide essential information for success in trap and handline fisheries. Fishers use their knowledge of the landscape in a TEK framework – fish are found on some features during certain seasons, and under certain conditions, but migrate to or past others at other times (see Manore and Van West, this volume). Some features are rich in larger fish, while others lack them. Knowledge of landmarks and underwater names permits fishers to share knowledge across generations and with community members, thereby permitting local people 'to accumulate a body of information larger than [they] alone would possess, reducing [their] dependency, real and imagined, on good fortune and success' (Butler 1983: 18). Butler notes that in the community he studied on the west coast of Newfoundland the landmarks have been constant for generations.[2]

Our research is providing invaluable information not only on fishing strategies and marine ecosystems but also on fishers' and the industry response to the combined effects of resource decline and related policy initiatives. The timing and impact of local technological changes, such as the introduction of sounders, new boat engines, and novel cod traps, the fishers' explanations for these changes, and their influence on fishing location have been tabulated (Neis et al. 1996). Interviews with fishers in the study region show that while total landings may have been increasing for some fishers between 1978 and 1990, there was a clear increase in size of boat, engine horsepower, number of gear units, frequency of net hauling, and spatial scale of fishing effort. Fishers attribute these increases in effort primarily to local resource decline. Thus, in the case of the gillnet fishers in vessels over thirty-five feet (11 metres), our data indicate a clear progression from boats under forty feet (12.2 metres) to those fifty feet (15.2 metres) and over. As boats were upgraded, the engine horsepower, amount of gear, fish-finding and communication capacity, and spatial scale of this fishery also increased. Fishing activity shifted from nearshore areas less than twenty miles (32 kilometres) from communities to areas such as the Virgin Rocks, 150 miles (241 kilometres) from shore and far from home communities on the Bonavista Peninsula (Figure 11.2). Information on such changes and their effect on catchability is essential to any stock assessment based on catch statistics (Pope 1977). This information is often available only from fishers. It should help us to arrive at a more definitive formulation of the events that led to the collapse of the northern cod stocks and provide a basis for better stock assessment and management in the future.

The concomitant development of fishing technology and large-scale fishing enterprises in the latter half of the twentieth century has increased the degree to which fishing effort is focused on high-density aggregations. The biological consequences are known to be substantial for pelagic schooling fish, and similar fishing practices are thought to have masked stock decline and encouraged the recent collapse of the Atlantic groundfishery (Hutchings, this volume). We are contributing to understanding in this area by interviewing participants in the offshore commercial fishery. In the late twentieth century there has been growing state intervention within fisheries management. Our research should help explain how such policies as the introduction of enterprise allocations and moratoriums on groundfish stocks affect fishing strategies (Neis et al. 1997).

Our research is also broadening existing knowledge of the ecosystem in this area. Fish assemblages are patterns of co-occurrence of species over time (Gomes 1993; Villagarcia et al., this volume). Assemblage distribution patterns have more biological meaning than standard Northwest Atlantic Fisheries

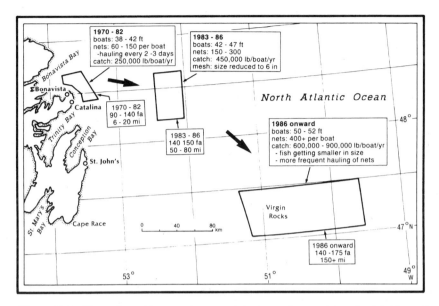

Figure 11.2   Upgrading of boats and gear and the increasing spatial scale of fishing off eastern Newfoundland after 1970

Organization (NAFO) statistical areas. They may also fit better with fishers' knowledge, which is influenced by their practical experience. Ocean scientists have used stock assessment data to document the location and composition of fish assemblages and changes that have taken place in these since 1971 for the offshore areas (Gomes 1993; Gomes and Haedrich 1992; Gomes, Haedrich, and Rice 1992; Gomes et al. 1995). However, such data are not available for coastal and bay portions of the ecosystem. Interview data should help us identify assemblages, as well as recent changes in them, for the coastal and bay areas. We are also attempting to match data on fishing strategies with that on fish assemblages and on changes in fishing location and effort (drawn from purchase slips and interviews).

A useful illustration of the benefits from this research can be found in indications from our interviews that overfishing of groundfish populations may have resulted in an increase in invertebrates, such as snow crab (*Chionoecetes opilio*), in the ecosystem off northeast Newfoundland. Fishers reported that large cod feed on smaller, female snow crab, which are not harvested in the commercial snow crab fishery. Snow crab populations have indeed grown, and many fishers report their expansion into new areas in recent years. These changes coincide with the fishing out of large cod in the 1980s, and observations from the TEK interviews suggest that this reduced predatory pressure on mature female crab is leading to replacement of cod by crab as a biomass dominant in benthic habitats on the continental shelf east of Newfoundland.

Our TEK interviews are also providing data on stock structure, as well as spawning times and locations for northern cod. There is considerable controversy about the stock structure of northern cod. In addition, there is some debate about when and where fish congregate, where and when spawning occurs, and the impact of fishing effort on local populations (Hutchings, Myers, and Lily 1993; Hutchings, this volume). Some fishers and scientists maintain, for example, that specific bay stocks may have played an important role in stock recruitment during particular years and in the success of inshore, fixed-gear fisheries in some areas (Wroblewski, Bailey, and Howse 1994). Fishers' knowledge could provide another source of information for defining stock and population structures and identifying local spawning areas, where such exist (Neis et al. 1997).

From a social science perspective, our interviews should allow us to identify fishers' knowledge frameworks – the assumptions, terminology, and form of argument connected to strategies for understanding marine ecosystems and the fishery (Pinch 1986). Because we are carrying out similar interviews with scientists and analysing scientific reports, older, archival data, and DFO memos and correspondence related to stock assessment in the 1980s, this research should allow us to compare systematically the knowledge systems of fishers and scientists, possible changes in these over time (see, for example, Hutchings, Neis, and Ripley 1996), and the impact of both scientific and fishers' knowledge on management initiatives.

To the extent that these two sets of ecological knowledge are socially produced from the ecological, technological, and social contexts in which they are embedded, one might reasonably hypothesize that the knowledge systems of scientists would be influenced by their sensitivity to elements of the marine environment, the sources of information they draw on to develop their understanding of the marine ecosystem, the factors they identify as most threatening to marine resources, and the means they use to transmit ecological knowledge. Preliminary evidence suggests that the ongoing debates among fishers, among scientists, and between fishers and scientists concerning the ecological effects of different technologies, the state of the resource, and the causes of its decline are substantially affected by the ecological, technological, and social contexts of their work (Harris 1990; Neis 1992).

Conclusion

In his classic study of indigenous fishers in Palau, Johannes (1981) showed that gathering the ecological knowledge of such fishers can dramatically and quickly expand our often meagre understanding of marine ecosystems. Until recently, however, TEK in fisheries has been largely neglected by both natural scientists and social scientists. It has been marginalized by disciplinary boundaries that accord to science the study of nature and to social sciences the study of human societies. Scientists sometimes dismiss fishers' knowledge as 'anecdotal' or 'unsubstantiated,' despite the historical depth and richness of experience on which it is often based (Johannes 1993; Neis 1992).

There are at least two important consequences of excluding resource users from meaningful participation in scientific research and management: the management process does not incorporate the experience and knowledge of those physically closest to the resource; and acceptance of management initiatives and compliance with them are jeopardized (Freeman and Carbyn 1988, 1992; Johannes 1981; Johnson 1992; Pinkerton 1988, and this volume). As argued by Berkes (1988), 'the ability to use resources in a sustainable manner cannot be accomplished merely by the possession of appropriate ecological knowledge and social institutions. There also has to be an environmental ethic to keep exploitive abilities in check and to provide "ground rules" by which relations between humans and nature may be regulated' (8). Development of an environmental ethic is probably more likely where such an ethic fits with locally accepted systems of management and local views about nature and sustainability.

TEK can also help people to develop critical sensitivity towards the dominant knowledge system within science. Freeman (1989b) echoes other writers in the sociology of scientific knowledge in arguing that Western science, like other knowledge systems, is not immune to social forces that shape the problems it identifies, as well as its findings and the presentation of those findings. Both TEK and Western scientific fisheries knowledge need to be regarded as 'knowledge frameworks' (Pinch 1986).

We see at least four areas in which TEK can be incorporated into fisheries management. The first is in defining management goals. TEK includes goals and values in the organization of knowledge. Eliciting and articulating these values and incorporating them into management reduce one source of conflict and misunderstanding. More specifically, a series of biological targets (F_{max}, F_{osy}, $F_0.1$) have been used in fisheries management to set total allowable catches. These targets are global (computed for entire stocks), rather than local. If we can articulate the values underlying TEK, we might be able to develop a set of biological targets that take into account local variation in both fish and fishing activity in ways that do not result in one group's 'borrowing' or intercepting the catch of another.

A second and related area where TEK can be incorporated into management is in allocation of the resource, once a total catch, free of hidden borrowing, has been set as a management target. It might seem that allocation is a political matter, divorced from TEK. However, the participants in a fishery might choose to include some of the values implicit in the organization of TEK, such as conservation or continuity of stock. For example, they might find that there is agreement, at least in principle, on an allocation earmarked for conservation through, for example, such methods as holding back 10 per cent, or closing spawning grounds, or establishing marine protected areas (on the latter see Pauly, this volume).

In a third area, TEK can also be integrated into the development of ecological concepts that can then be tested within the framework of formal science, with its emphasis on repeatability, standardized observational formats, and objective assessment of evidence and its acceptance. Examples of this approach already exist in scientific research on the Newfoundland cod and capelin fisheries (Rose and Leggett 1993; Schneider and Methven 1988). Hypotheses tested within the framework of standard science are necessarily based on substantial amounts of observation. TEK offers a way to gather large amounts of information, winnowed over several generations, to be used for developing testable hypotheses within the organizational framework of scientific knowledge.

A fourth area where TEK can be integrated into fisheries science is in permitting changes in fishing behaviour to be incorporated into assessment of stock size (see Usher and Tough, this volume). Standard assessment methods such as virtual population analysis (VPA) assume constant catchability and hence are vulnerable to overestimates of stock size if fishing skill and fishing capacity are on the rise (see also Pinkerton, this volume). TEK, with its focus on the behaviour of individual fishing enterprises at local scales, can be used to evaluate such changes (Neis et al. 1996). Another promising means of integration is to use interviews to determine factors that change the spatial extent over which fishers gather the data they use to make decisions about where and when to fish and the number of people they involve when forming their fishing plans. Integration of research on traditional ecological knowledge into fisheries science and management has already begun (Neis et al. 1996), and we see considerable promise for further integration.

NOTES

1 'Dirty water' is in fact clear, cold water driven into areas by winds that blow parallel to the coast. The 'dirty' effect results from fouling of nets by greenish/black slime, consisting largely of the discarded mucus houses of two species of the zooplanktonic appendicularian *Oikopleura* and the mucus nets of a pteropod mollusc, *Limacina helicina*, known locally as 'blackberry' (Deibel 1987).

2 Butler (1983) uses the concept 'cognitive map' to analyse inshore fishers' knowledge. Eythorsson (1993) uses a similar concept, 'mental map,' in his discussion of the ecological knowledge of Sami fishers.

WORKS CITED

Babbie, Earl. 1989. *The Practice of Social Research*. Belmont, Calif.: Wadsworth.
Berkes, Fikret. 1993. 'Traditional Ecological Knowledge in Perspective.' In Julian T. Inglis, ed., *Traditional Ecological Knowledge: Concepts and Cases*, 1–10. Ottawa: International Development Research Centre.
Berkes, Fikret, ed. 1989. *Common Property Resources: Ecology and Community-Based Sustainable Development*. London: Belhaven Press.
Bishop, Claude A., C.F. Murphy, M.B. Davis, J.W. Baird, and G.A. Rose. 1993. 'An Assessment of the Cod Stock in NAFO Divisions 2J and 3KL.' NAFO SRC Doc. 93/86, Ser. no. 2271.
Butler, Gary. 1983. 'Culture, Cognition, and Communication: Fishermen's Location-Finding in L'anse-a-Canards, Newfoundland.' *Canadian Folklore canadien* 5, nos. 1–2: 7–21.
Cushing, David H. 1988. *The Provident Sea*. Cambridge: Cambridge University Press.
Deibel, Don. 1987. 'Slub: Science Looks at Fishermen's Torment.' *Fo'c'sle* 6, no. 2: 7–11.
Dewees, Christopher M., and Glenn R. Hawkes. 1988. 'Technical Innovation in the Pacific Coast Trawl Fishery: The Effects of Fishermen's Characteristics and Perceptions on Adoption Behaviour.' *Human Organization* 47: 224–34.
Ellerston, B., P. Fossum, P. Solemdal, and S. Sundby. 1989. 'Relation between Temperature and Survival of Eggs and First-feeding Larvae of Northeast Arctic Cod (*Gadus morhua* L.).' In *Report and Verbal Proceedings of the Meetings of the International Commission for Exploration of the Sea* 191: 202–19.
Eythorsson, Einar. 1993. 'Sami Fjord Fishermen and the State: Traditional Knowledge and Resource Management in Northern Norway.' In Julian T. Inglis, ed., *Traditional Ecological Knowledge: Concepts and Cases*. Ottawa: IDRC.
Felt, Lawrence F. 1994. 'Two Tales of a Fish: The Social Construction of Indigenous Knowledge among Atlantic Canadian Salmon Fishers.' In C. Dyer and J. McGoodwin,

eds., *Folk Management of World Fisheries*, 251–86. Boulder: University of Colorado Press.

Finlayson, A. Christopher. 1994. *Fishing for Truth: A Sociological Analysis of Northern Cod Stock Assessment from 1977 to 1990*. St John's: ISER, Memorial University.

Fisheries Resource Conservation Council (FRCC). 1993. *1994 Conservation Requirements for Atlantic Groundfish*. Report to the Minister of Fisheries and Oceans. Ottawa.

Food and Agricultural Organization (FAO). 1989. *Review of the State of World Fishery Resources*. Rome: United Nations.

Freeman, Milton. 1989a. 'The Alaska Eskimo Whaling Commission: Successful Co-Management under Extreme Conditions.' In Evelyn Pinkerton, ed. *Co-operative Management of Local Fisheries: New Directions for Improved Management and Community Development*, 137–53. Vancouver: University of British Columbia Press.

– 1989b. 'Graphs and Gaffs.' In Fikret Berkes, ed., *Common Property Resources: Ecology and Community-Based Sustainable Development*, 92–109. London: Belhaven Press.

– 1992. 'The Nature and Utility of Traditional Ecological Knowledge.' *Northern Perspectives* 20, no. 1: 9–12.

Freeman, Milton, M.R., and Ludwig N. Carbyn, eds. 1988. *Traditional Knowledge and Renewable Resource Management in Northern Regions*. Occasional Publication no. 23. Edmonton: Boreal Institute for Northern Studies.

Gadgil, M., F. Berkes, and C. Folke. 1993. 'Indigenous Knowledge for Biodiversity of Conservation.' *Ambio* 22, nos. 2–3: 151–6.

Gaston, A.J. 1984. *Guide to the Seabirds of Eastern Canada*. Ottawa: Canadian Wildlife Service.

Gomes, Manuel C. 1993. *Predictions under Uncertainty*. St John's: ISER, Memorial University.

Gomes, Manuel C., and R. Haedrich. 1992. 'Predicting Community Dynamics from Food Web Structure.' In G.T. Rowe and V. Pariente, eds., *Deep-Sea Food Chains and the Global Carbon Cycle*, 277–93. Dordrecht: Kluwer Academic Publishers.

Gomes, Manuel C., R. Haedrich, and J. Rice. 1992. 'Biogeography of Groundfish Assemblages on the Grand Bank of Newfoundland.' *Journal of Northwest Atlantic Fisheries Science* 14: 13–27.

Gomes, Manuel C., R. Haedrich, Richard L. Villagarcia, and M.G Villagarcia. 1995. 'Spatial and Temporal Changes in the Groundfish Assemblages in the Northwest Atlantic, 1978–1991.' *Fisheries Quarterly* 4: 88–101.

Goonatilake, S. 1984. *Aborted Discovery: Science and Creativity in the Third World*. London: Zed Books.

Graham, Robert. 1993. 'Customary Users Near Shore and Coastal Inventory.' Waterloo: Department of Recreation and Leisure Studies, University of Waterloo.

Gunn, Anne, Goo Arlooktoo, and David Kaomayok. 1988. 'The Contribution of the Ecological Knowledge of Inuit to Wildlife Management in the Northwest Territories.' In Milton M.R. Freeman and Ludwig Carbyn, eds., *Traditional Knowledge and Renewable Resource Management in Northern Regions*, 22–30. Edmonton: Boreal Institute for Northern Studies.

Haedrich, Richard L. 1995. 'Structure over Time of an Exploited Deep Water Fish Assemblage.' In A.G. Hopper, ed., *Deep Water Fisheries of the North Atlantic Oceanic Slope*, 35–58. Dordrecht: Kluwer Academic Publishers.

Hanna, Susan S., and Courtland L. Smith. 1993. 'Attitudes of Trawl Vessel Captains about Work, Resource Use, and Fishery Management.' *North American Journal of Fisheries Management* 13: 367–75.

Harris, Leslie. 1990. *Independent Review of the State of the Northern Cod Stock*. Ottawa: Department of Fisheries and Oceans.

Holling, Carl S. 1993. 'Investing in Research for Sustainability.' *Ecological Applications* 3, no. 4: 552–5.

Hrenchuk, Carl. 1993. 'Native Land Use and Common Property: Whose Common?' In Julian T. Inglis, ed., *Traditional Ecological Knowledge: Concepts and Cases*, 99–109. Ottawa: IDRC.

Hutchings, Jeffrey A., and R.A. Myers. 1994. 'What Can be Learned from the Collapse of a Renewable Resource? Atlantic Cod, *Gadus morhua*, of Newfoundland and Labrador.' *Canadian Journal of Fisheries and Aquatic Sciences* 51, no. 9: 2126–46.

– 1993. 'Effect of Age on the Seasonality of Maturation and Spawning of Atlantic Cod, *Gadus morhua*, in the Northwest Atlantic.' *Canadian Journal of Fisheries and Aquatic Sciences* 50, no. 11: 2468–74.

Hutchings, Jeffrey A., R.A. Myers, and G.R. Lily. 1993. 'Geographic Variation in the Spawning of Atlantic Cod, *Gadus morhua*, in the Northwest Atlantic.' *Canadian Journal of Fisheries and Aquatic Sciences* 50, no. 11: 2457–67.

Hutchings, Jeffrey A., B. Neis, and P. Ripley. 1997. 'The "Nature" of Cod (*Gadus morhua*): Perceptions of Stock Structure and Cod Behaviour by Fishermen, "Experts" and Scientists from the Nineteenth Century to the Present.' In D. Vickers, ed., *Proceedings of the Marine Resources and Human Societies in the North Atlantic since 1500 Conference*, 123–88. St John's: ISER, Memorial University.

Johannes, R.E. 1978. 'Traditional Marine Conservation Methods in Oceania and Their Demise.' *Annual Review of Ecological Systems* 9: 349–64.

– 1981. *Words of the Lagoon: Fishing and Marine Lore in the Palau District of Micronesia*. Berkeley: University of California Press.

– 1993. 'Integrating Traditional Ecological Knowledge and Management with Environmental Assessment.' In J.T. Inglis, ed., *Traditional Knowledge: Concepts and Cases*, 33–40. Ottawa: IDRC.

Johnson, Martha, ed. 1992. *Lore: Capturing Traditional Environmental Knowledge*. Hay River, NWT: Dene Cultural Institute, International Development Research Centre.

Kearney, John. 1993. 'Diversity of Labour Process, Household Forms, and Political Practice: A Social Approach to the Inshore Fisheries of Clare, Digby Neck and the Islands.' PhD dissertation, Laval University.

Kloppenburg, Jack. 1991. 'Social Theory and the De/Reconstruction of Agricultural Science: Local Knowledge for an Alternative Agriculture.' *Rural Sociology* 56, no. 4: 519–48.

Lear, W. Henry, and L.S. Parsons. 1993. 'History and Management of the Fishery for Northern Cod in NAFO Divisions 2J, 3K and 3L.' In L.S. Parsons and W.H. Lear, eds., *Perspectives on Canadian Marine Fisheries Management. Canadian Bulletin of Fisheries and Aquatic Sciences* 226: 55–89.

Ludwig, Donald, Ray Hilborn, and Carl Walters. 1993. 'Uncertainty, Resource Exploitation, and Conservation: Lessons from History.' *Ecological Applications* 3, no. 4: 547–9.

McCay, Bonnie J., and J. Acheson, eds. 1987. *The Question of the Commons.* Tucson: University of Arizona Press.

McGoodwin, James R. 1990. *Crisis in the World's Fisheries: People, Problems and Policies.* Palo Alto, Calif.: Stanford University Press.

Mailhot, José. 1993. *Traditional Ecological Knowledge: The Diversity of Knowledge Systems and Their Study.* Montreal: Great Whale Public Review Support Office.

Martin, Cabot. 1992. *No Fish and Our Lives: Some Survival Notes for Newfoundland.* St John's: Creative Publishers.

Martin, Kent. 1973. 'The Law in St John's Says ...' MA thesis, Memorial University of Newfoundland.

Munn, William A. 1922. 'Annual Migration of Codfish in Newfoundland Waters.' *Newfoundland Trade Review*, 23 Dec.: 21–4.

Myers, Ransom A., G. Mertz, and N.J. Barrowman. 1995. 'Spatial Scales of Variability in Cod Recruitment in the North Atlantic.' *Canadian Journal of Fisheries and Aquatic Sciences* 52, no. 9: 1849–62.

Myers, Ransom A., K.F. Drinkwater, N.J. Barrowman, and J.W. Baird. 1993. 'Salinity and Recruitment of Atlantic Cod (*Gadus morhua*) in the Newfoundland Region.' *Canadian Journal of Fisheries and Aquatic Sciences* 50, no. 8: 1599–1609.

Nadel-Klein, Jane, and Donna Lee Davis. 1988. *To Work and to Weep: Women in Fishing Economies.* St John's: ISER, Memorial University.

Nakashima, Douglas J. 1993. 'Astute Observers on the Sea Ice Edge: Inuit Knowledge as a Basis for Arctic Co-Management.' In Julian T. Inglis, ed., *Traditional Ecological Knowledge: Concepts and Cases*, 99–109. Ottawa: IDRC.

Neis, Barbara. 1992. 'Fishers' Ecological Knowledge and Stock Assessment in Newfoundland and Labrador.' *Newfoundland Studies* 8, no. 2: 155–78.

Neis, Barbara, L. Felt, D.C. Schneider, R. Haedrich, J. Hutchings, and J. Fischer. 1996. 'Northern Cod Stock Assessment: What Can Be Learned from Interviewing Resource Users?' NAFO SCR Doc. 96/45, 22.

Neis, B., D.C. Schneider, L. Felt, R.L. Haedrich, J. Fischer, and J.A. Hutchings. 1997. 'Stock Assessment: What Can Be Learned from Resource Users?' Occasional Paper. St John's: Eco-Research Project, Memorial University.

Oceans, Ltd. 1990. 'Reproductive Success in Atlantic Cod: (*Gadus Morhua* L.): The Potential Impact of Trawling.' Report submitted to the Newfoundland Inshore Fisheries Association. St John's.

Pethon, Per. 1985. 'Aschehougs Store Fiskebok: Alle Norske Fisker I Farger.' Stockholm: H. Aschehoug and Co.

Pinch, Trevor. 1986. *Confronting Nature: The Sociology of Solar-Neutrino Detection.* Dordrecht: D. Reidel Publishing.

Pinkerton, Evelyn, ed. 1988. *Co-operative Management of Local Fisheries: New Directions for Improved Management and Community Development.* Vancouver: University of British Columbia Press.

Pope, John G. 1977. 'Estimation of Fishing Mortality, Its Precision and Implications for the Management of Fisheries.' In J.H. Steele, ed., *Fisheries Mathematics*, 63–76. London: Academic Press.

Richards, Paul. 1985. *Indigenous Agricultural Revolution: Ecology and Food Production in West Africa.* Boulder, Col: Westview Press.

Rose, George A. 1992. 'Indices of Total Stock Biomass in the "Northern" and Gulf of St Lawrence Atlantic Cod (*Gadus morhua*) Stocks Derived from Time Series Analyses of Fixed Gear (Trap) Catches.' *Canadian Journal of Fisheries and Aquatic Sciences* 49: 202–9.

– 'Cod spawning on a Migration Highway in the North-west Atlantic.' *Nature* 366: 458–61.

Rose, George A., and W.C. Leggett. 1989. 'Predicting Variability in Catch-per-Effort in Atlantic Cod, *Gadus morhua*, Trap and Gillnet Fisheries.' *Journal of Fish Biology* 35 Supplement A: 155–61.

– 1991. 'Effects of Biomass–Range Interactions and Catchability of Migratory Demersal Fish by Mobile Fisheries: An Example of Atlantic Cod (*Gadus morhua*).' *Canadian Journal of Fisheries and Aquatic Sciences* 48: 843–8.

– 1993. 'Use of Oceanographic Forecasts and Echosounders to Guide and Enhance an Inshore Gillnet Fishery for Atlantic Cod (*Gadus morhua*).' *Canadian Journal of Fisheries and Aquatic Sciences* 50, no. 10: 2129–36.

Rosenburg, A.A., M.J. Fogarty, M.P. Sissenwine, J.R. Beddington, and J.G. Shepherd. 1993. 'Achieving Sustainable Use of Renewable Resources.' *Science* 262: 828–9.

Schneider, D.C., and D.A. Methven. 1988. 'Response of Capelin to Wind-Induced Thermal Events in the Southern Labrador Current.' *Journal of Marine Research* 46: 105–18.

Scott, W.B., and M.G. Scott. 1988. *Atlantic Fishes of Canada.* Toronto: University of Toronto Press.

Steele, Donald H., R. Andersen, and J.M. Green. 1992. 'The Managed Commercial Annihilation of Northern Cod.' *Newfoundland Studies* 8, no. 1: 34–68.

Story, George M., W.J. Kirwin, and J.D.A. Widdowson. 1990. *Dictionary of Newfoundland English.* Toronto: University of Toronto Press.

Valerio, P.F., S.V. Goddard, M.H. Kao, and G.L. Fletcher. 1992. 'Survival of Northern Atlantic Cod Eggs and Larvae Exposed to Ice and Low Temperature.' *Canadian Journal of Fisheries and Aquatic Sciences* 49: 2588–95.

Wilson, James, J.M. Acheson, M. Metcalfe, and P. Kleban. 1994. 'Chaos, Complexity and Community Management of Fisheries.' *Marine Policy* 18, no. 4: 291–305.

Windsor, Fred. 1994. 'In Search of a Sustainable Northern Cod Fishery.' Paper presented at, the Fifth International Symposium on Society and Resource Management, Fort Collins, Col., June.

Wroblewski, J.S., W.L. Bailey, and K.A. Howse. 1994. 'Observations of Adult Atlantic Cod (*Gadus morhua*) Overwintering in Nearshore Waters of Trinity Bay, Newfoundland.' *Canadian Journal of Fisheries and Aquatic Sciences* 51, no. 1: 142–50.

12

Groundfish Assemblages of Eastern Canada Examined over Two Decades

MARIMAR G. VILLAGARCÍA, RICHARD L. HAEDRICH, AND JOHANNE FISCHER

Fishing grounds of the world ocean are surveyed at least yearly, mainly for the purpose of assessing changes at the levels of individual commercial species or stocks. But very few have been approached from a multispecies perspective to detect possible variations at the community scale. In many cases the groundfish survey's trawling techniques are not selective with regard to the target species, and the result is a catch of mixed species that can be quite useful for general scientific and ecological study. Such is the case for the continental shelves off eastern Canada, where sampling by non-selective trawls brings up a good mixture of species that appears to be quite representative of the community present at that particular moment in this large fisheries ecosystem.

In the 1970s and 1980s, there developed a general interest in applying multivariate techniques to identify and study fish assemblages on continental shelves of the Pacific and Atlantic oceans (Colvocoresses and Musick 1984; Gabriel and Tyler 1980; Mahon and Smith 1989; Overholtz and Tyler 1985; Rogers and Pikitch 1992; Tyler et al. 1982). Assemblages were reported to be persistent in time and within defined geographical contours for the temporal and spatial scales used in each study. Gomes (1993) and Villagarcía (1995) have followed on in this tradition with their studies of the Grand Banks and the northeast Newfoundland shelf.

The Grand Banks of Newfoundland and the eastern Newfoundland-Labrador continental shelves have been among the best fishing grounds in the world, as other chapters in this volume (Cadigan, Hutchings) discuss in greater detail. Focus in this region has been primarily on the Atlantic cod, *Gadus morhua*, the dominant and by far the most important commercial species. In some areas, the biomass of cod represented up to 60 per cent of the total catch. But in the post-Second World War period, within a time-span of about thirty years, the adult biomass of this fish declined by 99 per cent, leading the Canadian government in 1992 to declare a total ban on all cod fishing in the Newfoundland area (Sinclair, Squires, and Downton, Hutchings, this volume). Even with the moratorium in effect, however, scientific surveys continue to record declines; the estimated spawning biomass in 1994 was little more than ten thousand tonnes, down two orders of magnitude from the 1.6 million tonnes estimated in 1963. The socioeconomic effects resulting from the major declines of this species have shown themselves to be very critical for both Canada and Europe (Safina 1995; Sinclair, Squires, and Downton, this volume).

The Newfoundland region (NAFO divisions 2 and 3) has been sampled regularly on a year-to-year basis since 1971 on the Grand Banks and, since 1978, on the eastern Newfoundland–Labrador shelf. The species of commercial interest in the surveys are analysed, together with information from the commercial catch, so that a total allowable catch (TAC) for the fishery in the following year can be set.

Two series of sampling data were made available to Memorial University of Newfoundland by the Canadian Department of Fisheries and Oceans (DFO) for these important shelf areas. The series are for the Grand Banks and for the northeast Newfoundland shelf and include data on species taken in survey trawls and their numbers and aggregate weights. Their analysis provides insight into the groundfish community structure of the zones that may have some implications for multispecies management; the ability of the fish assemblages to reflect community-level changes can complement information at the species level. Study from an ecological community perspective also allows analysis of variability in the community structure induced by disturbances that could be of either anthropogenic or environmental origin. This chapter describes changes in the fish community observed over the twenty-year time series.

Material and Methods

DFO research cruises conducted surveys over sixteen years for Northwest Atlantic Fisheries Organization (NAFO) subdivisions 3LNO in spring and for NAFO 2J3K3L over fourteen years in fall. Each year the cruises sampled a different number of stations, according to a stratified random technique that allocated a minimum of two stations per stratum; tows lasted for thirty minutes at a speed of 2.5 knots (Atkinson 1994). Each species in every catch was identified, and its abundance and biomass were recorded. For the commercial species, other, more specific data of interest were also obtained. Depth and surface and bottom temperature at each station were recorded as well.

The multivariate analysis that we conducted was based on the biomass of the groundfish in the sample, including in the analysis only those species whose biomass represented at least 0.1 per cent of the total catch in a given spring or fall survey. Before starting the procedure, we log-transformed the catch of each species by station to avoid having the most abundant species dominate the results of the multispecies analysis. Gauch (1982) found that differences in population abundance tend to be of an exponential nature and that the reduction of information after logarithmic transformation does not alter the results significantly.

The first step was to construct, for each year, a two-way matrix of biomass (kg) by station. We applied two methods to identify groups – the clusters or fish assemblages that were present each year in the study area. We used the CLUSTAN package (Wishart 1978) to define the clusters of species; the semimetric distance of Bray and Curtis was the measure of proximity (dissimilarity), and stations were clustered using two agglomerative polythetic methods – group average (Sokal and Mitchener 1958) and error sum of squares (Ward 1963).

The validity of the clustering technique was confirmed via a divisive hierarchical software package named TWINSPAN (Hill 1979) – an ordination technique identified as a 'TWo-way INdicator SPecies ANalysis.' The method makes an initial ordination of species and stations using correspondence analysis; a rough division of samples is done based on scores on the first factorial axis, and 'differential' species are identified by their preference for either side of the dichotomy; a second ordination is done employing 'differential' species, and the new ordination is further continually divided to obtain a final dichotomy. Each side of the dichotomy proceeds in the same way until final results emerge that are displayed in a two-way table that fulfills requirements of non-exclusivity; the table permits recognition of biological features such as differential species, presence or absence of species in a cluster, and anomalies in the number of species present in each cluster. We checked the geographical continuity of the similar stations grouped by the above methods by plotting the stations in their exact latitude and longitude; we checked stations with a non-clear preference for an assemblage using the ALLOCATE procedure in the TWINSPAN analysis.

A temperature study (Haedrich, Fischer, and Chernova 1997) considered the immediate and direct effects of ocean temperature on distribution and abundance. We used data from the surveys off northeast Newfoundland to examine the relationship between abundances of species at individual stations and the bottom temperature in the period 1978–93. We studied the twenty-seven most abundant groundfish species in the area. We established seven temperature classes, ranging from −2 to >4°C, and determined the number of stations and the abundance of each species falling in each category. We then compared the distribution of each species with the distribution of stations, and differences

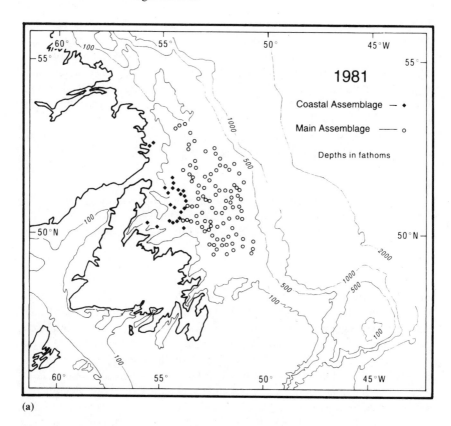

(a)

Figure 12.1   Map in (a) 1981 and (b) 1991 (on facing page): approximate positions
occupied by coastal assemblages (closed squares) and main assemblages (open circles)
on the northeast Newfoundland shelf

between expected (E) and observed (O) values were expressed as gamma-
values (gamma = [O − E]/E). We used the distribution of gamma-values, which
by their departure from zero indicated either preference for, or avoidance of,
temperature classes, to assign each species to a preferred temperature range.

Results

Four groundfish assemblages (*sensu* Underwood 1986) that persisted over the
years were found both on the Grand Banks, where thirty species were analysed
(Gomes 1993), and on the northeast Newfoundland shelf, where thirty-five were
included (Gomes, Haedrich, and Villagarcía 1995; Villagarcía 1995). The
assemblages have a homogeneous species composition and relative abundance

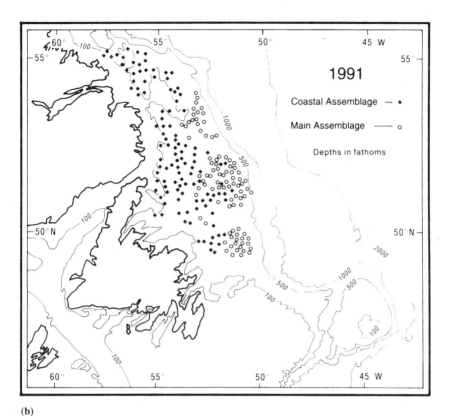

(b)

is fairly stable. The geographical limits of each assemblage were relatively fixed for both regions until 1987, when changes in the geographical position of the assemblages occurred across the shelf. This change coincided with the decline of the main species present in most of the assemblages, especially the commercial cod (Atlantic cod, *Gadus morhua*) and flounders (for example, Greenland halibut, *Reinhardtius hippoglossoides*). Species present in lesser abundance also declined, usually before the dominant species, which may have had a time lag of at least two years. This finding indicates that, whatever the reasons for the changes, they are affecting the structure of the whole fish community.

Mante, Durbec, and Dauvin (1997) found that dominant species reflected very well the temporal changes of the community, while rare species provided information concerning the qualitative evolution of the community and also translated certain aspects of the main characteristics of general community evolution. Thus less abundant species should be looked at carefully in a community analysis. Over time, the Newfoundland assemblages shifted mainly towards the east (Figure 12.1) at the same time as some individual species shifted their centre of abundance towards the south. These phenomena were detected at the community level earlier than they were at the individual species level in both commercial and non-commercial species. The Grand Banks was studied up until 1987, and, since it was at that time that depletion started to become evident, no sign of the coming decline appeared in the data analysed. Our further studies of the Grand Banks assemblages for the period from 1987 to the present show declines in biomass and abundance that are quite similar to those reported by Gomes, Haedrich, and Villagarcía (1995) on the northeast Newfoundland shelf (Figure 12.2).

The findings about community dynamics show the need for further sampling, probably seasonal, to find out if the same or alternative assemblages can be found at different times of the year in the same areas. The cruises to sample the NAFO areas took place once a year, and therefore the assemblages are representative only of the season of the year and the kind of trawl used; this fact imposes some restrictions when one is judging the data, and therefore there is a need for further investigation, especially if the aim is to relate oceanographic changes to possible variation in fish dynamics (Helbig, Mertz, and Pepin 1992). It was possible to categorize the twenty-seven common fish species studied according to preferred temperature ranges (I. warm = >3°C [max 6°C], II. intermediate = 0–3°C, III. cold = <0°C [min –2°C], and IV. = broad temperature range), as indicated in Table 12.1. Despite the identification of these preferred temperature ranges, changing sea temperatures over the survey period are not clearly reflected in patterns of changing fish abundances.

Discussion

Several factors could have produced the variability observed in the distribution of the fish assemblages as well as the changes in species abundance and biomass in NAFO subdivisions 2J3K. From the anthropogenic side, there has been an increase over the period of study in the commercial trawling effort, especially by mid-size trawlers (Figure 12.3), even though catch per unit of effort has decreased (Hutchings and Myers 1994). Overfishing, as reflected in this increased trawling effort, is without doubt the primary cause of the decline in fish biomass and abundance (Myers, Hutchings, and Barrowman 1996). But furthermore, beyond the observed declines and distributional shifts, there is a widespread pattern of reduced average size in many groundfish species (Haedrich and Barnes 1997). Whether the species was commercially fished or not was unimportant (for example, Figure 12.4), and species ranging in size from a few hundred grams to several kilograms responded in the same way. In fact, changes

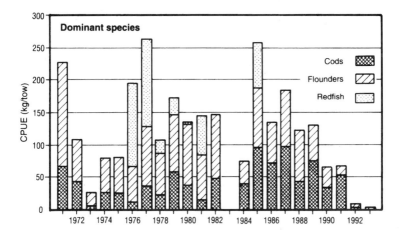

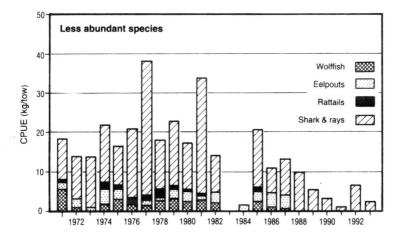

Figure 12.2   Survey CPUE (catch per unit effort) for (top) dominant (commercial) and (bottom) less abundant (mostly non-commercial) groundfish groups on the northeastern Grand Banks, 1972–93

in size were sometimes evident before declines in abundance became manifest, as was the case for Greenland halibut (Haedrich and Fischer 1996).

We interpret widespread reductions in average size as a sign that the community has lost the bigger-sized individuals and therefore many non-mature fish must now be a part of the catch. The change in size (and thus implied age)

Table 12.1 Percentage distribution in temperature classes, average number/tow, and total number for twenty-seven common Newfoundland groundfish species. Species are grouped according to temperature preferences, and preferred temperature categories for each species are in italics.

	Temperature categories (°C)							Average no./tow	Total number
	(−1.9)−(−1.0)	(−0.9)−0	0.1–1.0	1.1–2.0	2.1–3.0	3.1–4.0	>4.0		
Numbers of stations	581	1,019	840	840	1,275	1,049	79	812	5,683
Higher temperatures									
Coryphaenoides rupestris	0.0	0.0	0.0	0.0	0.3	*39.1*	*60.7*	9.6	31,042
Nezumia bairdii	0.0	0.0	0.1	0.5	4.0	*41.5*	*54.0*	0.9	3,475
Sebastes marinus	0.0	0.1	1.5	25.8	*41.2*	23.3	8.1	2.3	16,271
Glyptocephalus cynoglossus	0.1	0.8	2.3	9.5	27.2	32.7	*27.5*	5.4	30,937
Sebastes mentella	0.0	0.1	0.6	2.5	26.0	29.9	*40.9*	106.7	525,682
Macrourus berglax	0.0	0.8	5.3	10.2	14.3	30.8	*38.6*	3.6	17,194
Raja senta	0.0	0.2	1.6	9.7	*33.3*	32.2	22.9	0.5	2,863
Lycodes vahlii	1.1	2.3	6.8	15.8	*29.7*	21.9	22.5	1.9	11,415
Bathylagus euryops	0.0	0.0	0.0	0.0	0.0	11.5	*88.5*	0.1	158
Cottunculus microps	0.1	1.0	2.1	12.5	21.7	29.7	*32.9*	0.2	979
Anarhichas denticulatus	0.2	7.0	11.3	15.3	13.5	20.3	*32.4*	1.1	5,301
Intermediate temperatures									
Boreogadus saida	19.4	21.5	26.5	*26.7*	5.2	0.6	0.1	13.6	80,956
Anarhichas lupus	0.2	4.5	14.4	25.6	*26.7*	17.0	11.4	1.5	9,481
Agonus decagonus	16.2	13.9	21.9	*28.7*	15.7	2.3	1.4	0.3	1,609
Gadus ogac	7.4	29.2	*49.2*	12.6	1.5	0.0	0.0	0.0	183
Anarhichas minor	1.6	16.9	*22.2*	21.2	17.2	11.9	9.0	0.4	2,536

Table 12.1 (concluded)

	Temperature categories (°C)							Average no./tow	Total number
	(−1.9)–(−1.0)	(−0.9)–0	0.1–1.0	1.1–2.0	2.1–3.0	3.1–4.0	>4.0		
Numbers of stations	581	1,019	840	840	1,275	1,049	79	812	5,683
Artediellus sp.	4.4	7.1	13.7	26.4	24.1	14.2	10.1	0.2	1,011
Aspidophoroides monopterygius	4.9	20.4	11.8	28.4	18.1	9.2	7.2	0.0	319
Lower temperatures									
Triglops sp.	19.0	23.3	52.5	2.7	1.8	0.3	0.4	1.0	5,611
Lycodes reticulatus	31.3	27.0	18.6	14.4	5.6	1.5	1.7	1.5	8,629
Cyclopterus lumpus	16.1	15.3	32.1	17.8	10.8	6.9	0.9	0.2	1,287
Myoxocephalus scorpius	14.5	23.6	37.6	20.4	3.1	0.9	0.0	0.0	286
Broad range of temperatures									
Mallotus villosus	16.6	13.4	18.2	37.4	13.4	0.6	0.5	18.3	112,566
Reinhardtius hippoglossoides	1.3	5.1	9.9	16.9	36.1	16.3	14.4	32.8	212,836
Hippoglossoides platessoides	35.2	30.7	15.8	9.1	5.7	2.1	1.3	68.4	394,006
Raja radiata	21.2	17.4	16.7	16.2	14.9	7.7	5.9	5.9	34,976
Gadus morhua	4.5	10.4	17.4	19.6	21.0	20.7	6.4	38.5	251,141

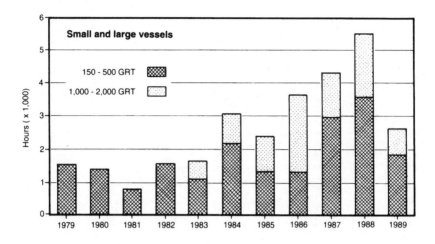

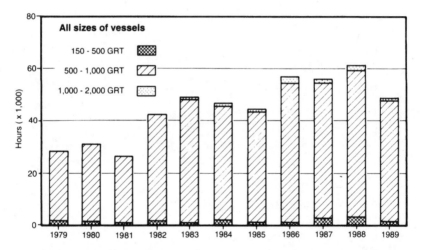

Figure 12.3   Hours of trawling, by size of vessel – (top) small and large, and (bottom) all sizes (GRT: gross registered tons) – in NAFO areas 2J3K3L, 1979–89

structure translates into fish populations possibly arriving at currently unstable points and therefore eventually moving to extinction (see Pauly, this volume). That a shift in age structure has actually happened is evident in one of the last detailed northern cod assessments (Bishop et al. 1994); its Table 26 and Figure 18 show no fish older than eight years in 1992, whereas previously, in 1982 for

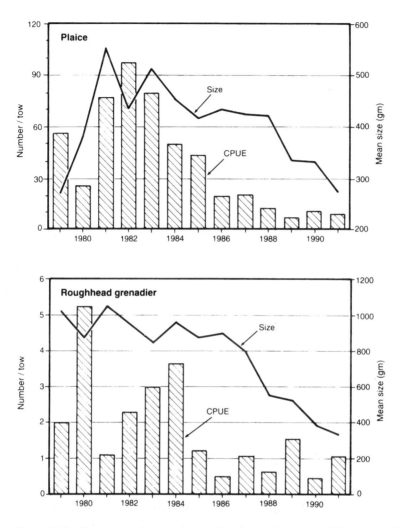

Figure 12.4 Survey abundance and mean size for (top) a commercial species (plaice, *Hippoglossoides platessoides*) and (bottom) a non-commercial one (roughhead grenadier, *Macrourus berglax*) on the northeast Newfoundland shelf, 1979–91

example, fish as old as twenty-two were found. The observed changes in average fish size for many species are certainly the result of massive removal of susceptible size classes, either netted for market or taken as by-catch and dumped over the side dead.

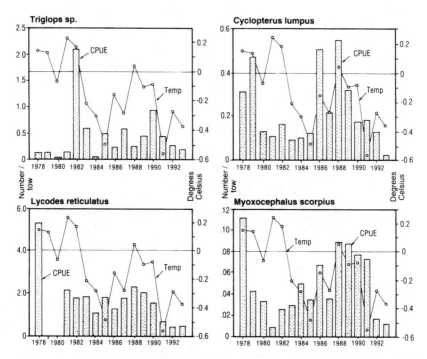

Figure 12.5   Average number/tow in fall surveys 1978–93 for four fish species that prefer lower temperatures, plus average yearly bottom temperature of station 27, near Cape Spear, Nfld. The horizontal line in the figure marks 0°C.

From the environmental side, fluctuations have also been detected in recent years, most especially ocean temperatures as seen in a long-term temporal series for ocean-monitoring station 27 near St John's, which is thought to be well-correlated with the temperature on the shelf generally (see Figure 12.5). The fish assemblage data for the northeast Newfoundland shelf also contain a series of data on bottom temperatures for the fourteen years analysed, though these reflect the temperature only at the time of fishing at each station. This temporal series shows that temperature in the fall was colder on average in the Coastal and North assemblages for the years 1982–5 (Villagarcía 1995), but in general there is little trend apparent in the data.

Temperature can affect fish in at least three ways: reduced recruitment levels because of higher mortality of early life stages (Ellertsen et al. 1989); starvation or migration, because of changes in abundances and distribution of food organisms (Ellertsen et al. 1987; Nilssen and Hopkins 1992); and increased mortality, migration, and/or concentration in small, favourable areas in response to physiological changes relating to growth, food requirements, nervous system function, swimming activity, and reduced immunity (Loeng 1989; Sissenwine 1984). Laevastu (1993) provides an excellent overall summary. Drinkwater and Mountain (1997) discuss the role of climatic conditions throughout the full life history in determining the year-class strength of fish.

The results of our temperature-preference analyses show distinct and coherent patterns of fish species distributions across the temperature gradient (Fischer and Haedrich 1997). This, however, does not necessarily imply that fish migrate to remote areas or perish during, for example, colder as opposed to warmer years. Equally likely, fish may search for a relatively more suitable spot in their immediate neighbourhood and concentrate there during unfavourable periods. Savvatimsky (1987) studied groundfish in NAFO areas 2J and 3K and found that in the colder years some of them were distributed at much greater depths than in years prior to the cooling of the water masses.

Cod – in contrast to most of the other fish species studied and also in contrast to widespread public opinion – showed no pronounced preference or avoidance patterns with respect to temperature above $-1°C$. Obviously the range of temperatures found off Newfoundland is well within the tolerance limits of this fish, and an average decline of temperature (meaning that areas of intermediate and warmer temperatures become less frequent) should not affect the abundance of cod there. This is not surprising, since cod, as a northern species, has been around for some ten million years and has therefore lived through the recent ice ages and certainly through much colder eras before that. Even in our times, there have been very cold periods which do not seem to have affected the fish at all (Hutchings and Myers 1994).

There are some commercial species, however, that are found mainly in higher temperatures off Newfoundland, including the roundnose grenadier (*Coryphaenoides rupestris*), redfish (*Sebastes marinus* or *S. mentella*), and witch flounder (*Glyptocephalus cynoglossus*) (see Table 12.1). The observed decline of these species may have been encouraged by colder temperatures, but such assertions are problematic. Mapping studies show, for example, that even in the cold years of the early 1990s the areal extent of suitably warm water still comprised tens of thousands of square kilometres on the Newfoundland shelf. There is moreover the converse situation found in the arctic eelpout (*Lycodes reticulatus*), a non-commercial species that prefers cold areas and strongly avoids warmer areas (>2°C; Table 12.1). It might therefore be expected that this eelpout should be more abundant during colder years, but this is not the case. A decline has also been observed in this species during the cold recent years (Figure 12.5).

Obviously a variety of other factors, including especially human predation, combine in influencing fish abundance and, in addition, may also act differently for each species. So, even having been able to show that temperature does influence the small-scale distribution of many fish species, the argument cannot be extended to imply that a small average change of temperature extrapolated across the entire huge area of the shelf has any measurable effect on the overall abundance of common Newfoundland marine fish species (Haedrich, Fisher, and Chernova 1997).

Ecological theory suggests that declines of one species in an ecosystem will be matched by increases in other species, the so-called compensatory effect. Sherman (1992) points out a number of major fisheries where compensation seems to have occurred: the Northeast Atlantic Shelf, where sand lance (*Ammodytes*) replaced herring (*Clupea*) and mackerel (*Scomber*); the Benguela Current systems, where horse mackerel (*Trachurus*) increased as hakes (*Merluccius*) declined, and Georges Bank (northwest Atlantic), where dogfish (*Squalus*) have filled the gap left by the much overfished gadids. But in the Newfoundland region, there appears to have been no compensation by other fish species, though crab and shrimp populations seem to be flourishing. In fact, too little time may have elapsed for a compensatory effect to be evident (Ross 1997). None the less, the common and relatively short-lived pelagic species such as capelin (*Mallotus villosus*), which might have been expected to take over from groundfish, began to decline in the 1990s (Lilly 1994).

There is a need for a global, holistic approach to examine linkages between Newfoundland's physical oceanography, which is dominated by the inner and outer branches of the Labrador Current, and the dynamics of fish populations. Since the area is so large and has such a complex topography, there is insufficient environmental information to relate fish and oceanography with respect to larval drift, preferred temperature, and habitat preferences. Biological oceanographic data are also very scarce, though they can be invaluable in relating the dynamics of zooplankton to those of fish. The global approach could lead to the production of a simple model to synthesize the current knowledge about the ecosystem, as in the ideas presented by McFarlane et al. (1997) for the Pacific. New physical-oceanographic (Helbig, Mertz, and Pepin 1992) and phytoplankton-standing-stock models (Haedrich 1993) bear on questions related to fish production on the Grand Banks and Newfoundland shelf, but the general state of knowledge is very far from allowing an overall synthesis.

Though the local oceanographic climate has certainly affected fish populations over time, it is without doubt unrelenting overfishing (Sinclair and Murowski 1997), and arguably ideological management (Hutchings, Walters, and Haedrich 1997; Hutchings, this volume) that have brought the Newfound-

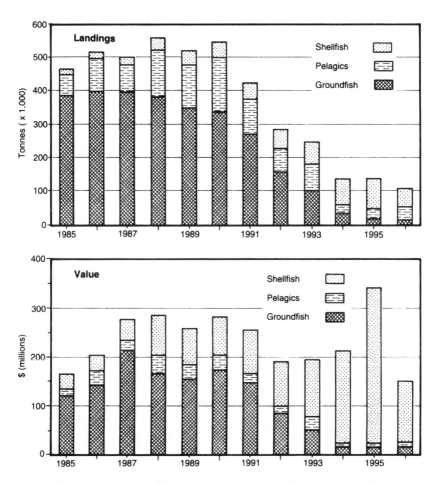

Figure 12.6   Commercial landings (top) and value (bottom) of shellfish (crab and shrimp), pelagic fish, and groundfish in Newfoundland, 1985–96. *Source*: Data from the Department of Food, Fisheries and Agriculture, Province of Newfoundland and Labrador

land stocks to the sorry state seen now in the late 1990s. As mentioned elsewhere in this volume, the social consequences of the fishery's collapse are severe and far-reaching (Hamilton, Duncan, and Flanders 1996; Sinclair 1996; Sinclair, Squires, and Downton, and Ommer, this volume). This is what would be expected when landings decline to the extent shown in the top graph of Figure 12.6. Catches of all Newfoundland fisheries have fallen markedly in recent years. The catch figure here follows a trajectory similar to that described generally by Deimling and Liss (1994), with the original groundfish fishery building and then, as it begins to decline, being somewhat offset by the buildup of a pelagic fishery based largely on less valuable species (such as capelin) that were the food of the depleted groundfish (i.e., cod). Then, as the pelagic fishery began to wane, there was a shift to the shellfish (crab and shrimp), which were also groundfish food and had become more abundant when their major predator, the cod, was decimated.

The catch figure is interesting in its own right but becomes even more so when expressed in terms of value rather than catch alone (Figure 12.6, bottom). Quite surprisingly, the figure shows that the Newfoundland fishery was worth more in 1995, in the midst of the 'fishery crisis,' than it has ever been in recent history. This value, however, is generated to a considerable degree from crab, which briefly enjoyed high prices in Newfoundland because of the closure of crab fisheries elsewhere (as in Alaska). As the Alaskan fishery is reopened, and Russia begins to exploit its stocks in earnest, the value of Newfoundland crab will drop – as it appeared to be starting to do in 1996. A similar scenario can be expected in northern shrimp, a relatively short-lived but widespread species that is therefore certain to vary greatly in annual local abundance.

The fundamental problem in the Newfoundland fishery is that stocks of whatever species are being fished simply cannot sustain recent rates of exploitation (Matthews 1995). But this ecological constraint, it can be argued, is overlain by a significant socioeconomic problem related both to shifts in the species exploited, as a consequence of the dominant traditional species having been overfished, and, most significantly, to the distribution of income generated from the fishery. It would seem that socioeconomic solutions, not biological ones, are required to set things right (see also Pauly, this volume). The role of natural scientists must be only to raise the ecological warning flags and to hope that managers and politicians are paying attention (but see Hutchings, Pinkerton, this volume).

Conclusion

Fish communities on the Grand Banks of Newfoundland and the northeast Newfoundland–Labrador shelf have tended to be relatively resilient to environmental perturbations at an ecological scale of tens of years (Gomes 1993; Villagarcía 1995). For long periods these areas were exposed to fishing exploitation, but even though overall biomass declined a lot, the main components of the assemblages did not seem to change their relative abundances very much. However, the increasing and heavy pressure exerted in recent years appears to have induced a disturbance that has produced perturbations large enough to exceed the resiliency of the system.

The northeast Newfoundland–Labrador shelf fish assemblages show some of the consequences of this disturbance. Until 1987 the assemblages showed little variation in their relative geographical boundaries, and the component species displayed about the same relative biomass, though some of the less abundant species started to decrease. From 1987 to 1989, a decline of some of the most commercially important species began, but it was only in the last years of study (1990 and 1991) that the collapse of the northern cod resulted in a loss of fidelity by some of the assemblages to their usual geographical areas (Gomes, Haedrich, and Villagarcía 1995).

From this work, it appears that studies of fish assemblages have the potential to allow detection of perturbations by effects on community structure perhaps earlier than the same perturbations might be apparent at the level of a single species. This potential may be especially useful across larger areas, at the scale of the continental shelf. From this point of view, examination of assemblages may provide a useful and more holistic framework for fisheries assessment and possibly even management (Haedrich, Villagarcía, and Gomes 1994; Langton and Haedrich 1997).

Overfishing with direct (for example, loss of biomass by removing the larger fish) and indirect (for example, loss of potential spawners) effects has caused most of the changes found in the groundfish community of the Newfoundland shelf (Hutchings and Myers 1994; Hutchings, this volume). Changes in environmental parameters are part of the natural variability with which the fish community of the area has had to cope for at least five thousand years. The fact that, beginning in 1991, the surface temperature decreased by about two Celsius degrees and ice conditions lasted longer than normal, thereby producing a delay of at least one month in the spring bloom, is not a new phenomenon in the Newfoundland area. Periods with similar environmental changes have been reported before, but without causing such strong changes in the groundfish biomass (Hutchings and Myers 1994).

The depletion of stocks and loss of the larger fish because of overfishing at the same time as inhospitable changes occurred in the ocean environment may have worsened the situation. If such a situation continues for too long, as may have already happened in Newfoundland, the groundfish stocks could lose any possibility of recovery following more and continued adverse conditions. Such harmful conditions could be either environmental – for example, another few years of very cold winters – or anthropogenic – for example, reopening of the fishery before the stocks have fully recovered. To judge from recent history, the latter course seems unfortunately to be the one that is being realized.

NOTE

The research on which this paper is based was supported by funding from the Ministerio de Educacion y Ciencia (Spain), a DFO/NSERC Subvention and an SSHRC Strategic Grant (Canada), and a German–Canadian Agreement on Co-operation in Scientific Research and Technological Development (Germany).

WORKS CITED

Atkinson, D.B. 1994. 'Some Observations on the Biomass and Abundance of Fish Captured during Stratified Random Bottom Trawl Surveys in NAFO Divisions 2J3KL, Fall 1981–1991.' In NAFO. *Scientific Council Studies* 21: 43–66.

Bishop, C.A., J. Anderson, E. Dalley, M.B. Davis, E.F. Murphy, G.A. Rose, D.E. Stansbury, C. Taggart, G. Winters, and D. Methven. 1994. An Assessment of the Cod Stock in NAFO Divisions 2J+3KL, NAFO SCR Doc. 94/40, Ser. no. N2410.

Colvocoresses, J.A., and J.A. Musick. 1984. 'Species Associations and Community Composition of Middle Atlantic Bight Continental Shelf Demersal Fishes.' *Fishery Bulletin* 82, no. 2: 95–313.

Deimling, E.A., and W.J. Liss. 1994. 'Fishery Development in the Eastern North Pacific: A Natural-cultural System Perspective.' *Fisheries Oceanography* 3: 60–77.

Drinkwater, K.F., and D.G. Mountain. 1997. 'Climate and Oceanography.' In J. Boreman, B.S. Nakashima, J.A. Wilson, R.L. Kendall, eds., *Northwest Atlantic Groundfish: Perspectives on a Fishery Collapse*, 3–25. Bethesda, Md.: American Fisheries Society.

Ellertsen, B., P. Fossum, P. Solemdal, and S. Sundby. 1989. 'Relations between Temperature and Survival of Eggs and First Feeding Larvae of Northeast Arctic Cod (*Gadus morhua* L.).' *ICES Rapports et Procès-Verbaux des Réunion du Conseil Internationale pour l'exploration de la Mer* 191: 209–19.

Ellertsen, B., P. Fossum, P. Solemdal, S. Sundby, and S. Tilseth. 1987. 'The Effect of Biological and Physical Factors on the Survival of Arcto-Norwegian Cod and the Influence on Recruitment Variability.' In H. Loeng, ed., *The Effect of Oceanographic Conditions on Distribution and Population Dynamics of Commercial Fish Stocks in the Barents Sea, Proceedings of the Third Soviet–Norwegian Symposium, Murmansk 26–28 May 1986*, 101–26. Bergen, Norway: Institute of Marine Research.

Fischer, J. and R.L. Haedrich. 1997. 'Common Marine Fish Species and Ocean Temperatures off Newfoundland.' Occasional Paper. St John's: Eco-research Project, Memorial University.

Gabriel, W.L., and A.V. Tyler. 1980. 'Preliminary Analysis of Pacific Coast Demersal Fish Assemblages.' *Marine Fisheries Review* March–April: 83–8.

Gauch, H.G. 1982. *Multivariate Analysis in Community Ecology*. Cambridge: Cambridge University Press.

Gomes, M.C. 1993. *Predictions under Uncertainty: Fish Assemblages and Food Webs on the Grand Banks of Newfoundland*. St John's: ISER, Memorial University.

Gomes, M.C., R.L. Haedrich, and M.G. Villagarcía. 1995. 'Spatial and Temporal Changes in the Groundfish Assemblages on the Northeast Newfoundland/Labrador Shelf, Northwest Atlantic. 1978–1991.' *Fisheries Oceanography* 4: 85–101.

Haedrich, R.L. 1993. 'Variability in Newfoundland's Ocean Environment.' In J. Hall and M. Wadleigh, eds., *The Scientific Challenge to Our Changing Environment*, 50–1. Canadian Global Change Program, Incidental Report Series IR93-2.

Haedrich, R.L., and S.M. Barnes. 1997. 'Changes over Time of the Size Structure in an Exploited Shelf Fish Community.' *Fisheries Research* 31: 229–39.

Haedrich, R.L., and J. Fischer. 1996. 'Stability and Change of Exploited Fish Communities in a Cold Ocean Continental Shelf Ecosystem.' In *Proceedings of the ELAWAT Workshop on 'The Concept of Ecosystems,' Wilhelmshaven, September 1994, Senckenbergiana maritima* 27, nos. 3/6: 237–43.

Haedrich, R.L., J. Fischer, and N. Chernova. 1997a. 'Ocean Temperatures and Demersal Fish Abundance on the Northeast Newfoundland Continental Shelf.' In L.E. Hawkins and S. Hutchinson, eds., *The Response of Marine Organisms to Their Environments: Proceedings of the Thirteenth European Marine Biology Symposium, Southampton, U.K., October 1995*, 211–22.

Haedrich, R.L., M.G. Villagarcía, and M.C. Gomes. 1994. 'Scale of Protected Areas on Newfoundland's Continental Shelf.' in N.L. Shackell and J.H.M. Willison, eds., *Marine Protected Areas and Sustainable Fisheries, Proceedings of the 2nd International Conference on Science and the Management of Protected Areas, Halifax, NS, May 1994*. 48–53.

Hamilton, L.C., C.M. Duncan, and N.E. Flanders. 1998. 'Management, Adaptation and Large-scale Environmental Change.' In D. Symes, ed., *Property Rights and Regulator Systems for Fisheries*, 17–33. Oxford: Blackwell Science.

Helbig, J.H., G. Mertz, and P. Pepin. 1992. 'Environmental Influences on the Recruitment of Newfoundland/Labrador Cod,' *Fisheries Oceanography* 1: 39–56.

Hill, M. 1979. *TWINSPAN, a FORTRAN Program for Arranging Multivariate Data in an Ordered Two-way Table by Classification of the Individuals and Attributes, Ecology and Systematics*. Ithaca, NY: Cornell University Press.

Hutchings, J.A., and R.A. Myers. 1994. 'What Can Be Learned from the Collapse of a Renewable Resource? Atlantic Cod, *Gadus morhua*, of Newfoundland and Labrador.' *Canadian Journal of Fisheries and Aquatic Sciences* 51, no. 9: 2126–46.

Hutchings, J.A., C.W. Walters, and R.L. Haedrich. 1997. 'Is Scientific Inquiry Incompatible with Government Information Control?' *Canadian Journal of Fisheries and Aquatic Sciences* 54: 1198–1210.

Laevastu, T. 1993. *Marine Climate, Weather and Fisheries: The Effects of Weather and Climatic Changes on Fisheries and Ocean Resources*, New York: Halsted Press.

Langton, R.W., and R.L. Haedrich. 1997. 'Ecosystem-based Management.' In J. Boreman, B.S. Nakashima, J.A. Wilson, and R.L. Kendall, eds., *Northwest Atlantic Groundfish: Perspectives on a Fishery Collapse*, 153–7. Bethesda, Md.: American Fisheries Society.

Lilly, G.R. 1994. 'Predation by Atlantic Cod on Capelin on the Southern Labrador and Northeast Newfoundland Shelves during a Period of Changing Spatial Distributions.' *ICES Marine Science Symposium* 198: 600–1.

Loeng, H. 1989. 'The Influence of Temperature on Some Fish Population Parameters in the Barents Sea.' *Journal of Northwestern Atlantic Fisheries Science* 9: 103–13.

McFarlane, G.A., D.M. Ware, R.E. Thomson, D.L. Mackas, and C.L.K. Robinson. 1997. 'Physical, Biological and Fisheries Oceanography of a Large Ecosystem (West Coast of Vancouver Island) and Implications for Management.' In J. Castel, ed., *Long Term Changes in Marine Ecosystems, Oceanologica Acta* 20, no. 1: 191–200.

Mahon, R., and W.R. Smith. 1989. 'Demersal Fish Assemblages on the Scotian Shelf, Northwest Atlantic: Spatial Distribution and Persistence.' *Canadian Journal of Fisheries and Aquatic Sciences* 46: 134–52.

Mante, C., J.P. Durbec, and J.C. Dauvin. 1997. 'Analysis of Temporal Changes in Macrobenthic Communities on the Basis of Probable Species Presence.' In J. Castel, ed., *Long Term Changes in Marine Ecosystems, Oceanologica Acta* 20, no. 1: 71–80.

Matthews, D.R. 1995. '"Commons versus open access." The Collapse of Canada's East Coast Fishery.' *Ecologist* 25, nos. 2/3: 86–96.

Myers, R.A., J.A. Hutchings, and N.J. Barrowman. 1996. 'Hypotheses for the Decline of Cod in the North Atlantic.' *Marine Ecology Progress Series* 138: 293–308.

Nilssen, E.M., and C.E. Hopkins. 1992. 'Regional Variability in Fish-Prawn Communities and Catches in the Barents Sea, and Their Relationship to the Environment.' *ICES Marine Science Symposium* 195: 331–48.

Overholtz, W.J., and A.V. Tyler. 1985. 'Long-Term Responses of the Demersal Fish Assemblages of Georges Bank.' *Fishery Bulletin* 83: 507–20.

Rogers, J.B., and J.B. Pikitch. 1992. 'Numerical Definition of Groundfish Assemblages Caught off the Coasts of Oregon and Washington Using Commercial Fishing Strategies.' *Canadian Journal of Fisheries and Aquatic Sciences* 49: 2648–56.

Ross, M.R. 1997. *Fisheries Conservation and Management*, Upper Saddle River, NJ: Prentice Hall.

Safina, C. 1995. 'The World's Imperiled Fish,' *Scientific American* 273, no. 5 (Nov.): 46–53.

Savvatimsky, P.I. 1987. 'Changes in Species Composition of Trawl Catches by Depth on the Continental Slope from Baffin Island to Northeastern Newfoundland. 1970–85.' *NAFO Scientific Council Studies* 11: 43–52.

Sherman, K.L. 1992. 'Monitoring and Assessment of Large Marine Ecosystems: A Global and Regional Perspective.' In D.H. McKenzie, D.E. Hyatt, and V.J. McDonald, eds., *Ecological Indicators*, vol. 2, 1041–74. Essex, England: Elsevier.

Sinclair, A.F., and S.A. Murawski. 1997. 'Why Have Groundfish Stocks Declined?' In J. Boreman, B.S. Nakashima, J.A. Wilson, and R.L. Kendall, eds., *Northwest Atlantic Groundfish: Perspectives on a Fishery Collapse*, 71–93. Bethesda, Md.: American Fisheries Society.

Sinclair, P.R. 1996. 'Sustainable Development in Fisheries Dependent Regions? Reflections on Newfoundland's Cod Fisheries.' *Sociologia Ruralis* 36, no. 2: 225–32.

Sissenwine, M.P. 1984. 'Why Do Fish Populations Vary?' In R.M. May, ed., *Exploitation of Marine Communities*, 59–94. Dahlem Konferenzen. Berlin: Springer Verlag.

Sokal, R., and C. Mitchener. 1958. 'A Statistical Method for Evaluating Systematic Relationships.' *University of Kansas Science Bulletin* 38: 1409–38.

Tyler, A.V., W.L. Gabriel, and W.J. Overholtz. 1982. 'Adaptive Management Based on Structure of Fish Assemblages of Northern Continental Shelves.' In M.C. Mercer, ed., *Multispecies Approaches to Fisheries Management Advice*, Special Publication of the *Canadian Journal of Fisheries and Aquatic Sciences* 74: 149–56.

Underwood, A.J. 1986. 'What Is a Community?' In D.M. Raup and D. Jablonski, eds., *Patterns and Processes in the History of Life*, Dahlem Konferenzen 1986. 351–67. Berlin: Springer-Verlag.

Villagarcía, M.G. 1995. 'Structure and Distribution of Fish Assemblages on the Northwest Newfoundland and Labrador Shelf.' MSc thesis, Memorial University of Newfoundland.

Ward, J. 1963. 'Hierarchical Grouping to Optimize an Objective Function,' *Journal of American Statistical Association* 58: 236–44.

Wishart, D. 1978, *CLUSAN User Manual*. 3rd ed. Edinburgh: Program Library Unit, University of Edinburgh.

13

The Biological Collapse of Newfoundland's Northern Cod

JEFFREY A. HUTCHINGS

The sustainability of communities from southeastern Labrador to southeastern Newfoundland is tied inextricably to the sustainability of the fish stock known as northern cod, *Gadus morhua*. This has been so since the early seventeenth century, when the English and French first established year-round settlements in Newfoundland (Cell 1982). When the moratorium on commercial fishing for northern cod was announced in 1992, an estimated 16 per cent of the total workforce – as much as 90 per cent in many communities – was directly employed by the fishery (Cashin 1993). However, since Canada's extension of its fisheries jurisdiction limit to 200 miles (322 kilometres) in 1977, these communities have not simply depended on the sustainability of northern cod. They have also been dependent on the management strategies established by the Canadian Department of Fisheries and Oceans (DFO), whose mandate includes the establishment of harvesting regulations designed to conserve commercial fish species and to ensure their long-term viability. Against these socioeconomic and institutional backgrounds, this chapter[1] assesses the strengths and weaknesses of the main scientific hypotheses that have been proposed to explain the collapse of northern cod.

The influence of social interactions within DFO on the interpretation of fish stock assessments has been recently examined by Finlayson (1994). Within this context, and based on personal experience with recent reviews of stock assessments, I comment on the process by which a consensual view is reached at DFO stock assessment meetings. While providing personal comment is unavoidably encumbered with personal bias, the difference between this approach and that of a formal research study lies in the nature of the bias, since the latter has its own methodological and researcher-induced biases.

For simplicity, and to be consistent with previous treatments of this topic (for example, Finlayson 1994), I refer to the assessment of stock status as 'science,' though some would argue it is not. Involving the formulation and testing of hypotheses, fisheries science embodies that body of work intended to provide information on the biology and population dynamics of harvested fishes. Such information can include data on age structure, movement and migration, vertical and horizontal patterns of abundance (i.e., distribution), relationships with other species, morphological, behavioural, and genetic distinctiveness, life history strategies (i.e., age-specific rates of survival and fecundity), and associations between physical environmental factors (such as temperature and salinity) and biological variables (such as growth rate and survival) (see Villagarcia et al., this volume). Stock assessment, in contrast, is the estimation of past, present, and future abundance (or biomass) of, and the past fishing mortality imposed on, a given fish stock. These estimates are based on commercial catch statistics and research surveys.

The Collapse

The Newfoundland fishery for Atlantic cod was once the largest (McGrath 1911) and the most productive (Thompson 1943) cod fishery in the world. Its 'northern' cod component has constituted upward of 70 per cent of all Newfoundland catches since 1954 and probably did so for most of the five hundred–year history of the fishery (Hutchings and Myers 1995). The geographical range of northern cod extends from southern Labrador southeast along the northeast Newfoundland shelf to include the northern half of the once biologically rich Grand Banks. For management purposes, northern cod are defined as those cod found in Northwest Atlantic Fishery Organization (NAFO) divisions 2J (southern Labrador), 3K (northeast Newfoundland shelf), and 3L (northern Grand Bank). Despite this geographical designation of the northern cod 'stock,' it is highly probable that several inshore and offshore populations of cod exist within these NAFO divisions (Hutchings, Myers, and Lilly 1993; Lear 1984; Ruzzante et al. 1996; Templeman 1966; Thompson 1943).

Historical data suggest that northern cod catches of 200,000 to 300,000 metric tonnes (1 t = 1,000 kg) were sustainable throughout the nineteenth and early twentieth centuries (Hutchings and Myers 1995). However, the enormous catches made possible by the introduction of the 'factory freezer' stern trawler, which facilitates the freezing and packaging of fish on board, in the late 1950s and early 1960s (the maximum *reported* catch for northern cod was 810,000 tonnes in 1968; Murphy et al. [1997]) may have changed for ever the ability of northern cod to sustain catches of the size that were apparently sustainable in the nineteenth century. By the time Canada extended its fisheries jurisdiction limit to 200 miles in 1977, it was clear that northern cod were in very serious trouble. Since 1962, the biomass of cod available for harvest had declined by 82 per cent from an estimated three million tonnes to 526,000 tonnes (data available in Bishop et al. 1994). The reproductive portion of the stock (cod seven years and older) had declined by 94 per cent from 1.6 million tonnes to 93,000 tonnes. Concomitant with this decline, recruitment (number of three-year-olds) decreased from an estimated one billion to 168 million individuals.

Imposition of the 200-mile limit resulted in an immediate and dramatic decrease in fishing mortality; most foreign trawler fleets were restricted from fishing, and Canada had yet to develop its own trawler fleet. This lull in trawler activity permitted a modest stock 'recovery' between 1977 and 1985, during which time harvestable biomass approximately doubled (Bishop et al. 1994). The stock declined thereafter until a moratorium on commercial exploitation of northern cod was announced on 2 July 1992. Similar moratoria were imposed on five other Atlantic Canadian cod stocks in September 1993. Comparing spawner biomass at the time the moratoria were announced with historical maxima for each of the six stocks, we find that northern cod experienced a decline considerably more severe than other cod stocks; spawner biomass in 1992 was only approximately 1 per cent of its historical maximum. By contrast, the size of the spawner biomass of other cod stocks when their fishing moratoria were announced averaged 13 per cent of their historical maxima (Myers, Hutchings, and Barrowman 1996).

Failure to Control Fishing Mortality

The primary cause of the collapse of northern cod was failure to maintain fishing mortality at a level that would allow the stock to sustain itself through time. The management strategy of DFO from 1977 to 1992 was based on, in technical terms, the $F_{0.1}$ strategy. The fishing mortality corresponding to $F_{0.1}$ can be estimated from the dome-shaped relationship presumed to exist between yield per recruit (plotted on the y-axis) and fishing mortality (plotted on the x-axis). The positive slope of such a curve is greatest near the origin (i.e., the point corresponding to x = 0 and y = 0). As fishing mortality increases, the rate at which yield per recruit rises per unit increase in fishing mortality (i.e., the slope of the curve) declines. The fishing mortality at $F_{0.1}$ corresponds to that point on the curve at which the slope is 10 per cent of the slope at the origin.

By comparison, the maximum sustainable yield (MSY) strategy would correspond approximately to an F_0 strategy – i.e., the point on the curve where the slope is equal to zero. The instantaneous fishing mortality (F) that corresponded to $F_{0.1}$ for northern cod was $F = 0.2$. This level of fishing mortality would allow approximately 18 per cent of the harvestable biomass to be caught by commercial fishing gear every year. To maintain harvest rates at this 18 per cent target, the stock was regulated on the basis of catch quotas, or total allowable catches (TACs). In contrast, changes in harvesting capacity were not monitored.

Estimating Stock Size

The success of a catch-quota management system depends on the reliability of the estimate of stock size and on the accuracy of the reported statistics on catches. Errors in these cornerstones of the catch-quota strategy will become manifest as errors in the setting of TACs at the $F_{0.1}$ level. There is considerable evidence of errors in stock-size estimation and catch statistics for the northern cod stock (Walters and Maguire 1996).

There are two primary means of estimating the size of a commercially fished stock: virtual (or sequential) population analysis (VPA) and research surveys. VPA provides abundance-at-age estimates for cohorts (i.e., fish born in the same year) that have passed through the fishery and for those still present. These estimates, used to predict stock size in the future and to estimate past and present levels of fishing mortality, are based on the following general model:

$$\begin{array}{ccccccc} \text{abundance} & & \text{abundance} & & \text{last year's} & & \text{those fish that died} \\ \text{last year} & = & \text{this year} & + & \text{catch} & + & \text{from natural causes} \\ & & & & & & \text{last year} \end{array}$$

The accuracy of VPA estimates of stock size and F depend on the validity of two primary assumptions – that commercial catch data are reported without error and that natural mortality is constant from one year to the next and does not vary with age. Research surveys have been conducted throughout the entire management unit of the northern cod stock since 1981. Annually, these surveys have consisted of 300 to 500 thirty-minute tows by a stern-hauled bottom trawl at randomly selected locations within each of seventy-five to eighty sampling areas, called 'strata.' The data from these random-stratified surveys provide the only available independent estimates of stock abundance and are typically reported as mean biomass or number per tow (see Villagarcia et al., this volume).

Between 1978 and 1988, catch rates from Canadian trawlers and the research surveys were used to describe trends in northern cod abundance. Catch rate was assumed to be directly proportional to abundance – i.e., a given increase in catch rate reflected a given increase in stock size, an assumption that now appears unjustified (Hutchings and Myers 1994; Walters and Pearse 1996). Between 1978 and 1985, the data on the commercial catch rate suggested that the northern cod stock had more than trebled in size, while the survey data indicated a 50 per cent increase at best. An arbitrary decision was made to use the mid-point of the two catch rate trends in the stock assessments (Baird, Bishop, and Murphy 1991), despite the very real possibility that the increase in commercial catch rate could be attributed to increased harvesting efficiency resulting from the continual introduction of new technology as well as from learning by the novice Canadian trawler fleet. The use of data on the commercial catch rate to describe trends in fish abundance contributed to the overestimation of stock size in the 1980s.

The main consequence of this overestimation was that realized fishing mortalities exceeded the targeted $F_{0.1}$ level more than two-fold between 1978 and 1983 and more than three-fold between 1984 and 1989 (Hutchings and Myers 1994; Murphy et al. 1997). By the late 1980s and early 1990s, VPA estimates of fishing mortality indicated that fishing was removing 60 to 80 per cent of the harvestable stock biomass every year (recall that the target, and presumed sustainable, harvest rate was 18 per cent).

However, since one assumption of VPA is that natural mortality is constant from one year to the next, the VPA-based fishing mortalities estimated for the late 1980s and early 1990s may have reflected a dramatic, short-term increase in natural mortality rather than an increase in fishing mortality. One recent analysis suggests that a biologically significant increase in natural mortality of northern cod did not occur prior to the stock's collapse (Myers and Cadigan 1995), while additional analyses report that estimates of fishing mortality determined from mark-recapture experiments (the recaptures are tag returns from fishers) are consistent with those estimated by VPA, indicating that the rapid increase in fishing mortality in the late 1980s and early 1990s was real (Myers, Hutchings, and Barrowman 1996; Myers et al. 1996; Myers, Barrowman, and Hutchings 1997).

Why Did Northern Cod Collapse?

Hypotheses for the collapse of northern cod characterize the increase in mortality of cod in the 1980s and early 1990s as being largely a function of increasing natural mortality, increasing fishing mortality, or some combination thereof.

Decline of Non-targeted Fish Species

The concomitant decline of some fish species not directly targeted by the cod fishery has been cited as evidence that ecosystem change helped produce the collapse of northern cod (Atkinson 1993; Atkinson and Bennett 1994; DFO 1995; FRCC 1995). Some researchers have identified bycatch fishing mortality as a dominant cause of changes of abundance in non-commercial species on the northeast Newfoundland shelf (Gomes, Haedrich, and Villagarcía 1995; Haedrich 1995; Villagarcia et al., this volume). Others disagree, particularly with regard to the decline of American plaice (*Hippoglossoides platessoides*) in division 2J. Brodie, Morgan, and Bowering (1995), for example, argue in effect that changes in the environment caused the natural mortality of plaice to rise dramatically in the 1980s.

Data from research surveys and commercial catches from 1981 to 1992 suggest that the mortality of American plaice caught as by-catch in the northern cod and other fisheries may have been considerable (Hutchings 1996). First, by-catch of plaice with cod in division 2J appears to be unavoidable: on average, 95 per cent of all research-survey trawls that caught cod between 1981 and 1992 also took American plaice. Second, there is a statistically significant positive relationship between plaice and cod survey biomass in most years prior to 1991; that is, the amount of plaice in a given survey tow increased with the amount of cod in that tow. Linear regressions between plaice and cod biomass, used to predict annual by-catches of plaice in the northern cod fishery in 2J, have resulted in predicted catches of plaice generally more than ten times greater than the reported catches (Hutchings 1996). Notwithstanding the considerable estimation errors associated with these predicted catches, this analysis does suggest that the attribution of the decline of 2J plaice primarily to non-fishing causes may be unwarranted.

Environmental and Ecosystem Change

Northern cod, and other Northwest Atlantic cod stocks, were closed because of a scarcity of individuals of reproductive age – i.e., low spawner biomass. It has been suggested that an increase in the natural mortality of cod aged one to three years in the mid-1980s helped produce this relative absence of spawners (Atkinson and Bennett 1994; Mann and Drinkwater 1994). This 'poor recruitment' hypothesis has recently been examined for the six Canadian cod stocks for which fishing moratoria currently exist. Using research survey data, Myers, Hutchings, and Barrowman (1996, 1997) tested the null hypothesis that recruitment of those year classes that would have dominated the spawner biomass at the time of the moratoria did not differ significantly from the average annual recruitment preceding those year classes. This hypothesis could not be rejected for northern cod. Of related interest, analyses that previously linked poor recruitment of Atlantic cod to environmental factors such as cold water temperature (deYoung and Rose 1993) and low salinity (Myers et al. 1993) have not been supported on re-examination of the data (Hutchings and Myers 1994; Ouellet 1997).

Hypotheses linking the collapse of northern cod to environmental or ecosystem change bear the implicit assumption that environmental conditions in the late 1980s and early 1990s were temporally anomalous, given that collapses of the magnitude documented in 1992 have never been previously recorded. However, a comparison of various environmental indices on a decadal and on a century time scale indicates that the environmental conditions for northern cod since the late 1980s have clearly been experienced by the stock in the past.

Contrary to public perception (for example, FRCC 1995), water temperatures on the northeast Newfoundland shelf in the early 1990s were not temporally anomalous. Though low, the depth-averaged water temperatures recorded at a hydrographic station east of St John's (station 27) in 1991 did not differ statistically from those in 1972–4 and 1984–5; bottom and near-bottom temperatures there were in fact lower in these two previous periods than in 1991 (Hutchings and Myers 1994). In addition, the volume of water at less than 0°C (the cold intermediate layer, or CIL) off southern Labrador in 1990 and 1991 was actually less than that of the CIL in 1972 and in 1985; the volume of the 1990–1 CIL farther south off Cape Bonavista was also less than the CIL volume in 1985 and equal to that recorded in 1972 (Colbourne 1995; see also Villagarcia et al., this volume).

Data available over a considerably longer time scale, such as water temperatures on Grand Banks since 1910, data on ice clearance from Labrador since 1800, and air temperature data from St John's since 1874 (the latter two significantly associated with water temperature at station 27) all suggest that northern cod catches of the magnitude that proved unsustainable in the 1980s were sustainable in the nineteenth and early twentieth centuries in an environment that was on average considerably colder (Hutchings and Myers 1994). Therefore the description of oceanographic conditions off northeast Newfoundland in the early 1990s as temporally anomalous is simply not supported by empirical data.

There is to date no direct evidence linking the mortality of northern cod to changes in the environment or ecosystem: any apparent direct effect of temperature or salinity on recruitment can be rejected on statistical grounds (Hutchings and Myers 1994). The hypothesis that cod shifted their distribution southward in 1989 or 1990 – one possible response to environmental and/or ecosystem change (deYoung and Rose 1993; Rose et al. 1994) – is not supported by consideration of research survey data, age-specific growth data, and the results of tagging experiments (Hutchings 1996). The hypothesis that a severe decline in prey abundance spurred the cod collapse (Atkinson and Bennett 1994) has been rejected by Lilly (1995), who reported that there is no evidence that average stomach fullness of cod decreased before, or in parallel with, the decline in northern cod.

Interaction between Environmental Change and Fishing

Some have argued that fishing pressure was indeed too high on northern cod – i.e., actual F exceeded the $F_{0.1}$ target – but not high enough to effect a stock collapse. For example, Shelton (1995) has argued that unusually poor recruitment (i.e., low production of offspring and/or survival of cod to age three) and declining weight-at-age may also have furthered the stock's decline. However, his estimates of recruitment are based on VPA. These data indicate that recruitment declined in the mid-1980s to levels that would have resulted in very low spawner biomass at the time when the fishery was closed in 1992 (northern cod typically attain maturity at age five to seven years). However, as indicated above, the reliability of VPA abundance estimates declines with increased unreliability in the reporting of catch statistics. Research survey data, in contrast, indicate no such decline in recruitment during the same period (neither for northern cod nor for the other cod stocks closed in 1993; Myers, Hutchings, and Barrowman 1997).

The conclusion that unusually poor recruitment contributed heavily to the collapse of east coast cod is problematic. One reason for this is that underreporting, misreporting, and discarding of catches (see Palmer and Sinclair 1996 for one example) negatively bias VPA estimates of past abundance (refer to the equation above). Thus an increased tendency to underreport actual catches will cause a VPA to indicate a declining trend in past abundance when none had actually occurred. Thus the VPA downward trend in recruitment in the mid-1980s can be attributed to increased discarding of undersized cod of recruitment age. The observation that research survey estimates of recruitment do not show the decline in abundance of three-year-olds suggested by VPA also suggests that the collapse cannot be attributed to poor recruitment (Myers, Hutchings, and Barrowman 1996, 1997). Finally, the idea that the VPA decline in recruitment really reflects increased rates of discarding rather than poor recruitment is consistent with research survey–based estimates of mortality of three-year-old cod in the late 1980s and early 1990s (Myers, Hutchings, and Barrowman. 1997) and with statements by fishermen (as in Palmer and Sinclair 1996; J.A. Hutchings and M. Ferguson, unpublished data).

The argument that the collapse can be attributed to an interaction between fishing mortality and a deteriorating environment, reflected by declining age-specific body sizes, suffers somewhat from Pauly's (1995) 'shifting baseline syndrome of fisheries.' The time series over which the declining trend in cod weights-at-age is described begins in the late 1970s or early 1980s (for example, Bishop et al. 1994; Taggart et al. 1994). If this time series is extended to include age-specific body sizes over all years for which such data are available, it is apparent that it is the large age-specific body sizes of the late 1970s and early 1980s that are anomalous (Fleming 1952; Hutchings 1996).

The decline in age-specific body sizes observed through the 1980s is undoubtedly a consequence partly of water temperature but also of the increased rate of harvesting of the fastest growers within each cohort (Hutchings 1996). Consistent with this hypothesis is the observation that the high age-specific body sizes measured in the late 1970s and early 1980s coincided with the period during which cod experienced the lowest offshore fishing mortality since the early 1960s (Bishop et al. 1994; Myers et al. 1996). In addition, the physical oceanographic environment of the late 1970s and early 1980s (for example, water temperature; Hutchings and Myers 1994), was not notably favourable for fast growth. The decline in weights-at-age may also be a consequence of the observed trend towards earlier age at maturity (Trippel et al. 1997), which reduces body size at later ages (Hutchings 1993).

Was the Decline of Northern Cod Gradual or Sudden?

It has been suggested that the collapse of northern cod occurred suddenly over a single year, the spring of 1991 being the favoured time (Atkinson and Bennett 1994; Lear and Parsons 1993). This perception may be partly the result of use of trends in arithmetic mean biomass as a primary indicator of stock abundance. Estimators of temporal trends in stock biomass that are less sensitive to infrequent high-biomass tows (for example, geometric mean biomass, a rank-based biomass trend) indicate that the decline of northern cod was gradual rather than abrupt (Hutchings 1996). Changes in the density composition of the stock, as reflected by research survey data, coupled with spatial-temporal variation in stock biomass, also indicate that northern cod were declining throughout the 1980s (Hutchings 1996). The inshore fishery provides further evidence of a gradual stock decrease. Declining fixed-gear catch rates, evident from the mid-1980s on, suggest that the stock had been decreasing since at least 1985 (Hutchings and Myers 1994). The dramatic spatial shift in the gillnet fishery in the mid-1980s from inshore to offshore waters in response to declining catch rates is further evidence of a decline prior to the 1990s (Hutchings and Myers 1994, 1995; Neis et al. 1996).

There is thus considerable evidence that overfishing was the primary significant cause of the collapse of northern cod and of other Northwest Atlantic groundfish stocks (Sinclair and Murawski, in press). The decline in northern cod through the mid-1980s occurred at the same time that offshore (Hutchings and Myers 1994) and inshore (J.A. Hutchings and M. Ferguson, unpublished data; Neis et al. 1996) fishing effort was increasing (a description of temporal changes in gear technology in the northern cod fishery is provided by Hutchings and Myers 1995).

Northern cod were apparently not sustainable at the age-specific rates of survival and fecundity experienced between 1985 and 1992 (Hutchings and Myers 1994). Statistical analyses of catch-at-age data for the species provide no evidence of a rise in natural mortality in 1991 (Myers and Cadigan 1995). One of the strongest sources of evidence of excessive fishing mortality are the estimates of fishing mortality derived from tagging studies, which indicate that fishing mortality on northern cod had exceeded $F = 1.0$ by the late 1980s (Myers, Barrowman, and Hutchings 1996, 1997). The primary role of overfishing has been accepted as an explanation for the collapse of southern Gulf of St Lawrence cod, for which trends in environmental factors and seal abundance were inconsistent with trends in cod mortality (Sinclair et al. 1995). More recently, Canadian government officials have also acknowledged that overfishing and poor fishing practices were the primary cause of the collapse of Atlantic cod throughout eastern Canada (Doubleday, Atkinson, and Baird 1997).

A Comment on Stock Assessment Review

Management consists of establishing and implementing harvesting regulations as means of controlling fishing mortality. These regulations can take the form of catch quotas, such as total allowable catches (TACs), and individual transferable quotas (ITQs); effort restrictions, such as minimum mesh sizes and maximum vessel sizes; and seasonal closures.

Prior to the collapse of northern cod, science was often poorly represented in DFO's fishery management decisions. Uncertainties in data quality and abundance estimates, and the existence of alternative hypotheses for observed changes in fish stocks, were rarely evident in the stock-assessment documents on which management, in consultation with industry, based its decisions. For example, rather than including confidence intervals on abundance metrics or demonstrating how perception of stock status depends on the validity of different sets of assumptions, stock assessments usually offered a single perception of the health of a stock and little or no indication of the variability associated with the parameters used in abundance models or of the robustness of such models to such variability. In addition, documents on stock assessment did not reflect the different opinions that often existed among scientists regarding the health of fish stocks. Because documents did not formally acknowledge the existence of differing opinions on the health of a fish stock, and did not report and quantify sources of variability and their effects on the robustness of abundance models, they provided fishery managers with input that was significantly inferior to that transferred among scientists (Hutchings, Walters, and Haedrich 1997).

Explicit identification of the sources of variability in stock assessment, and to a lesser extent differences in opinion among scientists, are now becoming increasingly common features of stock assessment documents prepared by DFO. There is, however, considerable room for strengthening the communication of fisheries science within DFO. Though this subject is dealt with more extensively elsewhere (Hutchings, Walters, and Haedrich 1997), I wish to comment briefly on what I perceive to be limitations in the means by which stock assessments are conducted and reviewed.

There tends to be a high degree of conceptual 'compartmentalization' in the stock assessments. Reviews evaluate the stock status of one species at a time, without consideration of the population dynamics of co-existing species, and rely overly on information enclosed in what I would identify as three 'boxes.' The first contains VPA estimates of fishing mortality and age-specific abundance – this is the 'fishing box.' The second contains survey-based trends in stock biomass, weight-at-age, and stock distribution – the 'biology box.' The third contains temporal trends in environmental variables such as temperature, ice coverage, and salinity – the 'environment box.' If trends in the biology box cannot be accounted for by the fishing box, they must, it is often argued, be the result of changes in the environment box.

Despite its considerable deficiencies (Myers, Hutchings, and Barrowman 1997; Walters and Maguire 1996), a surprising amount of confidence is still expressed in the results produced by VPA. This often leads many individuals to accept the data in the fishing box with little question. Ecological factors such as inter-specific interactions (competition, predation), density-dependent habitat selection, density-dependent migratory behaviour, and the effects of fishing on age-specific weights, distribution patterns, and age at maturity receive comparatively little attention. Such information has not been considered essential for stock assessment; such data have not historically been part of one of the boxes. In addition, most individuals involved in stock assessment reviews are neither ecologists nor population biologists.

These observations should be interpreted not as criticisms of those involved in stock assessment but as suggestions for improvement. Despite the problems associated with multispecies VPAs (Hilborn and Walters 1992) and the practical limitations of defining ecosystem-based management strategies, there is a clear and urgent need to include basic behaviour, ecology, population biology, and life history research in the stock assessments of commercially harvested fishes.

Conclusion

The observation that over 90 per cent of the world's fish stocks have been over-exploited (Alverson et al. 1994) underscores the fact that the effects of fishing are consistently underestimated and poorly understood by those charged with fisheries management (see also Pauly, this volume). What are the effects of fishing on the population biology of commercially harvested fishes? This is a central question facing fisheries science in the aftermath of the collapse of northern cod.

Fishing represents the cumulative mortality effected by multiple predators (i.e., various types of fishing gear). The different age- and size-specific mortality rates effected by these predators produce a multitude of biological responses that may appear to be the product of environmental change. For example, as discussed above, declines in age-specific weight, rather than being a consequence of a lack of prey or increased physiological stress, can result from increasing fish mortality on the fastest-growing individuals of age classes partially recruited to the fishery. Similarly, spatial changes in distribution, rather than reflecting ecosystem change, may reflect density-dependent changes in habitat selection (for example, Swain and Wade 1993). This does not mean that the environment does not influence fish biology and behaviour. But any legitimate examination of temporal changes in northern cod, or in any other fish species, prior to the northern cod fishing moratorium is incomplete if it fails to account fully for the potential influence of the single factor known to effect the greatest mortality of commercially harvested species – fishing.

NOTE

The author's fisheries research is supported by a NSERC research grant, by a DFO/ NSERC subvention, and by Memorial University of Newfoundland's Tri-Council Eco-Research Project. I thank Rosemary Ommer for the invitation to undertake this work and Daniel Pauly for comments on a previous version of the manuscript.

WORKS CITED

Alverson, D.L., M.H. Freeberg, J.G. Pope, and S.A. Murawski 1994. 'A Global Assessment of Fisheries Bycatch and Discards.' *FAO Fisheries Technical Paper No.* 339.
Atkinson, D.B. 1993. 'Some Observations on the Biomass and Abundance of Fish Cap-

tured during Stratified Random Bottom Trawl Surveys in NAFO Divisions 2J3KL, Fall 1981–1991.' *Northwest-Atlantic Fisheries Organization Scientific Study Document* 93/29, Ser. No. N2209.

Atkinson, D.B., and B. Bennett. 1994. 'Proceedings of a Northern Cod Workshop Held in St. John's, Newfoundland, Canada, January 27–29, 1993.' *Canadian Technical Report Fisheries Aquatic Sciences.*

Baird, J.W., C.A. Bishop, and E.F. Murphy. 1991. 'Sudden Changes in the Perception of Stock Size and Reference Catch Levels for Cod in North-Eastern Newfoundland Shelves.' *NAFO Science Council Studies* 16: 111–19.

Bishop, C.A., J. Anderson, E. Dalley, M.B. Davis, E.F. Murphy, G.A. Rose, D.E. Stansbury, C. Taggart, G. Winters, and D. Methven. 1994. 'An Assessment of the Cod Stock in NAFO Divisions 2J+3KL.' *NAFO Science Council Research Document* 94/40.

Brodie, W., J. Morgan, and W.R. Bowering. 1995. 'An Update of the Status of the Stock of American Plaice in Subarea 2 + Div. 3K.' *DFO Atlantic Fisheries Research Document* 95/35.

Cashin, R. 1993. *Charting a New Course: Towards the Fishery of the Future.* Ottawa: Communications Directorate, Fisheries and Oceans.

Cell, G.T. 1982. *Newfoundland Discovered: English Attempts at Colonisation, 1610– 1630.* London: Hakluyt Society.

Colbourne, E. 1995. 'Environmental Conditions in Atlantic Canada, Summer 1995, With comparisons to the 1961–1990 Average.' *DFO Atlantic Fisheries Research Document* 95/98.

Department of Fisheries and Oceans (DFO). 1995. 'Report on the Status of Canadian Managed Groundfish Stocks of the Newfoundland Region.' *DFO Atlantic Fisheries Stock Status Report* 95/4E.

deYoung, B., and G.A. Rose. 1993. 'On Recruitment and Distribution of Atlantic Cod (*Gadus morhua*) off Newfoundland.' *Canadian Journal of Fisheries, and Aquatic Sciences.* 50: 2729–41.

Doubleday, W.G., D.B. Atkinson, and J. Baird. 1997. 'Comment: Scientific Inquiry and Fish Stock Assessment in the Canadian Department of Fisheries and Oceans.' *Canadian Journal of Fisheries and Aquatic Sciences* 54: 1422–26.

Finlayson, A.C. 1994. *Fishing for Truth: A Sociological Analysis of Northern Cod Stock Assessment from 1977 to 1990.* St John's: ISER, Memorial University.

Fisheries Resource Conservation Council (FRCC). 1995. *Conservation: Come Aboard. 1996 Conservation Requirements for Atlantic Groundfish. Report to the Minister of Fisheries and Oceans.* Ottawa.

Fleming, A.M. 1952. 'A Study of the Age and Growth of the Cod (*Gadus callarias L.*) in the Newfoundland Area.' MSc thesis, University of Toronto.

Gomes, M.C., R.L. Haedrich, and M.G. Villagarcia. 1995. 'Spatial and Temporal

Changes in the Groundfish Assemblages on the North-East Newfoundland/Labrador Shelf, North-West Atlantic, 1978–1991.' *Fisheries Oceanography.* 4: 85–101.

Haedrich, R.L. 1995. 'Structure over Time of an Exploited Deep-Water Fish Assemblage.' In A.G. Hopper, ed., *Deep-water Fisheries of the North Atlantic Oceanic Slope*, 27–50. Dordrecht, Netherlands: Kluwer Academic.

Hilborn, R., and C.J. Walters. 1992. *Quantitative Fisheries Stock Assessment.* New York: Chapman and Hall.

Hutchings, J.A. 1983. 'Adaptive Life Histories Effected by Age-Specific Survival and Growth Rate.' *Ecology* 74: 673–84.

– 1996. 'Spatial and Temporal Variation in the Density of Northern Cod and a Review of Hypotheses for the Stock's Collapse.' *Canadian Journal of Fisheries and Aquatic Sciences* 53: 943–62.

Hutchings, J.A., and R.A. Myers. 1994. 'What Can Be Learned from the Collapse of a Renewable Resource? Atlantic Cod, *Gadus morhua*, of Newfoundland and Labrador.' *Canadian Journal of Fisheries and Aquatic Sciences* 51: 2126–46.

– 'The Biological Collapse of Atlantic Cod off Newfoundland and Labrador: An Exploration of Historical Changes in Exploitation, Harvesting Technology, and Management.' In R. Arnason and L. Felt, eds., *The North Atlantic Fisheries: Successes, Failures, and Challenges*, 37–93. (Charlottetown: Institute of Island Studies.

Hutchings, J.A., Myers, R.A., and G.R. Lilly. 1993. 'Geographic Variation in the Spawning of Atlantic Cod, *Gadus morhua*, in the Northwest Atlantic.' *Canadian Journal of Fisheries and Aquatic Sciences* 50 (1993): 2457–67.

Hutchings, J.A., C. Walters, and R.L. Haedrich. 1997. 'Is Scientific Inquiry Incompatible with Government Information Control?' *Canadian Journal of Fisheries and Aquatic Sciences* 54: 1198–1210.

Lear, W.H. 1984. 'Discrimination of the Stock Complex of Atlantic Cod (*Gadus morhua*) off Southern Labrador and Eastern Newfoundland, as Inferred from Tagging Studies.' *Journal of Northwest Atlantic Fisheries Science* 5: 143–59.

Lear W.H., and L.S. Parsons. 1993. 'History and Management of the Fishery for Northern Cod in NAFO Divisions 2J, 3K and 3L.' In L.S. Parsons and W.H. Lear, eds., *Perspectives on Canadian Marine Fisheries Management Canadian Bulletin of Fisheries and Aquatic Sciences* 226, 55–89.

Lilly, G.R. 1995. 'Did the Feeding Level of the Cod off Southern Labrador and Eastern Newfoundland Decline in the 1990s?' *DFO Atlantic Fisheries Research Document* 95/74.

McGrath, P.T. 1911. *Newfoundland in 1911.* London: Whitehead, Morris, and Co.

Mann, K.H., and K.F. Drinkwater. 1994. 'Environmental Influences on Fish and Shellfish Production in the Northwest Atlantic.' *Environmental Research* 2: 16–32.

Murphy, E.F., D.E. Stansbury, P.A. Shelton, J. Brattey, and G.R. Lilly. 1997. 'A Stock

Status Update for NAFO Divisions 2J + 3KL Cod.' *NAFO Science Council Research Doc.* 97/59.

Myers, R.A., N.J. Barrowman, and J.A. Hutchings. 1997a. 'Inshore Exploitation of Newfoundland Atlantic Cod (*Gadus morhua*) since 1948 as Estimated from Mark-Recapture Data.' *Canadian Journal of Fisheries and Aquatic Sciences* 54: 224–35.

– 1997b. 'Why Do Fish Stocks Collapse? The Example of Cod in Atlantic Canada.' 7: 91–106.

Myers, R.A., and N.G. Cadigan. 1995. 'Was an Increase in Natural Mortality Responsible for the Collapse of Northern Cod?' *Canadian Journal of Fisheries and Aquatic Sciences* 52: 1274–85.

Myers, R.A., K.F. Drinkwater, N.J. Barrowman, and J.W. Baird. 1993. 'Salinity and the Recruitment of Atlantic Cod (*Gadus morhua*) in the Newfoundland Region.' *Canadian Journal of Fisheries and Aquatic Sciences* 50: 1599–1609.

Myers, R.A., J.A. Hutchings, and N.J. Barrowman, N.J. 1996. 'Hypotheses for the Decline of Cod in the North Atlantic.' *Marine Ecology Progress Series* 138: 293–308.

Myers, R.A., N.J. Barrowman, J.M. Hoenig, and Z. Qu. 1996. 'The Collapse of Cod in Eastern Canada: The Evidence From Tagging Data.' *International Council for the Exploration of the Sea* (ICES) *Journal of Marine Science* 53: 629–40.

Neis, B., L. Felt, D.C. Schneider, R.L. Haedrich, J.A. Hutchings, and J. Fischer. 1996. 'Northern Cod Stock Assessment: What Can Be Learned from Interviewing Resource Users?' *DFO Atlantic Fisheries Research Document* 96/45.

Ouellet, P. 1997. 'Characteristics and Vertical Distribution of Atlantic Cod (*Gadus morhua*) Eggs in the Northern Gulf of St. Lawrence, and the Possible Effect of Cold Water Temperature on Recruitment.' *Canadian Journal of Fisheries and Aquatic Sciences* 54: 211–23.

Palmer, C.T., and P.R. Sinclair. 1996. 'Perceptions of a Fishery in Crisis: Dragger Skippers on the Gulf of St. Lawrence Cod Moratorium.' *Society and Natural Resources* 9: 267–79.

Pauly, D. 1995. 'Anecdotes and the Shifting Baseline Syndrome of Fisheries.' *Trends Ecol. Evol.* 10: 430.

Rose, G.A., B.A. Atkinson, J. Baird, C.A. Bishop, and D.W. Kulka. 1994. 'Changes in Distribution of Atlantic Cod and Thermal Variations in Newfoundland Waters, 1980–1992.' *ICES Marine Sciences Symposium* 198: 542–52.

Ruzzante, D.E., C.T. Taggart, D. Cook, and S. Goddard. 1996. 'Genetic Differentiation between Inshore and Offshore Atlantic Cod (*Gadus morhua* L.) off Newfoundland: Microsatellite DNA Variation and Antifreeze Level.' *Canadian Journal of Fisheries and Aquatic Sciences* 53: 634–45.

Shelton, P.A. 1995. 'Analysis of Replacement in Eight Northwest Atlantic Cod Stocks.' *DFO Atlantic Fisheries Research Document* 95/75.

Sinclair, A., G. Chouinard, D. Swain, G. Nielsen, M. Hanson, L. Currie, T. Hurlbut, and

R. Hébert. 1995. 'Assessment of the Southern Gulf of St. Lawrence Cod Stock, March 1995.' *DFO Atlantic Fisheries Research Document* 95/39.

Sinclair, A.F., and S.A. Murawski. In press. 'Why Have Groundfish Stocks Declined in the Northwest Atlantic?' In J. Bormenn, B. Nakashima, H. Powles, J. Wilson, and R. Kendall, eds., *Perspectives on a Fishery Collapse* Bethesda, Md.: American Fisheries Society.

Swain, D.P. and E.J. Wade. 1993. 'Density-Dependent Geographic Distribution of Atlantic Cod (*Gadus morhua*) in the Southern Gulf of St. Lawrence.' *Canadian Journal of Fisheries and Aquatic Sciences* 50: 725–33.

Taggart, C.T., J. Anderson, C. Bishop, E. Colbourne, J. Hutchings, G. Lilly, J. Morgan, E. Murphy, R. Myers, G. Rose, and P. Shelton. 1994. 'Overview of Cod Stocks, Biology, and Environment in the Northwest Atlantic Region of Newfoundland with Emphasis on Northern Cod.' *ICES Marine Sciences Symposium* 198: 147–57.

Templeman, W. 1966. *Marine Resources of Newfoundland. Bulletin of the Fisheries Research Board of Canada.* 154. Ottawa.

Thompson, H. 1943. *A Biological and Economic Study of Cod (Gadus callarias, L.)* Department of Natural Resources, Research Bulletin no. 14. St John's: Government of Newfoundland and Labrador.

Trippel, E.A., M.J. Morgan, A. Fréchet, C. Rollet, A. Sinclair, C. Annand, D. Beanlands, and L. Brown. 1997. 'Changes in Age and Length at Sexual Maturity of Northwest Atlantic Cod, Haddock and Pollock Stocks, 1972–1995.' *Canadian Technical Report of Fisheries and Aquatic Sciences* no. 2157.

Walters, C., and J.-J. Maguire. 1996. 'Lessons for Stock Assessment from the Northern Cod Collapse.' *Reviews in Fish Biology and Fisheries* 6: 125–37.

Walters, C., and P.H. Pearse. 1996. 'Stock Information Requirements for Quota Management Systems in Commercial Fisheries.' *Reviews in Fish Biology and Fisheries* 6: 21–42.

14

Tying It Together along the BC Coast

PATRICIA GALLAUGHER AND
KELLY M. VODDEN

Everywhere along the coast of British Columbia there is unprecedented concern about the future of fishery-dependent communities. It is not uncommon to hear comments such as 'Our future, our communities are dying – my town is dying,' which is reminiscent of the strong feelings of east coast peoples about the collapsed east coast northern cod fishery. In 1995–6 this sense of crisis led to a series of public forums in Vancouver and fishery-dependent towns and villages along the BC coast. The meetings brought together people who were concerned about the west coast fishery to share knowledge about fish stocks and habitat conditions, to exchange information about changes in the fishing industry and fisheries management, and, it was hoped, to develop ways to sustain the fisheries and communities that depend on them to support their traditional way of life.[1]

Background

Salmon, historically the most abundant and valuable fisheries resource on the Pacific coast, was and continues to be the backbone of the BC commercial fishing industry. These anadromous fish begin their lives in freshwater rivers or streams, migrate to the sea where they mature, and return several years later to spawn in their ancestral waters. The six species of Pacific salmon (*Oncoryhnchus*), known by the common names sockeye (*O. nerka*), coho (*O. kisutch*), chinook (*O. tshawytscha*), pink (*O.gorbuscha*), chum (*O. keta*), and steelhead (recently reclassified as *O. mykiss*), differ in a number of ways, including size, length of time spent in fresh and seawater habitats, and spawning conditions, locations, and seasons. There are major commercial fisheries in Pacific herring (*Clupea pallasii*) and halibut (*Hippoglossus stenolepis*).

Though 'wild' salmon and halibut remain significant fisheries, the commer-

cial herring-reduction fishery ended in the late 1960s because of stock depletion. Fishing for herring roe products (roe herring and herring spawn-on-kelp),
however, began in the 1970s and is a new, lucrative, licensed commercial operation (see Newell, this volume). Other vertebrate species fished in a minor way
in the past have of late become increasingly targeted as salmon stocks decline
and the number of licences for salmon fishing is reduced. Sablefish or black cod
(*Anaplopoma fimbria*), rockfish (*Sebastes mystinus*), sole (*Soleus* sp.), dogfish
(*Scyliorhinus stellaris*), skate (*Raja* sp.), and Pacific hake (*merluccius
productus*), for example. The same is true for many species of marine invertebrates, such as shrimp, crab, oysters, clams, and abalone, all traditionally fished
by Aboriginal people. Finally, BC salmon farming grew tremendously in the
1980s to become in the 1990s the largest agricultural sector in the province.
Between 1995 and 1997, because of perceived environmental problems that the
farms pose to wild stocks (for example, the development of hybrid species and
diseases) and the tidal foreshore (marine pollution), a BC government moratorium pending an environmental review stalled continued expansion in this
controversial industry. The review, completed in the autumn of 1997, recommended removing the moratorium and proceeding, with caution, with the development of this industry.

The statutory obligation to conserve and protect the BC fish resource rests
with the government of Canada and is met by its Department of Fisheries and
Oceans (DFO). The provincial and federal governments share responsibility in
some areas such as fish habitat. DFO exerts much of its direction of the fishery
through input control, as in licensing policies, though it has increasingly
favoured output control; and many fisheries (for example, sablefish, groundfish, and herring spawn-on-kelp) are now managed by production quotas. As
well, resource management decisions, such as design of annual fishing plans
and setting of the total allowable catch (TAC) limits and times of fisheries
openings, are DFO's responsibility. Over the years its ability to make good
decisions has been jeopardized by funding cutbacks and organizational restructuring (Pearse 1982). Cutbacks have also severely curtailed DFO's ability to
perform essential functions such as stock assessment, habitat monitoring, and
enforcement.

DFO's inability to manage the fishery properly was evident in 1992 with the
'missing fish' on the salmon-spawning grounds of the Fraser River and its tributaries. Several 'independent' investigations into the causes of the shortfall led
to a series of recommendations for corrective measures (for the DFO-sponsored
investigation, see Pearse 1992). Despite Pearse's recommendations of greater
enforcement capacity for DFO, better consultation and communication, and
more careful planning, there were even more 'missing fish' in 1994. This situa-

tion led to yet another inquiry, the Fraser River Sockeye Public Review Board (see Fraser 1995), and a number of public forums – the subject of this discussion.

Getting the Missing Fish Story Straight

There was an enormous amount of public interest in the 'case of the missing fish,' interest which grew as information heard at the public inquiries was reported. Much of the evidence was conflicting, and the public found it difficult to know what the facts were. These events also raised the fear in many that salmon may be going the same way as the Newfoundland cod. As a resident of Prince Rupert would express it at a community meeting, 'Do not kid yourself. We are headed down the same road as Newfoundland.'

In an attempt to get at the facts, Simon Fraser University (SFU), in partnership with the Steelhead Society of BC, organized a public forum, with principal funding from Environment Canada, 'Getting the Missing Fish Story Straight. The East Coast Fishery Crisis and Pacific Coast Salmon Fisheries: Facts and Suggestions.' This forum was held on 31 March 1995. The idea was to provide an objective platform for open discussion and information-sharing on fisheries-related issues.

This was a landmark meeting in the fishing industry; for the first time representatives of all the 'stakeholders' met together with the public in one room. With the aim both of following up on the findings of the Fraser River Sockeye Public Review Board and assessing the Pacific coast fishery in the context of both global overfishing and the collapse of the Newfoundland cod fishery, panels discussed stock and habitat status, harvesting, and fisheries management. Contributing to the high level of audience participation were the comprehensive briefing books that each participant received in advance. These contained background information on the issues, facts about stocks status, recommendations of the Fraser Review Board and DFO's response, findings of previous commissions and task forces, and media coverage from different perspectives.

The result was animated and knowledgeable discussion involving many different industry stakeholders. One representative of the Native Brotherhood of BC summed it up: 'I feel very strongly that getting together a group of the kind that you represent is really important to the survival and future of our resource, and I feel that this kind of event should happen on a regular basis and should be part of a process of developing management style and management programs.' Representatives of coastal communities who attended this forum invited SFU to extend this process to fishing communities coastwide. The result was a series of coastal community forums.

Coastal Communities Building the Future Forum Series

SFU organized a small pilot forum in the Vancouver Island community of Ucluelet in late June 1995, just before the opening of the commercial salmon fishing season there. Guest speakers and members of the community engaged in dialogue on the east coast fishery crisis, the proposed Pacific Fisheries Conservation Council, the concept of owner-operator licensing, and current fisheries management policy (see also Fraser 1995).

The 1995 salmon fishing season turned out to be much the same as that in 1994, with more 'missing fish' on the Fraser, no agreement with the United States on the U.S.–Canadian Pacific Salmon Treaty, allegations of poaching and estimation errors, fluctuating estimates of run size, and calls for major changes in fisheries management (see Routledge 1995). This crisis, together with the knowledge that 1996 was going to be an 'off' (low) cycle year for salmon runs on the principal river, the Fraser, led to further public and industry concern about the future of the BC fishing industry and confirmed the need for a series of coastal forums.

SFU, in partnership with the Coastal Community Network, organized forums held in eight fishing communities from fall 1995 through spring 1996. Environment Canada continued to provide the main funding, with further support coming from governmental, private-sector, and non-profit partners. The forum process was community-driven, with the project coordinators helping community members with organization and dissemination of information. In most cases, a representative of Continuing Studies in Science at SFU went out to forum locations in advance to meet with members of community committees to plan details of the program, local sponsorships, venue, and advertising. Some towns and villages had committees already in place. In others, a group of community stakeholders formed ad hoc committees for the purpose of planning and hosting the forum. The formation of local committees proved to be an invaluable aspect of the process, because several of these committees would live on after the forums concluded.

The same basic program was used in each community and was designed to encourage as much dialogue as possible. At an evening orientation session the issues of most concern to the community were identified in small group discussions, reported back to all participants, and 'prioritized.' Next came a full-day program, which included presentations by invited speakers, first-hand reports of news from other coastal communities, and small group discussions about possible solutions for issues of concern raised the previous evening. At the conclusion, all participants came together to report back on the proposed solutions and to identify the next steps for community action. Also, as part of

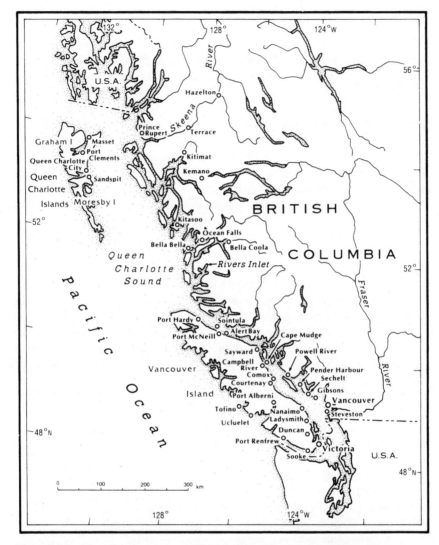

Figure 14.1   Map of the west coast of British Columbia, showing communities
involved in the Community Forums Series, 1995–6. *Source*: Simon Fraser University,
Continuing Studies

the overall process, several resource people visited one or more local schools
to meet with students and teachers, share knowledge, and discuss fisheries-
related issues, such as biodiversity, salmon enhancement, and habitat, and pos-

sibilities for forming communication links with schools in other coastal communities.

The organizers gave feedback a high priority. Participants received forum minutes within weeks of a forum's being held. An interim status report was prepared in January 1996 and circulated widely along the coast and throughout the communities. Video profiles were prepared for each community, aired coastwide on community cable television, and sent to forum committees, municipalities, and regional districts.

Issues and Solutions

During the forum process, the frustration most commonly expressed along the coast was that communities do not have a voice in decision-making in fisheries management: 'For too long, decisions that affect the well-being of our communities have been made by people who do not live here, and who have little or no personal interest in the future of the Islands' ('Consensus Document' 1996). As a result, community health suffers as the benefits obtained from adjacent marine resources continue to decline. Opportunities for fishing decrease as stocks decline, areas and times for fishing are restricted, and licence ownership leaves the community. Centralization of processing for the sake of economic efficiency and introduction of technological capabilities, such as on-board freezing and coastwide fishing, have aggravated the situation. Even the need for boats and crews to return from the fishing grounds to gas up, obtain supplies and ice, mend nets, or make repairs in fishing communities has diminished. Instead, the fleet tends to head for major processing centres.

As for the involvement of coastal people in fisheries management, three themes emerged. First, coastal people have the right to share in benefits obtained from the resources adjacent to them and to participate in decisions that affect these resources and benefits. The situation was well-described by a Masset-area fisher: 'We have an enormous amount of marine wealth around these islands, but we have no say in how the resource is harvested and no access. We're going to be allowed to wither away while this vast wealth is harvested right from our doorstep.' Second, the importance and validity of local knowledge must be taken into account as a useful tool in fisheries management. Third, coastal communities have stewardship capabilities that cannot be cost-effectively matched by government, not only in data collection but also in enhancement, habitat protection, and even enforcement.

Each of these points, participants agreed, could be taken into account under a new, decentralized system of management that would include the voice of communities. Representatives from Alaska and the Philippines described ocean ranching and municipal fisheries boards, respectively, as examples of fisheries regimes that include communities as major stakeholders in their regions. This vision for the BC fishery was referred to as co-management (see Pinkerton, this volume).

Community members realized that in order to have an effective voice they must be organized. Several multi-stakeholder community and regional/watershed committees addressing fisheries issues were already in place before the forum process began. These were offered as models for other communities to consider. Prominent among these was the Skeena Watershed Committee; formed in 1992, it was composed of five equal partners – north coast commercial fishing interests, First Nations, the 'recreational' (sports) fishing sector, and the federal and provincial governments. Owing to the inability to reach consensus over harvesting issues, this committee disbanded just before the 1997 Salmon-fishing season began (see Pinkerton, this volume). Other such groups included the Westcoast Sustainability Association of Ucluelet, the Campbell River and Community Fisheries Committee, and the Coastal Community Network. The last-named, formed in 1992, is an organization of representatives from regional districts along the coast that encourages the sustainability of coastal communities by spreading information, advising on fisheries policy, and organizing annual conferences.

Several new committees were launched at this time, in part the result of the forum process: the Coastal Communities Conservation Society, in Steveston; the Sunshine Coast Community Working Group, in Pender Harbour; and a fisheries subcommittee of the Masset Community Adjustment Committee, in the Queen Charlotte Islands (Haida Gwaii). (Later, the mayor of Prince Rupert appointed a local advisory committee on fisheries and the Fisheries Association of Alert Bay held its first meeting in June 1996.) These were promising beginnings for the organizational infrastructure needed to facilitate community involvement in fisheries management coastwide.

Community members targeted harvest allocation as one area for the decentralizing of management responsibility. The issue of allocation among competing sectors has become increasingly complex and hotly contested. Demands for allocation come repeatedly from the commercial industry, itself comprised of a diverse range of gear types; from First Nations – as modern treaty negotiations (comprehensive claims) between First Nations and governments proceed, larger Aboriginal resource claims, with aquatic resources often heading the list, are being taken into account (see Newell, this volume) – and from a rapidly-expanding sport-fishing industry.

Forum participants claimed that negotiations over allocation now occur behind closed doors and that the winners are those with the most money to spend or the most political influence to exercise in lobbying Ottawa. Through a community-based management system, they argued, people at a local level, with a stake in the overall well-being of their communities, could work out distribution. It is at this level, as one Nanaimo community adviser pointed out, that individual members of the various interest groups routinely interact: 'People living in a community understand one another by a different set of rules, a local set of rules.'

With regard to the commercial fisheries, there was surprisingly little discussion in these forums about allocation among types of gear in the commercial salmon fishery (i.e., troll, gillnet, and seine), perhaps because the issue was too controversial or because coastal community fishers instead united in speaking out against other threats to their communities and the commercial fishery as a whole. However, suggestions surfaced about use of more selective types of gear, such as in-river traps and weirs and brailing techniques, and some fishers expressed concern about the catch efficiency of large seine boats and the product quality of gillnet- and seine-caught fish. Participants cared about the state of the stocks.

For BC First Nations, their claims to allocation of fish rest on their Aboriginal rights. For centuries Aboriginal peoples along the coast and the major river systems structured their communities and lives around the fisheries resources in their traditional territories, where they developed rites and customs to govern access to fish and conducted communal and family-based enterprises. They were the great traders of Aboriginal Canada, and fish constituted a prime item of trade among villages and regions and later with early Europeans in British Columbia. Indian men and women entered the commercial fishery as fishers and plantworkers for the expanding cannery industry in the late nineteenth century, while also defending and exercising their traditional fisheries.

BC Aboriginal peoples' rights to harvest and use the resource have been regulated by the federal Fisheries Act, which was first applied to the province in 1878. By the 1920s this right had become narrowly defined as a licensed 'food fishery,' or 'Indian fishery,' for personal consumption only. Recently, this position was overturned in *Regina v. Sparrow* (1990). In this case, the Supreme Court of Canada ruled that there was an unextinguished Aboriginal right to fish for food, broadly defined to include societal and ceremonial purposes, and established that First Nations would receive priority for fish after conservation needs had been met.

In 1992 DFO established the Aboriginal Fisheries Strategy (AFS), which led to riverwide agreements on the Skeena and Fraser with respect to cooperative fisheries management and to the controversial pilot salmon-sales project allowing for commercial sale of 'food fish' by Indians bands of the lower Fraser,

Somass (Alberni), and Skeena rivers. The Strategy has met heated opposition from some non-Native fishing interests. It was clear from the dialogue during the coastal forums, however, that the AFS has not been universally accepted by BC First Nations, either. One Indian from the Sunshine Coast stated: 'No one asked for the AFS. We didn't see the differences between us before this. Now we size each other up when we're walking down on the dock. It's crazy!' Another from the same area agreed: 'Why should I defend the AFS? It was dumped in the laps of the Natives – to manage (the) Natives.'

In 1994 and 1995, the Aboriginal fishery accounted for only about 3 per cent of the total allowable catch of salmon in British Columbia. However, First Nations and DFO expected this allocation to increase. For example, part of the recent (February 1996), unprecedented agreement-in-principle reached in Nisga'a treaty negotiations, will entitle the Nisga'a to take, among other harvest entitlements, up to 26 per cent of the total allowable Nass River salmon run.

In the sport-fishing industry, there have been increasing expectations of success in catching salmon, particularly the valuable chinook and coho (and now, with the decline in these species, sockeye). Given more efficient equipment, it has come to the point where one second-generation owner of a sport-fishing lodge remarked, 'Sports fishing should be a total experience. Catching one or two salmon should be a big deal (it is anywhere else in the world). It's time to put the sport back into sports fishing and restrict the gear.' Historically the west coast sport fishery has been virtually unchecked and unregulated. Some observers maintain that 'commercialized sportfishing operations ... those commercial ventures that derive their revenue from selling fishing opportunities (fishing lodges, resorts and charter boats) ... represent a multi-million dollar industry that is growing at the expense of both commercial fishermen and resident anglers' (Cruickshank 1995).

Nowhere was the dispute over allocation to the commercalized sports-fishing industry more heated than in the village of Masset, where community members claimed that the commercial recreational sport-fishing lodges were a direct threat to their survival. A local commercial fisher stated, 'We've seen any chinooks gained by our conservation-based restrictions, plus a significant chunk of our agreed upon quota, ruthlessly handed over to the sports sector ... The catch of sports chinook has rocketed up 5,720 per cent.' Further, 'those two areas [lodges] generated conservatively $30 million or more likely nearer to $40 million in 1995. We [the community] received no benefits.' In Prince Rupert, a member of a small boat charter association proposed a solution for the sport-fishing industry: 'Area licensing [of the sport fishery] would reduce uncontrolled growth and would provide a means to preserve resources for the community.'

Will future stocks support such competing demands for allocations of fish?

The gradual settlement of the BC coast over the past century has led to increasing demands for the once-bountiful marine and freshwater resources and to major interference with the habitats that support them. Thus in recent years many of the natural resources along the coast, including the valuable and cherished salmon runs, have declined significantly; most are seriously threatened, some have disappeared, and others are being rebuilt. In the early part of this century there was a significant pilchard (a large sardine, *Sardinops sagax*) fishery, but pilchard disappeared from the BC coast in the 1940s. The prized abalone (*Haliotus* sp.) fishery was closed in 1990 after being fished to commercial extinction under individual-quota (IQ) licensing. But lingcod (*Ophiodon elongatus*) stocks fished to near-extinction twenty years ago are currently being rebuilt.

During the coastal forums, each community told its own stories about streams once full of salmon or about once-rich shellfish beds where viable populations no longer exist – of habitat destruction and overfishing. An account frequently heard from the older generation was, 'You could walk across the rivers on the backs of salmon.' Or as one long-time fisher from Ucluelet put it, 'People just wouldn't believe how much there was.' Another from Fort Rupert was quoted as saying, 'When the fish disappeared before, I said "don't worry, they'll be back." I don't say that now.'

Healthy communities, it was widely recognized, depend on healthy resources. 'The resource is our saviour ... If we look after the resource the resource will look after us,' declared a long-time Prince Rupert fisher and city councillor. 'The fish must come first,' said a member of the Skeena Watershed Committee. Fishers all along the coast confirmed their commitment to conservation, and a fisher from Masset noted, 'Every fishing organization on the coast has been shouting at DFO to put conservation and fish first, even if it means closing down everything.' Fishers joined other concerned residents in calling for a moratorium on fishing where stocks are in peril, along with a renewed commitment to enhancement.

Efforts at stock enhancement, it was suggested, should focus not just on hatcheries but also on protection and restoration of fish habitat and even control of (non-human) predators. Predators of concern were primarily seals and sea lions, but also mackerel. Large-scale hatchery operations came under criticism by some for artificially increasing large runs to the detriment of the multitude of smaller runs that are key to maintaining the genetic diversity of Pacific salmon. Enhancement is most effective, local experts claimed, when used to boost smaller stocks. In many such cases, once a stream has been restored, the hatchery has the capacity to move on to yet another stream. Education and community involvement were cited as valuable benefits of salmon enhancement. One hatchery manager proudly described the impact of enhancement work on school children: 'Two weeks ago we had forty-two five-year-olds up on site. They were responsible for the lives of 500–800 pink salmon fry. You could see a lot of pride when they came back.'

Adequate, accurate, and unbiased data, however, are required for proper assessment of stock status (see Hutchings, this volume). After years of cutbacks in DFO's field staff, there is little confidence that the information currently available is sufficient. 'One of the most harmful decisions taken by DFO's senior management was to dismantle the strong, effective Fishery Officer field organization that had evolved in the Pacific Region over the years since the late 1920s,' observed a former fisheries officer and DFO director general. Data collection, it was suggested, must be supplemented with information gathered from fishers and other citizens. 'After all, if somebody from one of our community groups wades a river that hasn't been waded by a fisheries officer for ten years they are going to know a lot more about it than the fisheries officer that wasn't there, or a scientist in a laboratory,' stated Canada's ambassador for the environment at the Vancouver wrap-up forum.

Uncertainty exists regarding the overall health of BC salmon stocks and, for that matter, all stocks. The development of 'new fisheries' such as red and green sea urchin (*Strongylocentrotus*), sea cucumber (*Stichopus californicus*), horse clams (*Schizothaerus*), squid, and zooplankton has caused particular concern about potential effects on the ocean ecosystem that might result from lack of information on the status of and interconnections among these stocks.

A number of First Nations reported on their own stock assessments for marine invertebrates. A fisheries biologist from Port Alberni commented, 'The Tribal Council takes a different approach to management ... an ecosystem perspective, not stock by stock.' A member of the Sechelt Indian Band supported this view: 'The Band has taken their own initiative by getting hard data of what's left in terms of the stocks, and the biodiversity of the stocks, and the food chains that are within our territory. People want to go after the krill, but that's part of the food chain. People want to go after the Venus clams, the shellfish, but they're part of the food chain ... Before we take one herring, or one scoop full of *Euphausiis*, we have to know what's in the ecosystem.'

Participants suggested that the public, which has demonstrated that it cares about the state of the BC fishery, receives mixed messages about the health of fish stocks. The resulting confusion has furthered the commonly expressed feelings of public mistrust in the DFO. A fisheries biologist cited numerous examples of nearly extinct salmon populations on the west coast of Vancouver Island, and a biostatistician described the Rivers Inlet sockeye population, which for over twenty years saw annual returns of over one million sockeye, 'as a minor remnant' of what it once was. At the same time, with fisheries managers concentrating on the two main rivers, some larger runs of the Fraser and the Skeena have returned in record numbers in recent years. Still others, however, have been greatly reduced. The question 'Are BC salmon stocks healthy?' evokes complex answers.

To monitor data quality and adequacy, and to make this information available to decision-makers and the people of Canada, forum participants called for an independent fact-finding council – a public watchdog – as recommended by the Fraser River Sockeye Public Review Board (Fraser 1995). 'A Pacific Fisheries Conservation Council would provide an annual report to Ministers and the public on the status of stocks – what is known and what is not known ... There could be no escape from the glare of public exposure of serious deficiencies in conservation and protection programs,' explained the former DFO director general. In 1998 both Ottawa and Victoria committed themselves to the formation of this independent fact-finding council.

The most commonly cited reasons for declines of BC fish stock were habitat loss, degradation, and overfishing. As chief of the Nanaimo Indian Band reminded the audience so powerfully, 'Our people, all the people here, all members of the coastal communities, we must accept shared ownership of this problem which we have created ... We haven't really accepted the shame, the disgrace, for the harm that each and every one of us has caused ... We have taken it for granted that these resources would never die out, would never go away.' Among the threats listed by forum participants were logging, urban development, construction of hydro-electric dams, predation, changing ocean conditions, and industrial and municipal pollution, including runoff, sewage outfall, mining, and pulp-and-paper effluent. 'The Campbell River is [one of] the most famous salmon rivers in the world ... You have a mine operating in the upper watershed ... In the last forty years there have been three hydro projects. There is heavy industry in the estuary and a sewage outfall proposed,' complained a Campbell River environmentalist.

One of the most severe problems has arisen from the rapid expansion of the logging industry along the coast. The development of logging roads in watersheds, and the vast tracts of land clearcut earlier in this century, have led to subsequent slides, blockages of small streams and creeks, siltation, and pollution. For many waters this situation has severely impeded both the ability of salmon to return to spawn and the quality of their spawning beds, and so salmon runs in creeks and streams have been lost or are near extinction all along the coast. 'There were a hundred rivers and streams in Barkley Sound that used to have coho. Now there are none,' remarked an elder of the Ucluelet fishing community. A representative of the Native Brotherhood of BC told a similarly disturbing story about another place: 'The logging company moved in there and decided to move their logs and the cheapest way ... was to have the river dragged. And they completely destroyed the habitat.'

A steadily growing population and consequent development and urbanization, particularly on the south coast, in the Fraser River estuary, and on the east coast of Vancouver Island, have also helped decimate salmon populations as small creeks and streams have become polluted, diverted, and filled. For one concerned citizen at the Steveston forum, 'there are just too many people in these extremely sensitive areas.'

Concerns also found expression about the building of hydro-electric dams as perhaps the most severe threat to fish habitat in the future. The example of the destruction of the once-largest chinook-producing system in the world because of dams built on the Columbia River was cited, and fears were expressed about similar projects being on line for British Columbia.

Finally, participants spoke about the interception of Canadian fish by U.S. fishers, and some called for action by the Canadian government. The anadromous nature and migratory patterns of Pacific salmon mean that during their spawning runs Canadian fish often pass through American waters and thus are subject to U.S. interception (and vice versa). This circumstance has historically been a source of friction and management problems, particularly with recent increases in the American catch of fish of Fraser River salmon and in Alaskan interception of weakened BC chinook stocks. Since 1994, the Pacific Salmon Treaty, a management scheme designed in 1985 to address the conflicting interests of the two countries (see Pinkerton, this volume), has broken down because of lack of agreement. The treaty specified that each country is entitled to salmon harvests in relation to the production of salmon originating in its rivers. 'The overall figures show that in the last number of years, American interceptions of our fish, relative to our interceptions of their fish, have increased by 50 per cent, and ours have decreased about 20 to 25 per cent,' said John Fraser, the special envoy appointed to resolve the dispute.

BC salmon harvests have thus evolved into extremely complex fisheries that are exceedingly difficult to manage. The entire system of fisheries management for British Columbia was called into question by almost all communities visited. Forum participants asserted that coastal residents have lost confidence in DFO's ability to manage the fish resource or to fulfill its conservation mandate. Further, they reported that people no longer trust DFO, see it as lacking vision, and generally view it as not listening. 'We in the communities, the people who live there, know and understand the industry very well. Listen to the people,' said one participant from Prince Rupert – a typical remark. A guest speaker from Newfoundland remarked at the wrap-up forum in Vancouver, 'Communi-

ties [in Newfoundland] are calling for action ... But they are not even being listened to [by Ottawa].' The perceived weakness of DFO is one of the principal
reasons why local fisheries groups have been formed along the BC coast; communities dependent on effective fisheries management are attempting to fill the
void: 'We've been written off. DFO left us three years ago. They closed the
office and departed ... DFO has abandoned the coast. Well, we claim salvage
rights. We are taking back the management of our fishery,' stated a community
leader from Alert Bay.

While DFO field staff in the regions were generally viewed as vital components of an effective management and conservation system, both their input and
their numbers are shrinking. 'When I was young there used to be sixteen guardians in the fall looking after Barkley Sound. Then they took them off one by
one. Probably thirty years ago they were down to one out of Port Alberni with a
boat and the rest was left wide open,' reported an old-timer from Ucluelet. 'In
fact, we don't have the time and resources to even count the salmon in a lot of
these streams anymore,' said a fisheries biologist from the same region. Participants described a situation where politically motivated management decisions
are made either in Ottawa by politicians and high-level bureaucrats or in BC
urban centres. These decisions usually contradict the advice of personnel who
are closest to coastal communities and the stocks themselves. 'I'm kind of a
believer in the DFO,' said a plant manager from the north; 'unfortunately [the
department is] politically driven and the people in the field that have the knowledge have no say.' There were numerous remarks about expansion in the DFO
administration despite cutbacks on the ground. 'In a year when patrol boats are
tied up at the docks because they haven't got enough money for fuel they hire
four new Regional Directors. There's just no bloody logic in the department at
all,' said a village councillor from Ucluelet who chaired the Coastal Community Network.

Despite many problems, DFO, it was generally felt, should remain part of a
new co-management scheme, and it should continue to meet and maintain its
conservation mandate. The department's community-involvement initiative,
however limited, was popular, and some speakers proposed that its mandate be
expanded beyond salmon enhancement. As one DFO employee suggested at a
forum, 'There is no bureaucracy between the community adviser and headquar-
ters in Vancouver. I hope that the department will consider the community-
involvement program as a model for the management of fisheries.'

Nowhere was dissatisfaction with DFO more apparent than in discussions
about licensing policy. Participants feared corporate concentration of the fishing
fleet and made references to a similar process that had already taken place in the
BC logging industry, resulting in small, local wood processors' being left short

of wood and a handful of large companies gaining control of the industry. While companies have sold off much of their small boat fleet, many owner-operators are financed by the fishing companies. Financing arrangements stipulate that fish caught are delivered to the firm on its terms. Many felt that the BC fishery was heading in the same direction as forestry and that concentration of fleet ownership would complement the centralization of management and processing that had already occurred, leaving coastal communities with nothing. 'I'm seeing my community on the brink of extinction, to be pretty blunt about it,' said a Ucluelet troller.

The pattern of shorter fishing seasons and restricted areas for fishing that started after 1917 escalated after 1945, especially after the introduction of the Davis Plan (licence-limitation) in the late 1960s, as fleet capacity and value increased. As one fisher from the 'Sunshine Coast' described it, 'We used to fish five days a week, from April to October, and now we fish only one to two days a week, if that, over a much shorter season.' 'Twenty-five years ago most of the fishing was done in camps and rural areas ... Now it is more centralized,' explained a fisher from Nanaimo. The result has been limited access to the resource, with no distinction being made for local users. One crab fisher from Masset described a five-fold increase in catching capacity in his area over the past five years, because of the entry of outside users. It used to be that local people fished in their own 'backyard.' Today local crab fishers are calling for licence-limitation and a reduction in catch capacity as fishers had done with the halibut fishery of the past and are continuing to do with salmon.

Transferable licence-limitation schemes have historically failed, however, to protect the small-boat fisher and coastal communities where they live and to reduce catch capacity in the long term. Licence-limitation has instead contributed to overcapitalization, because owners try to get more from their investment, and to leasing and speculation, because fishing licences become worth trading. In fact, licence-limitation has in many cases either forced locals out of fishing or dramatically reduced their fishing incomes. As a Haida chief from Masset informed us, 'We had the largest fleet of boats here at one time and around 1969 we began to get legislated out.' Many participants saw solutions in a licensing scheme that would protect the owner-operator and small-boat fisher. 'The first thing that's going to have to be addressed to put people back to work is the area of licensing. If they don't fish it, they're not going to own it. It's got to be owner-operator. And do away with this leasing, where people who don't even have any stake in, or ties to, the fishing industry are getting wealthier every year,' declared a shoreworker and union leader from Prince Rupert.

We must not separate what has happened to fishing and fishing communities from trends in the processing industry. Many factors have changed the BC industry since the first canneries were built along the coast, beginning at the mouth of the Fraser River in the early 1870s. Several hundred cannery sites were developed over the years, with construction peaking in the early twentieth century. In the 1940s and 1950s, twenty canneries still operated on the BC coast, supporting thriving communities during summer and thousands of shore-workers and fishers of Aboriginal, Japanese, Chinese, and European descent.

But after the introduction of the Davis Plan in the 1960s the industry entered a new period of consolidation. Consolidation, together with unionization and increased mechanization of the canning process, which began earlier, improved refrigeration technology, and the development of ground transport facilities, brought an end to remote and isolated plant operations, and those that survived became year-round, diversified operations. Today, processing of salmon and other fish products takes place primarily in Prince Rupert and Vancouver, with much smaller fish-packing operations going on in communities such as Campbell River, Masset, Port Hardy, Sointula, and Ucluelet.

Forum participants expressed concern about the job loss that this pattern of restructuring has caused and continues to cause in remote communities. A plant owner from the central coast remarked, 'I had 120 to 130 people working for me two years ago. I have one now ... Each pay cheque I signed every two weeks supported an entire family ... That was called prosperity. Prosperity that we haven't seen in that area since the canneries pulled out eighty years ago. That's gone now.'

In recent years, both greater world supply of salmon (because of the growth of salmon farming and the increase in Alaskan runs of wild salmon) and depleted BC stocks have hampered the profitability of the surviving BC processing operations, causing even further business concentration. One industry senior executive stated: 'The dramatic change really is that we're in a world market for our product. At one time Canada was a dominant player ... We're down to a point where we're probably three to five percent of the world's supply of salmon. We're followers on price. We're followers on everything. What we really have to do is restructure ourselves so we can compete in these world markets.' Alaskan plants cannot handle the volume produced, so they sell unprocessed salmon to BC processors. According to one processor in Prince Rupert, 'What's happening now is we're dependent more on Alaska to supply the raw product to this plant than BC.' But for another, from further south in Campbell River, 'There is so little product around that there won't be enough production coming across the factory floor to justify that plant being there.' In fact, in 1996 these factors resulted in a recently announced merger of the operations of the province's two largest processing companies.

One response to these changes in the salmon market, other than plant closures and business mergers, has been for processors to turn to other species. A Ucluelet plant worker observed: 'Over the last few years they've gotten out of salmon – a few years ago they decided to target hake.' Another solution, many say, is to concentrate on value-added processing, creating more jobs and revenue from each fish by capitalizing on specialty export markets such as Japan. Participants in the Prince Rupert forum suggested that products should also be handled, packaged, and marketed more effectively. BC fish producers, they suggested, should pay more attention to developing domestic markets and creating new, more sophisticated products.

Coastal Communities Taking Action, Vancouver, April 1996

To plan the final wrap-up forum in the series, we struck an ad hoc committee made up of representatives of each of the coastal communities visited, as well as special advisers. Its two-day workshop in late March, funded by the provincial government, moved forward with the issues and solutions identified in the community forums and planned the program for the final wrap-up forum in Vancouver. Attendance at the latter was broad-based and included over fifty members of coastal communities. Parts of the day's proceedings were aired nationally, an edited hour-long video report was produced and aired in British Columbia, and the forum proceedings were published (Gallaugher, Knight, and Vodden 1991) and circulated throughout British Columbia and across Canada.

Representatives wanted not only to point out the problems of the status quo but also to present a 'proactive' vision for the involvement of communities in fisheries management: 'Coastal communities envision co-managing resources in true partnerships with the federal, provincial, regional and municipal levels of government. Partnerships would also include resource companies and organizations who harvest, replenish and process community resources, as well as outside users. This co-management relationship would be governed by the principles of sustainability and the values of community and fairness in all aspects of its management mandate.'

The organizing committee described local co-management decision-making authorities as the vehicle through which coastal communities could realize this vision. These authorities could be built from local organizations already in place. One member explained the activities of a fisheries group in his community: 'In Campbell River what we've tried to accomplish is to get all sectors together in one room, at one table, to talk about common ground ... We decided as a group to get into our own fund-raising, to get into our own enhancement projects. In other words accepting personal responsibility for our own area, our own streams, and our own fish habitat.' Another, from Alert Bay, said: 'We are beginning the co-management process ... We've begun our co-management process by bringing local experts, they're called fishers, together ... We must do it together.'

Acknowledged was the need to define the roles of various members in these local bodies and to link them together through a coastwide organization comprised of members from each. These bodies, it was thought, should have a structure suitable to participants. The forum heard comments such as 'We'd like to see fishing being owner-operated, as a small business, that would support our community' and 'The one thing I believe in is that if the resource is on the doorstep of a community, that community has to get some benefit from that resource.' Participants wanted a common set of guiding principles, such as involvement of owner-operators and of adjacency. They recognized that such a structure cannot rely on government funding and suggested cost recovery measures such as taxes, redirected licence fees, and government and company partnership agreements facilitated by changes to the federal Fisheries Act.

An issue identified throughout the forum process as needing the ongoing involvement of higher governments was international negotiations. Concerns about U.S. interceptions and a breakdown in the Canadian–U.S. salmon treaty had heightened domestic allocation disputes in the province. Canada's ambassador for the environment agreed with participants: 'This is now a matter of very grave importance with respect to relations between us and the United States.'

The Vancouver forum and the publicity surrounding it signalled to all levels of government that BC coastal communities were joining together to work out solutions to their common concerns and to strengthen their political influence. They demanded to be heard. The frustration of not being listened to, expressed throughout the community forum process, was particularly evident. Only the day before the forum, the minister of fisheries and oceans, Fred Mifflin, had stated publicly on CBC radio that in today's age of modern communication technology, he did not need to go to British Columbia to discuss the vast changes recommended for the BC salmon fishery. Many saw this attitude as highly inappropriate for either a federal department or its political head.

The recently announced, highly controversial Salmon Revitalization Strategy, otherwise known as the 'Mifflin Plan' (29 March 1996), after the fisheries minister at the time, became a focus of the day's debate. The plan aimed to reduce the number of licences in the salmon fleet by 50 per cent through a voluntary buy-back program and area licensing system. Further, it provided for the purchase of additional licences and the 'stacking' of these licences on a single boat to allow vessels to fish more than one area.

Many coastal representatives feared that the plan – especially stacking – would devastate their small communities and so spoke out against it, hoping to inform the media and the public, politicians, and the civil servants present. As one Cape Mudge fisher described it, 'If you reduce the boats by 50 per cent, and you don't reduce the licences by 50 per cent, what have you gained? We know that 20 per cent of the fishing fleet takes 70 per cent of the fish.' Another argument was that small-boat operators unable to purchase second or third licences for stacking purposes would be pushed out of the industry. Their licences would be purchased by large operators, who would thereby increase their share of industry ownership, or by fishers from urban centres with sufficient assets to allow them to finance the investment. Small coastal communities, many of them comprised predominantly of First Nation peoples, would be hit the hardest, it was felt, and forced out of business.

Though not many forum participants favoured the Mifflin Plan, and criticized stacking, some industry proponents and DFO officials were supportive. Industry representatives argued that stacking was voluntary, that fishers would still be able to make a living with a single licence, and that area licensing would bring communities and fisheries management decisions closer together. DFO claimed that a reduction in the number of licence-holders would create more efficient and cost-effective management. Nevertheless, participants adopted a resolution calling for the prime minister and the minister, Fred Mifflin, to postpone implementation of the plan until its impact on coastal communities and salmon conservation could be assessed. It was clear from the interchanges that the consultative process that led to the design of the Mifflin Plan had been ineffective and the goals of the plan poorly communicated. 'We've been manipulated, misinformed, and mishandled, to say the least,' said one fisher from Steveston. 'I think the message I've heard very clearly is that there wasn't an effective communication of the coastal communities' attitudes, in respect to the plan,' responded the minister's representative.

The need for real community involvement in fisheries management was strongly expressed in the Vancouver forum. A local MP and representative of the minister made encouraging remarks: 'So when I hear the word co-management, it's the right buzz word.' The provincial minister responsible for fisheries commented, 'What's missing on the West Coast is a holistic or integrated approach to fisheries renewal.' Coastal representatives described their vision of a BC fishery that would sustain both the resource and their communities. 'We have a vision of streams that are now decimated being rebuilt, of stocks being rebuilt, of a healthy community that has an economic connection to those stocks, and that's what we are working towards,' declared a fisher from Ucluelet who is a director of the Westcoast Sustainability Association.

The vision described in the previous planning workshop was expanded in the Vancouver forum discussions with examples of co-management that are in practice in British Columbia and being discussed in Newfoundland. Participants also learned about recent initiatives taken by United Nations fisheries policy advisers to 'take into account traditional knowledge and interests of local communities ... in development and management programs.' An industry executive and one of the leaders of the Skeena Watershed Committee summed it up, 'You've got to have a vision of where you want to be ten years from now. Or, as the Native people we work with say, a vision for seven generations, which may be a better perspective.' The forum concluded with a remark by Canada's ambassador for the environment, himself a British Columbian: 'Ultimately it is up to us ... We have the capacity to make sure that the total fishing community, all of us, can look with some satisfaction and some confidence to the future ... Because this is our heritage, and we have an obligation to ensure that our children, and our children's children, inherit it.'

Epilogue

The momentum that increased among the communities during the forum series has continued. Multi-stakeholder groups within and among communities are working together to strengthen their voice and effectiveness. In Steveston, the committee formed to organize the forum has formed a society to address the issues identified as most important by community members, particularly stewardship of the Fraser River estuary. Alert Bay has taken the lead in developing a regional (northern Vancouver Island) initiative: the Inner Coast Natural Resource Centre. The centre plans a number of economic-development, education, and research projects, ranging from workshops and training programs on value-added forestry and ecotourism to a conference on the links between local ecological knowledge and Western science. New research and education initiatives are under way in Prince Rupert and on northern and western Vancouver Island to build local knowledge about local natural resources and ecosystems and economic-diversification opportunities. In addition there are a number of initiatives to involve BC coastal communities in fisheries management. These include the Groundfish Development Authority (founded 1997) and the Fisheries Renewal Board of BC (1997), a provincial crown corporation, to promote the protection, conservation, and enhancement of fish stocks and restore fish habitat. In 1998 Ottawa agreed to negotiate the terms of a regional fisheries-management board for west coast Vancouver Island in a tripartite relationship with British Columbia and the Nuu-chah-nulth Tribal Council. Regional boards are also being formed for the management of clams on BC's southern coast.

Fisheries Minister Mifflin finally met with representatives of coastal communities while in Vancouver (for another engagement). Despite receiving a clear message from the coastal communities that the Pacific Salmon Revitalization Plan was 'unacceptable,' the minister proceeded with its implementation. This action strengthened the perception that fisheries management decisions were being made in Ottawa without consideration of regional knowledge or concerns.

Meanwhile, after fears being expressed throughout the forum process that the 1996 salmon season would be the worst on record, particularly on the Fraser, returns as of late August, though the lowest in years, were over twice as large as what DFO had forecasted. And on the Skeena, returns were the best ever. But this type of news for salmon on the principal rivers, as many participants at the forums had mentioned, only helps to mask the essential poor health of many other systems and the need for reform of BC fisheries as a whole.

The forum process demonstrated the ongoing role for Simon Fraser University as a neutral facilitator for the exchange of knowledge and opinions in a field full of passion and conflicting interests. The university has since sponsored a number of follow-up workshops, including, in 1997, 'BC Salmon: A Fishery in Transition' and 'Exploring Cooperative Management in Fisheries,' in partnership with the University of British Columbia, and in 1998 'Speaking for Salmon.' According to workshop participants, this continuing dialogue brings Pacific coast peoples together to work for the sake of the fish and of fishing places and fishing people.

NOTES

1 We have chosen to quote liberally from the forum proceedings and related materials in order to represent fairly the opinions of participants in a process that is still ongoing. Unless cited otherwise, the quoted material is from the proceedings of these forums (Gallaugher 1995; Gallaugher, Knight, and Vodden 1996) or from the hour-long video Fish Story (1996). For full quotations and identities of the speakers, whose names we have intentionally withheld here, we refer readers to these sources.

WORKS CITED

'Consensus Document.' 1996. Queen Charlotte Islands: Islands Community Stability Institute, 31 Jan.
Cruickshank, Don. 1995. *The Fisherman's Report. A Commission of Inquiry into Licensing and Related Policies of the Department of Fisheries and Oceans*. 2nd ed. Vancouver.

Fish Story: BC Coastal Community Report. 1996. Video report. Burnaby, BC, and Vancouver: Simon Fraser University and Rogers Community 4. Aug.

Fraser, John. 1995. *Fraser River Sockeye 1994: Problems and Discrepancies.* Vancouver: Fraser River Sockeye Public Review Board.

Gallaugher, Patricia, ed. 1995. *Forum Proceedings. Getting the Missing Fish Story Straight. The East Coast Fishery Crisis and Pacific Coast Salmon Fisheries: Facts and Suggestions.* Burnaby, BC: Simon Fraser University, March.

Gallaugher, Patricia, ed., with Carol Knight and Kelly Vodden. 1996. *Forum Proceedings. A Year of Dialogue along the British Columbia Coast. Coastal Communities Taking Action.* Burnaby, BC: Simon Fraser University, April.

Pearse, Peter H. 1982. *Turning the Tide: A New Policy for Canada's Pacific Fisheries. Final Report of the Commission on Pacific Fisheries Policy* (Pearse Commission, 1981). Ottawa: Minister of Supply and Services.

– 1992. *Managing Salmon on the Fraser: Report to the Minister of Fisheries and Oceans on the Fraser River Salmon Investigation, with Scientific and Technical Advice from Peter A. Larkin.* Vancouver: Department of Fisheries and Oceans, Nov.

Regina v. Sparrow (1990), 4 WWR 410 (Supreme Court of Canada), affirming (1987) 36 DLR (4d) 246 (BC Court of Appeal), which reversed (1986) BEWLD 599 (County Court).

Routledge, Richard. 1995. *1995 Briefing Book. Coastal Communities: Building the Future. Fisheries Sustainability, Development and Conservation.* Burnaby, BC: Simon Fraser University.

PART THREE:
COMMUNITIES OF INTEREST – WHERE NOW?

15

'That's Not Right': Resistance to Enclosure in a Newfoundland Crab Fishery

BONNIE J. McCAY

One of the hallmarks of modernization is enclosure of the commons: the privatization and individualization of resources that were either held in common by members of a community or open to members of the public. This process has taken far longer at sea than on land. It was not until the 1970s that coastal states carved up the common resources of the ocean within 200 nautical miles (322 kilometres) of land – the 200-mile limit – into exclusive fisheries or economic zones. And only in the 1980s did governments and industries seriously consider creating private rights in the fish stocks, a system that has come to be known as 'individual transferable quotas,' or ITQs. ITQs carry privatization to the level of firms and individuals, who are allotted exclusive rights to shares of a quota. Economists argue that this is a more rational system of management because it does away with dangerous and costly races for fish on fishing grounds and reduces incentives for overcapitalization in the fisheries (Christy 1973; Neher, Arnason, and Mollett 1989; Squires, Kirkley, and Tisdell 1995). Moreover, to the extent that these shares are fully transferable, ITQs enable market mechanisms to take over much of the task of deciding who gets how much fish – the sticky 'allocation' side of fisheries management.

In Canada 'enterprise allocations' appeared in some of the offshore fisheries in the early 1980s. In this variant of private rights–based management, the handful of corporations engaged in the fisheries were assigned portions of the annual quota so that they could use their vessels and processing plants more economically than when they were competing for a limited quota (Crowley and Palsson 1992; Fraser 1986). Other Canadian rights-based systems were primarily 'boat quotas,' in which each boat received a portion of the quota. In most cases of boat quotas, the fishing rights could not be transferred apart from the boats; markets in them were limited, and the rights associated with one boat were not divisible.

New Zealand and Australia were the first countries deliberately to take the next step – to plan rights-based fishery management systems that involved fully transferable rights, enabling market forces to participate in decision-making about allocation of rights and of capital and labour to fish stocks (see Boyd and Dewees 1992). Iceland followed suit in 1984, on a trial basis that became permanent in 1990 (Pálsson and Helgason 1995). Canada also soon did the same in a few fisheries, beginning with Atlantic herring and expanding to a wider number of fisheries, including the Scotia-Fundy groundfish fishery, prosecuted by small (less than sixty-five feet, 19.8 metres) otter-trawlers or draggers, and the Western Newfoundland small-dragger cod fishery. The first U.S. ITQ fishery also began in 1990, for surf clams and ocean quahogs on the Atlantic coast.

I have been working with colleagues on a comparative study of ITQs and their consequences in the U.S. and Canadian fisheries of the northwest Atlantic (Apostle, McCay, and Mikalsen 1994; McCay 1994; McCay and Creed 1994; McCay et al. 1995), with particular focus on the evaluative criteria of efficiency, equity, and stewardship (Young and McCay 1995).

There are intriguing similarities as well as differences among the cases, including whether concerns about distributional equity influence the design (McCay et al. 1995). In Nova Scotia and Newfoundland they do; in the U.S. case studied they do not, except in so far as equity is read broadly to mean respect for the rights of investors and owners of capital. For example, the U.S. sea clam ITQ system sets no limit on how much of the quota any one individual or firm may hold; in the Newfoundland and Nova Scotia ITQ cases that we have studied there are limits, and in Newfoundland (but not Nova Scotia) those limits seem to hold. In the U.S. sea clam system the new property rights created include the right to transfer and exchange ITQs freely; Nova Scotia's did not allow this – transfers could occur only within the fishing season – but in 1993 the system became one of fairly free transferability, though non-fishers need not apply. In the western Newfoundland dragger fishery accumulation was strictly limited and so was transferability; the past tense is appropriate because the fishery for cod had ended by 1994, when we began our study, because of the near-disappearance of said fishes and a moratorium on the fishery.[1]

Embeddedness and the Northeast Newfoundland Crab Fishery

The northeast Newfoundland crab fishery was on the brink of regulatory change, from competitive quotas, to individual or boat quotas, during the period of research, 1992–4. The various social units comprising and linked to this fishery seemed to be resisting the next step in enclosure of the commons – to fully transferable privatization – which would fundamentally change property rights. When looking at the crab fishery from the perspective of the comparative study of ITQs, we asked why the northeast Newfoundland crab fishery was not managed with individualized and transferable property rights. It has many of the criteria one might associate with acceptance for ITQs, including a relatively small set of actors, high values, and hence high stakes involved in management and allocation of the resource, and serious problems, some of which might be resolved by a system that allowed some people to buy out others.

I examine this question from the perspective of field research on Fogo Island, Newfoundland. The simple answer that I offer is that such a degree of privatization was beyond imagining, at least in 1994. In my 1994 interviews in Newfoundland, I found only one skipper in the fishery who seemed to recognize how an ITQ system would differ from boat quotas. The Fogo Islanders whom I interviewed saw transferability only in terms of loaning or leasing a licence if one's boat burned or sank. I noted one exception, a particularly entrepreneurial soul, who understood the language and logic of private property and capitalist accumulation. In 1995 the entrepreneurial perspective was more widespread, and boat quotas were established in part of the crab fishery. In 1996 boat quotas were finally agreed on for the entire offshore sector of that fishery.

The theoretical perspective that I bring to this analysis is that of economic anthropology, deeply marked by the concept of embeddedness. My very simple, testable explanation for the slow pace of and resistance to privatization in Newfoundland's crab fishery is that the coastal Newfoundland fishery (both inshore and offshore) continues to be, as the economic sociologists say, 'embedded' in the social networks, social structures, and culture of the local, geographically and historically defined community (Barber 1995; Barrett, Jentoft, and McCay 1995; cf. Granovetter and Swedberg 1992; McCay and Jentoft 1996). In situations where decision-making about the fishery takes place apart from the concerns of fishing families and larger communities, such as in distant (or local) boardrooms, or where crew members or fishplant workers are transients or strangers, it is easier to imagine situations in which market forces might allocate rights to fish. However, in the context of ecological and economic crisis, expressed in a moratorium on the traditional northern cod fishery and closure of other traditional fisheries, which began in 1992, the embeddedness of fishing has loosened.

The Fogo Island Crab Fishery

Fogo Island, about nine miles (fourteen kilometres) from the northeast coast of Newfoundland and on the edge of the Labrador Current and offshore fishing banks, has an important crab fishery. In 1994 there were eight 'full-time' crab licences, used primarily with large 'longliner' vessels (fifty-five to sixty-five feet, or 16.8 to 19.8 metres)[2] and twenty-nine 'supplementary' crab licences, used with a wide variety of boats, ranging from the small (thirty-five feet, or 10.7 metres, and over) open, wooden skiffs of the traditional trap fishery to a wide range of 'longliners.' About 190 fishers participate in this fishery, out of a total island fishing population of about three hundred.

The crabs are fished in grounds that extend from nearshore areas to as far as 200 miles (322 kilometres) offshore in waters as deep as 200 fathoms. They are caught in conical metal and webbing pots, similar to those used in the Pacific northwest for crabs. The crabs are 'Queen crabs' or 'snow crabs' in the markets, *Chionecetes opilio* in Latin, and they are highly prized in U.S. and foreign markets. The Fogo Island cooperative runs a large, modern crab-processing plant, which in recent years has hired as many as three hundred workers at a time, mostly women, and therefore touches most of the households on the island.

There being far more supplementary than full-time licences is the general pattern in Newfoundland's crab fishery. In 1993 in the larger region ('area 3K,' or 'the northeast coast') there were thirty-one full-time crab licences and 210 supplementary licences. The nomenclature is misleading; neither type of licence-holder actually crabs full-time because fishing is very seasonal on this coast, which is ice-bound and stormy from November to May. The distinction between 'full-time' and 'supplementary' evolved from responses to trouble in the groundfish fishery in the 1980s. Crab fishing began as a sharply restricted, limited-entry fishery, both for fishers and for processing plants. In a draw, a small handful of Fogo Island fishers won the right to fish for crabs in an 'experimental' fishery, in 1981/82 ('everybody threw into a hat and there was three licences picked'; R.M. interview, 2 Aug. 1994). Others remained with the groundfish fishery, either as inshore, small-boat fishers using traps, handlines, gillnets, and so on, or as nearshore, large-boat fishers embodying large fleets of gillnets. Northern cod and turbot ('Greenland halibut') were the major species caught.

The groundfish supported fish filleting and freezing plants run by the cooperative, and, as crab landings and prices increased, the co-op won a licence for processing crab as well. When the limited-licence crab fishers did well, some jealousy and back-biting followed, particularly as a few crabbers were able to invest in very expensive new vessels and homes. However, with one or two exceptions, the crabbers remained full members of the community, and class divisions were muted by inconspicuous consumption, an intensely egalitarian, communal ideology, and the fact that crabbers too had their troubles.

In the 1980s groundfish landings began to decrease again (after major decline in the early 1970s), and in 1986, in response to demands for access to the comparatively healthy crab fishery, the federal Department of Fisheries and Oceans (DFO) created a class of 'supplementary' licences. Crab catches were intended to supplement cod, turbot, and other species for the nearshore, large-boat 'longliner' fleet. DFO divided the annual quota between the full-time and the supplementary licencees and by Northwest Atlantic Fisheries Organization (NAFO) areas. Area 3K was the major site of the northeast-coast crab fishery, but some licence-holders fished to the north, in the waters off the northern tip of Newfoundland and southern Labrador (areas 3L, 2J; 4R).

The supplementary crab fishery had open access – within limits – and open access was a major contributor to the problems generated. However, it was also under limited licencing and had the kinds of equity-related restrictions that are taken for granted in Newfoundland: those eligible had to be 'full-time,' in the sense of being dependent on fishing for a living and resident in the area for which the licence was granted. Because the supplementary quota and season were intended to help the longliner fleet, there was also a restriction to owners or operators of vessels in the regular 'longliner' class – between thirty-five and sixty-five feet long – which excluded small-boat fishers. Also increasingly taken for granted, though less enthusiastically applauded, was the government's policy of controlled-entry licencing: 'I don't know if I can put a foot in the Atlantic Ocean without a license' (P.T. interview, 15 Aug. 1994).

In the late 1980s the crab fishery declined because of poor markets and landings. Accordingly, after the first 'flush' of entry into the supplementary fishery, participation slacked off a bit.[3] However, the number of licenced supplementary crab boats began to rise rapidly after 1990 – in 1990 there were 139; by 1993 there were 210 in 3K, plus another eight fishing waters of 2J and 3L, for an increase of over 50 per cent. This expansion was linked to a dramatic increase in landings, and the supplementary fleet began to represent over half of the total landings, partly because of the political strength in its numbers and hence its ability to get larger shares of larger quotas.[4]

Pressure to reform the system came from the fact that the gains of the early 1990s were spread among large numbers of boats and people who were increasingly dependent on the crab fishery. The supplementary crab boats rarely brought in more than about 25,000 kg of round crab; the 'full-timers,' in sharp contrast, averaged up to 140,000 kg in the period from 1990 on (and had averaged as much as 180,000 kg in 1983).

The problem was that by 1990–1, and definitely in 1992, the supplementary crabbers had little left to supplement. The cod fishery was in sharp decline and was closed in July 1992, and other groundfish were also scarce, especially turbot. Accordingly, the full-time/supplementary distinction created a 'have'/'have not' dichotomy, and annual meetings between crab fishers and DFO representatives became intense battles over 'carving up the pie' of quotas set for conservation and market reasons.

Though quotas and landings increased, the seasons became shorter, as more people entered the fishery in the early 1990s. The supplementary quotas were taken up very fast, within a few weeks (four or five weeks in 1994). This situation was problematic for fishers, fishplant workers, and the fish plants. The Fogo Island Co-op's crab plant had developed good markets for specialized and high quality crab products in Japan; short-term gluts would only hurt those markets. Consequently, weekly limits were instituted for supplementary crab fishers (in 1994 the weekly limit was 30,000 lb., or 13.6 metric tonnes, per vessel), and there were trip limits for full-time crabbers. However, the co-op and local fishers are only part of a larger system and 3K quota, so the quotas were filled rapidly, no matter what.

With short crab-fishing seasons and a decline in groundfish landings, fishers and fishplant workers found it harder to work enough weeks at the fishery to qualify for unemployment insurance (UI). UI, including special provisions for otherwise self-employed seasonal fishers, is a fact of life and has become a critical part of household income in the region.

In 1992 this problem was both diluted and compounded by the federal government's moratorium on northern cod landings. The program included a benefits package for fishers and fishplant workers dependent on cod. In theory, it solved the problem for people who could not work the minimum number of weeks in the crab fishery. However, it also reduced the set of options available for those who did not qualify for the moratorium package. It also provoked personal and institutional conflicts, such as between those who were content to stay home to receive 'the package' and the managers of the crab plant, who needed capable workers. Moreover, people who had specialized in work at the crab plant were at a severe disadvantage, because they would not qualify for the benefits. In addition, their chances of working long enough to qualify for regular UI benefits were reduced by the shorter crab-fishing seasons, which in turn resulted from increases in crab-fishing effort because of the closure of other fisheries.

The Fogo Island cooperative tried to help individuals caught in these binds, but its dependence on the crab fishery had increased dramatically. Whether or not weekly crab limits were in place, increased numbers of boats racing for limited quotas meant shorter seasons.

Relief arrived with the opening of new grounds for the fishery in 1993 and 1994, particularly in 1994, when DFO decided to let the fishers do the exploratory work in offshore waters suspected of having good populations of crabs. However, even this new fishery did not extend the season much beyond four or five weeks.

In addition, the new fisheries create very dangerous conditions, particularly for many of the 'supplementary' crab fishers with smaller, older vessels. In 1994, under pressure to increase quotas for full-time and supplementary crab fishers, given the continued moratorium on cod fishing and a strong market for crab meat, the government opened up exploratory areas off the northern coast of Newfoundland, extending south to the northern tip of the Grand Bank. The larger boats began crabbing in an area about seventy miles (113 kilometres) from port, already beyond the old 'offshore,' and then went even farther, to NAFO subareas 3Kb, 3Kc, and 3Kg. The area of 3Kg fished, known as 'Tobin's Point,' is the northern tip of the Grand Banks, about 130 miles (209 kilometres) east southeast of Fogo Island. Some of the small (i.e., fifty-foot) longliners went out there, along with the bigger boats, very few of which were equipped with appropriate survival and life-saving equipment, and some 'really drove to their limits,' beyond what insurers allowed.[5] As one of the small-boat skippers admitted, it was 'too far for we small boats' (P.T. interview, 15 Aug. 1994).[6]

Proposals for Change

The situation is arguably one that would benefit from individual quotas. IQs would allow boat owners and plant managers to make agreements that would avoid gluts and would help crews and plant workers 'get their stamps' or UI. With secure shares of the total quota, a crab-boat skipper could make fewer or shorter trips per week and fish more weeks. If these quotas were transferable, then those with less quota than what they required to have a viable fishery could either sell out or buy in.

In the rest of the chapter I discuss the issues as they emerged in eighteen interviews done with crab-licence holders on Fogo Island in the summers of 1994 and 1995. I show that while the notion of IQs appealed to most, in the form of 'boat quotas,' allocation of those quotas equitably proved problematic, and the notion of using transferability to bring a market mechanism into the picture seemed 'out of the question' – beyond the realm of the possible.

Boat Quotas

Given the problems created by the race for a limited quota, there have long been proposals to rationalize the crab fishery by creating individual boat quotas. The crews would know how much they were allowed to catch and could arrange with plant owners when and how much they should catch at a time. An important issue is dealing with breakdowns and bad weather in a competitive situation. As one skipper said, a boat quota would allow 'me to catch it when I like, when I got trouble I can get it fixed and can catch my quota' (C.R. interview, 15 Aug. 1994).

The full-time crab skippers were particularly keen on boat quotas. They operated under trip limits, not weekly limits; this placed very heavy pressure on them to get in as many trips as possible. One of the skippers (J.G. interview, 17 Aug. 1994) explained, 'We sleep with one eye open all summer,' never really getting a rest, a dangerous practice that interviewees suggested might help to explain the sinking of one of the crab boats in the summer of 1994. 'It's hard going, come back, unload and go on again.' However, 'We had to do it, to keep up with the next fellow'; as they all raced to catch as many of the 'full time' quota as they could before it was reached, and thus the fishery would close. With boat quotas, they could make just two trips a week and be home on Sundays. That would 'give a feller a chance' to work safely and sanely.

The obstacle to boat quotas was the problem of how to divide up a quota in a way that seemed fair, given variations in experience, skills, capacity, and expectations. The issue is less challenging for the small fleet of 'full-time' crab fishers, most of whom have similar investments, experience, and expectations; they actively pursued a boat quota, which came into effect in 1995.

It is more difficult for the 'supplementary' fleet to come to an agreement because of the tremendous variability within the fleet, as well as the large number of vessels in it. The owners of large 'supplementary' boats wanted to get enough quota to cover their mortgages and other expenses. Some suggested that the small-boat owners, having much lower expenses, were much better off than they are: 'A forty-footer don't need so much as a fifty-footer ... and I know heart and soul [X] needs more than me' (C.R. interview, 15 Aug. 1994). Some of the small-boat fishers involved in the 1994 crab fishery were said to have 'made a fortune, and they took it all home,' meaning that they had relatively few expenses. This left owners of the larger, supplementary boats discontented. Owners of small boats, however, wanted at least a chance to share in the wealth of nature, which should, by right, belong to everyone.

The general consensus was that boat quotas should be allocated according to some measure of the relative expenses of each boat. A possibility under discussion was using a simple measure of vessel cubic capacity (length x breadth x depth). DFO uses cubic capacity to control replacements of vessels in the crab fishery. In the U.S. sea clam fishery, cubic capacity served as a proxy for investment in the formula developed for allocation of initial ITQs.

In the interviews, I noted that people talked about expenses, with scarcely a word being offered about the principle of making allocations on the basis of track record or past performance. This mission seemed strange because I knew that elsewhere historical performance was a major determinant of allocations. I suspect that avoiding the criterion of performance relates to the embeddedness of the fisheries in Newfoundland outport life, which values a semblance of egalitarianism and avoidance of overt conflict. As shown in the U.S. sea clam case (McCay and Creed 1990), the use of 'history' in determining the size of an IQ can become extremely controversial and divisive if participants believe that misreporting – breaking rules about technology, seasons, and sizes – was taking place. Such cheating was allegedly rampant in Newfoundland's west coast dragger fishery and the east coast crab fishery.[7] It thus seems believable that Fogo Islanders, aware of the potential for conflict and divisiveness, would search for criteria other than individual performance.

Classes of Boat Quotas

One way around the dilemma of allocation is to create large classes of boats, each with a different quota, so that the larger boats get larger quotas, and the smaller ones, smaller. Fogo Islanders have heard about this system, which has been tried in other areas of the Newfoundland crab fishery (i.e., the Avalon area) and generally fits the pattern of Canadian fisheries management – subdividing licence privileges by area, gear sector, vessel-size class sector, and so on. This approach avoids directly pitting one vessel owner or crew against another, and it creates small, potentially workable groups of common rights holders, who can negotiate quotas and changes of method in resource management. Competition remains, but it occurs within smaller 'commons.'

Two obstacles to creating quotas for vessel-size class sectors were raised in the interviews. One, discussed more below, is the continued problem of there being too many boats participating to ensure that everyone can have a viable quota. The second is concern about the practical and social consequences of creating or exacerbating socioeconomic differences within the fishery. Such a system may provide incentives to owners to upgrade boats, to get into the larger size-class because 'you have to get your tonnage' (P.T. interview, 15 Aug. 1994), thereby intensifying overcapitalization. The fact that some would get more than others is even more troublesome to Fogo Islanders: 'That's not right: Why should my crew member make $1,000 and another crew member on another boat make $2,000?' (P.T. interview, 15 Aug. 1994).

Full-time versus Supplementary

Underlying many of the specific problems envisioned with a boat quota system, particularly one that rewards different classes of fishers, is deep concern about class itself, about what one fisher described as 'creating different categories, one fellow won't speak to the other fella passing on the wharf ... There's enough categories, with this fisheries management system, as it is' (P.T. interview, 15 Aug. 1994). One of the more highly regarded, and the wealthiest, full-time crab fishers depicted his social position with a similar image: 'Oh yes, out there on the wharf, everybody is friendly, but when you turn your back ...' (J.G. interview, 17 Aug. 1994).

This brings the conversation back to the problematic difference in position and wealth between the full-time and supplementary crab licence-holders, beside which the issue of allocating boat licences among highly variable supplementary crabbers is minor indeed. It is fair to ask: 'Why should twenty-nine licencees in Notre Dame Bay all get rich?' (P.T. interview, 15 Aug. 1994). Rarely brought up as a real possibility but underlying what was said in many of the interviews was the idea of starting over, reallocating the entire quota so as to erase the historical distinction between those who obtained their licences from draws and purchases back in the early 1980s and those who came in on the supplementary program, both of which groups have become similarly dependent on crab fishing.

However, this idea is not only politically chancy; it is obviously divisive. On Fogo Island there are two classes of full-timers – those who land crabs locally and sell to the local cooperative, and those who do not. The former – the majority – are well integrated into their outport neighborhoods and communities, and, though subject to jealousy, they are generally respected as decent, hard-working people. Their crab fisheries have also been a godsend for fishplant workers, particularly for hundreds of women who found paying work for the first time as this fishery developed.

Like most other 'longliner' fishing enterprises, the full-time crabbers do their share for the community by hiring so many people. For a not unusual example, one has a crew of eight (the skipper and his brothers, son, in-laws, and other, more distant relations), each supporting a household.[8] The sheer volume of the crabbers' production, as well as the limited number of their licences and hence less competition among them for the quota, makes their season longer than that of the large supplementary crab fleet. Full-time crab boats therefore make a huge difference to crab plant workers and whether they will get the income and UI that they need. Fogo Island has about 3,700 people living in nine small communities; during the crabbing season, any one there is likely to know when and whether a particular full-time crab boat is expected to land. That landing will affect the pattern of work for dock workers, dockside monitors, crab plant workers, and babysitters.

Equity and the Problem of Small-Boat Crab Fishing

The relationship between full-time and supplementary fishers is at the heart of the difficulty in developing a 'rational' system of management for the crab fishery. It is also ambivalent, variable, and complicated by the cross-cutting issue of how those with large, very seaworthy vessels relate to those with small boats designed for the inshore fishery. Both types are found in the supplementary fishery, and in the early 1980s both were also in the full-time fishery. As the fishery developed during the 1980s, both large and small boats were competing for crabs in the same areas, fairly close to Fogo Island. It did not take long, however, for local fishers to work out an arrangement whereby the larger boats would restrict themselves to offshore crab-fishing grounds, and the government legitimized the agreement by fixing a line of latitude, at 50°10′ N, demarcating inshore from offshore crab-fishing territory.

I could say much more about this system, including its fragility and its strength as an instance of local-level governance of the commons, but let me turn to the newer 'inshore' issue: the entry of boats under thirty-five feet (10.7 metres) into the crab fishery. The chances of developing an even more restrictive boat quota system were further diminished in 1994 by pressure from inshore fishers to be allowed into the crab fishery. The original rule was that boats under thirty-five feet in length could not participate, a rule intended to keep this important resource for the longliner fleet, which was in particular trouble in the 1980s. The line at 50°10′ N was drawn to help the smaller longliners – thirty-five feet to forty-five feet (13.7 metres) perhaps – not this inshore fleet of mostly open skiffs that were used for cod trap, handline, gillnet, and seine fishing. With the moratorium on northern cod fishing, closure of other major inshore fisheries (including salmon), and signs of tremendous abundance of crabs – perhaps filling a niche emptied by cod, a theory widely discussed by local fishermen and women – the small-boat fishers wanted in.

Pressure to change this rule mounted during the early 1990s, and the topic was brought up at industry–DFO meetings in 1994. As might be expected, there were differences of opinion, particularly since both full-time and supplementary fishers were fighting to get larger shares of the quota. One of the large-longliner skippers, struggling to keep his enterprise (which supports five households) viable in the context of fishery decline while not possessing a 'full-time' crab licence (these are either unavailable or prohibitively expensive), said, about bringing boats of under thirty-five feet into the crab fishery: 'It's a big job, if they starts that, there'll be a bigger problem than there is now. ... The quota is too small to share out' (C.R. interview, 15 Aug. 1994).

None the less, no one interviewed was entirely opposed, recognizing how desperate the situation was for the cod fishers. This again shows embeddedness: on Fogo Island one simply cannot ignore what is happening to others in the community, like it or not. Consequently, crabbers came up with the idea of a special area 'inside the 50°10′ N line' for the very small boats. One longliner skipper, himself an inshore fisher until 1990, identified as an appropriate ground a place where Fogo Island crabbing began and that had not been fished much for many years because of higher returns from more distant and deeper waters (P.T. interview, 15 Aug. 1994). In 1995, when this small-boat crab fishery did take place (after a 'draw' for about a dozen permits), that very ground was used, the area from Little Fogo Islands out about eight miles (thirteen kilometres). Another concern was that crab-pot fishing was too difficult and dangerous for the small-boat fishers. Experimentation in 1994 and experience and observation over a longer time span removed suspicion that small-boat fishers could not handle crab pots. One skipper recalled what he had observed while working as a seaman on coastal trade boats on the Labrador coast – three or four fellows in a punt or a speedboat working crab pots (P.T. interview, 15 Aug. 1994).

However, up to and during the August 1995 small-boat crab fishery, which happily turned out to be without loss of boat or life, people talked about how dangerous this fishery would be. (In retrospect, a few months later, sceptics gave credit to the unusually fine weather. In the summer of 1995 the risks of crabbing in small boats in poor weather were a popular topic of conversation, and many volunteered that the fishery was designed to fail, in part because people were confined to the boats they held when they put in for the draw, and many of these were really too small and unseaworthy for the task of handling unwieldy crab pots.)

Transferability and Morality

There is much more that can and should be said about how Fogo Islanders think and talk about ways to reform crab fishery management. I deal with one last topic: transferability – a critical variable, determining the extent to which market forces can be involved in allocation. I note the extent to which this topic was beyond the pale and obliquely suggest why this is the case, by showing the perspective of one of the full-time crab fishers, who was 'disembedded' from the community.

A powerful obstacle to both boat quotas and sector-based quotas was realization that, given the large and growing number of boats in the supplementary crab fishery, any attempt to divide the quota among them would result in far less than what any one of them would need to be able to survive (make boat payments, provide a living wage to crew, contribute to UI, make repairs, and so on). 'It won't work ...' said one of the skippers interviewed in 1994 (P.T. interview, 15 Aug. 1994). He went back to the argument about boat quotas, and the fact that if the quota were divided up among all the participants in the supplementary fishery, there would be only 25,000 lb. (11.4 metric tonnes) per boat, not enough for a boat (given the fact that the weekly limit is 30,000 lb. (13.6 metric tonnes), this would be less than a week's worth of crabbing). The only reason it was viable in 1994 for the big boats was because they could venture outside, offshore, in the exploratory areas; otherwise, he insisted, the fishery was not viable.

It was at this point in my structured interview that I asked about the possibility of transferable quotas, in which some fishers could buy up the quota allocations of others, allowing some to survive in the fishery and others to leave 'with money in their pockets,' as advocates of market-based fisheries management often say. And it was at this point that I came to realize that some things might be beyond imagining. Everyone I talked with either sidestepped the question or transformed it into something else. C.R. (interview, 15 Aug. 1994), for example, who seemed to be leading up to this in our conversation, dodged it right away by saying, 'There are enough crabs out there [for everyone], no need for that.'

A more specific and realistic concern about transferable quotas came from one of the small-boat crab fishers, one of the handful who had been in the cod fishery but had held supplementary crab licences and were told, 'Use it or lose it,' so went crabbing in earnest in 1994. He pointed out that 'the fear of [transferability] is boats coming in from other areas' after they had fished out quotas from those areas, buying up local boat quotas or licences and hence taking part of the local quota (M.L. interview, 15 Aug. 1994). Note the principle, reinforced to a large extent by licencing practice and policy, that area quotas belong to people in adjacent local communities.

Just about everyone else (the other exception is discussed below) recast the question in terms of a very specific, narrow issue: whether someone who holds a licence can transfer it to someone else during the fishing season. That is not what I was talking about, but it was what they were seeing and preferred to discuss. The case that they had in mind concerned a local fisher, highly trained and experienced, who had neither a boat nor a crab licence. The arrangement was that he fished someone else's boat, using someone else's licence and taking on some of the crew of the lessor – a common practice that also reflects the embeddedness of this fishery. In exchange, he gave the owner–licence-holder a share of the proceeds. He also sold the crabs to a dealer located off the island.

The moral outrage was directed mostly at the licence-holder, also a Fogo Islander. For personal reasons, he had chosen not to fish his crab licence. One term used to describe this kind of licence was 'arse pocket licence,' analogous in meaning to 'coupon clipper' (someone who makes money only from invest-ing in stocks) for a factory worker. Everyone agreed that there was no problem in leasing a licence to someone else if your boat broke down, so that you could not fish yourself, but beyond that ... : 'If a feller's into the crab fishery he should have a boat' (P.T. interview, 15 Aug. 1994).

Even a discussion that links transferability with concentration of ownership gets around to the same moral issue, which sidesteps the concentration-monopoly question: 'We talked about [transferability] last winter [at the DFO crab meetings],' said one of the skippers (J.G. interview, 17 Aug. 1994), but he and others were not enthused about it. They suspected that the man who brought it up had already lined up a few licences. J.G. followed this report with a moral commentary: 'It's not right, a feller with no boat, he can give someone his quota to catch, and he probably has another job on the side. That isn't right.'

This moral discussion is grounded and embedded in the harsh reality of high unemployment, chronic poverty, and rural Newfoundland's cultural expecta-tions that if jobs are scarce they should be spread around as broadly as possible. And the owners of capital should work it as well.

The issue of concentration of ownership was not brought up by Fogo Island-ers interviewed in 1994 and 1995. Concentrated power was – one of the reasons why people were upset about a particular instance of licence-leasing is that an outside crab-buying company was involved in the deal and allegedly used its 'pull' with the government to get the licence transfer approved. But the possi-bility that transferability of boat quotas would lead to concentration of owner-ship (for example, in companies that would soon own, or at least control, large numbers of boats, as has happened in the U.S. and Nova Scotia cases), did not even surface, even though it is the major topic in debates about transferability of IQs in the rest of North America. That degree of transferability – a reality in the Nova Scotia and U.S. ITQ fisheries – is hard to imagine when the burning issue is whether someone whose boat is out of commission one season can lease his licence to someone else during that season. It was unimaginable on Fogo Island in 1994 and 1995, I suggest, because it was too profoundly threatening to the coastal fisheries, which are still thoroughly embedded in the coastal, fishery-dependent communities.

Conclusion

In closing I offer an exception to the rule that I have tried to define: a crab fisher who understands what I mean about transferability, at least to a point, and who thinks that it is a good thing.

The 'full-time' crab fishers include some who are highly respected, even if begrudgingly, and others who are not. More problematic, from a social perspective, are the full-time crab licensees who no longer sell their crabs locally because of old or long-standing problems with the cooperative and/or because one of the costs of obtaining a full-time crab licence is becoming indebted to a crab plant elsewhere in the region. No one interviewed had developed this fact into an argument against transferable boat quotas – but, as I noted above, the case of full-fledged ITQs seemed outside the realm of discourse at the time.

The following quotation from an interview with one of these crabbers suggests the difference between a business that is embedded in the local community and one that is not, a distinction that is causally linked to the social isolation this fisher and his family experience in the local community: '[BMc: Is that an issue, where you would sell your fish?] 'No, a lot of it, too, your hands are more or less tied, too, in a lot of cases. You got to get somewhere to get financing, like that boat there [a large, expensive steel-hulled vessel], you go to [a large fishery company in Newfoundland] to get financing, [then] you wouldn't expect to sell your fish somewhere else. Wherever you negotiate the best deal, that's where you got to go. It's not like a small business, one time, you could do this and that and something else; you're more or less, we're in the big business and we got to treat it as such' (R.M. interview, 2 Aug. 1994).

This particular crab fisher (actually a 'shore skipper' these days; his sons run the two boats of the family fleet) is also the only one who responded with enthusiasm to the notion of really transferable IQs in the fishery. I mentioned the system as it existed in western Newfoundland. He said that he thought it a very good idea,

... especially now the way the quotas are set up, so low ... You got your quota and the other guy don't use his boat or he loses his boat, and I gets his quota, well I got enough to live with, and if you're an aggressive fisherman, you can live with that.
BMc: Yeah, you can then buy up quota from other people.
R.M.: Yeah, that's right, if that's what it takes, if you're going to stay in.
BMc: This gives more flexibility than boat quotas?
R.M.: 'Yeah, I'm not interested in boat quotas, you eventually gets down to paying to catch it, it gets to a point where it's not feasible to be at it.
BMc: But ITQs would be different, you could operate as a business, more flexibility.
R.M.: 'Yes, that's true ... If I had a grocery store up here and there's another one down there, and if I can buy that place for a reasonable price and close it down, well sure, I'm going to take all the business, that's it, that's the way it's done. One trucker goes out, the other guy gets a new truck. That's the way it is. (R.M. interview, 2 Aug. 1994).

Transferability of licences, in the sense of buying and selling and leasing, bothers him, however, in so far as it might offer 'unfair' advantages to those who lease, in contrast with those, like him, who slowly built up capacity with both boats and licences: 'I can see a guy losing his boat, and he's trying to get a boat built the next year, if he's hung up on financing, I can see them letting him lease the licence, but [not otherwise, not if he just doesn't want to fish], no, the way things went the year, they go and lease somebody else's boat to fish the crab. If you want to go fishing, you go to the bank and negotiate a loan like I did, 1.2 million bucks, have that hanging over your head the rest of your days. You know, that's the facts. If you're going to be leasing licences, leasing boats, all this old stuff, how can I compete with that, when I got a million-dollar loan? You can't do it; it just can't be done.'

This hard-nosed, capitalist perspective is not unique on Fogo Island, but it is rarely expressed so baldly. Another semi-retired skipper, who in contrast is highly esteemed and thoroughly embedded in the community, took a more 'scientific' perspective, reflecting his long association with scientists and administrators. He echoed the favourite phrase of fishery managers: 'Too many fishers going after too few fish,' or the tragedy-of-the-commons analysis, which does not dwell on issues such as too many of what kind of fishers, using what kind of gear. The cure is to 'downsize' the fishery, so that those who fish can make a good living without subsidization from the government. But what will the others do if they cannot fish? In an interview he said, 'If people knew I was saying this I might get shot tonight' (McIlroy 1995).

The moral economy of Fogo Island insists on the embeddedness of the fisheries in the life of the community, but this takes work, the work of critical analysis and persuasion. A journalist, Anne McIlroy, visited Fogo Island and published an article in the *Toronto Star* (McIlroy 1995); it began: 'There are crab sheiks in the town of Seldom Come By, fishers who are becoming millionaires while their neighbours cash government cheques and wait for the cod to return.' She tapped into some of the conflict surrounding the development of sharp differences in class in the crab fishery.

A copy of the article was reprinted in a local newsletter in January 1996, and the editor added her own comments, appealing to islanders to work together rather than let jealousy and envy destroy a cherished way of life. She wrote that people in Toronto 'are watching us in fascination to see how we are handling ourselves in a crisis. What they are seeing is not pretty. The way our greed is growing, we won't have much crab to squabble over in next couple of years. Last year the Co-op bought 3,500,000 pounds, and we're still crying for bigger quotas. Every man, woman, child and dog wants a licence!. ... It's the inshore against the offshore, the full-time crabbers against the supplementary, TAGS against non-TAGS, TAGS against UI, plantworker against plantworker. Will

Bosnia come to Fogo Island? We've made a mess of the cod fishery, let's not do the same with the crab' (see McIlroy 1995).[9]

The stakes are high, making it very difficult to come to acceptable social terms with the question of who is privileged to reap the wealth of the sea – or what is left of it, the snow crabs. But the task has not yet been given over to impersonal market forces. 'It's hard going,' another full-time crab skipper said, with reference to competition with the 'supplementary fleet' over quotas and his desire to keep his own privileged position in it. 'I want [i.e., need] so much crab to survive,' he said, justifying the very high boat quotas awarded to the full-time crabbers. But he then paused and offered, 'They got to make a living too.' And this is where even the bold capitalists faltered, and the impulse to privatize hesitated.

Epilogue

I completed the original version of this chapter in 1995. In 1996 the crab fishers of the northeast coast of Newfoundland reached an agreement to do what had seemed so difficult before: allocate boat quotas to both the full-time and the supplementary fleets, based on a formula that recognized historical performance in the fishery in assigning boats to different classes, with equal allocation within the classes. Transferability is sharply limited as before, and the quotas cover specific crab-fishing grounds. Weekly catch limits and other regulations circumscribe the ability of crabbers to use their quotas in the 'free market' mode envisioned by the economists who argued for individual quota systems. However, crabbers are freer than before to decide when and where to fish. In addition, part of the quota was allocated to the small-boat fishing sector, which had had an experimental fishery in 1995. In a determined effort to salvage a little of the traditional moral economy of the inshore fishing communities, the small-boat fishers in area 3K, including Fogo Island, settled on equal division of the quota among all fishers. Moreover – and most surprising, given the overall trend towards limited licencing – they insisted that this fishery should be open to all who met the criteria rather than using a lottery or an arbitrary rule to limit numbers. It remains to be seen whether they will be able to continue that policy.

NOTES

1 I presented the original version of this chapter at Meetings of the International Association for the Study of Common Property, Bodö, Norway, 24–8 May 1995. Research for it has extended over many years and many sources of support, including the Rut-

gers University Grants program. Support for the interviews reported here came from the U.S. National Science Foundation, the New Jersey Sea Grant College Program, and the New Jersey Agricultural Experiment Station. The acronyms for persons interviewed are pseudonymous; I have tried to protect individual identities. My thanks to the people of Fogo Island for their cooperation and inspiration; any errors in fact or interpretation are mine alone.

2 One of the licences was not actually used from island ports; its holder had moved his vessel elsewhere. Data on licences are courtesy of the local DFO fisheries officers.

3 In 1983 and 1984 the number of full-time 3K licences increased sharply (from eighteen to twenty-seven, then twenty-seven to thirty-six), and in 1985 there were 115! The figure is an artefact of bookkeeping and lag in creating new constructs: in 1985, under pressure from longliner fishers hurting from decline in traditional groundfish species, DFO created a special season and opened up the fishery. However, the full-time licences retained special standing and in 1986 resumed that status in distinction to the newly constructed class of holders of supplementary licences.

4 By 1994 the government had an explicit policy that any increases in quotas should go to the 'supplementary' crab fleet.

5 Even the bigger boats came close to violating their insurance agreements. The typical limit is 120 miles (193 kilometres), beyond which the skipper must be a class-2 fishing master; the Fogo Island fleet was saved by the interpretation that the distance is from the nearest point; the 3Kg boats were only 80 miles (129 kilometres) from the nearest point of land, Cape Bonavista, though 130 miles (209 kilometres) from Fogo Island.

6 Tobin's Point was almost too far for this anthropologist, too, who found herself out at sea for four days with no EPIRBS or survival suit and little potable water!

7 Following an official investigation of cheating, dockside monitoring was instituted in the crab fishery in 1994. This system, financed by fishers, ensured an objective, 'arm's length' report on what was actually landed per boat. The dockside monitoring system of the Nova Scotia dragger fishery, which was part of the key to the relative success of its ITQ system, was influential to the use of dockside monitoring in Newfoundland.

8 Moreover, even though on most such large, expensive vessels the 'share' system of payment has been changed in favour of the vessel owner, this one has not: 'I'm good to me crew,' says the skipper/owner, who keeps 55 per cent for 'the boat' and shares out 45 per cent among the crew. Most others have increased the amount 'for the boat' to 60 per cent, 70 per cent, or more. Some of the smaller longliner operations try to return the to the old system of dividing the entire catch, minus the expenses of the trip, among members of the crew once they have paid off the boat.

9 TAGS is The Atlantic Groundfish Strategy, the northern cod moratorium program compensating fishers and fishplant work (see Neis, Ommer, and Sinclair, Squires, and Downton, this volume).

WORKS CITED

Apostle, Richard, Bonnie McCay, and Knut Mikalsen. 1994. 'Centralization and Privatization: A Comparison of Fisheries Management Regimes in Atlantic Canada and Norway.' Paper presented at the Annual Meeting of the American Fisheries Society,
Halifax, 21–25 Aug.

Barber, Bernard. 1995. 'All Economies are "Embedded": The Career of a Concept, and
Beyond.' *Social Research* 62, no. 2: 387–413.

Barrett, Gene, Svein Jentoft, and Bonnie McCay. 1995. 'Global Fishing Villages in Crisis: Embeddedness and Community Development on the North Atlantic Rim.' Paper
presented to XVI Congress of the European Society for Rural Sociology, 31 July–
4 Aug. Prague, Czech Republic.

Boyd, Rick O., and Christopher M. Dewees. 1992. 'Putting Theory into Practice: Individual Transferable Quotas in New Zealand's Fisheries.' *Society and Natural
Resources* 5: 179–98.

Christy, Francis T., Jr. 1973. *Fishermen's Quotas: A Tentative Suggestion for Domestic
Management.* Occasional Papers 19. Kingston: University of Rhode Island, Law of
the Sea Institute.

Crowley, R.W. and H. Pálsson. 1992. 'Rights-based Fisheries Management in Canada.'
Marine Resource Economics 7, no. 2: 1–21.

Fraser, Cheryl. 1986. 'Enterprise Allocations in the Offshore Groundfish Fishery in
Atlantic Canada: 1982–1986.' In N. Mollett, ed., *Fishery Access Control Programs
Worldwide*, 207–13. Fairbanks: Alaska Sea Grant Program, University of Alaska.

Granovetter, Mark, and Richard Swedberg, eds. 1992. *The Sociology of Economic Life.*
Boulder, Col.: Westview Press.

McCay, Bonnie J. 1994. 'ITQ Case Study: Atlantic Surf Clam and Ocean Quahog Fishery.' In Karyn L. Gimbel, ed., *Limited Access Management: A Guidebook to Conservation*, 75–97. Washington, DC: Center for Marine Conservation and World Wildlife
Fund.

McCay, Bonnie J., Richard Apostle, Carolyn F. Creed, Alan C. Finalyson, and Knut
Mikalsen. 1995. 'Individual Transferable Quotas (ITQs) in Canadian and US Fisheries.' *Ocean and Coastal Management* 28, nos. 1–3: 85–115.

McCay, Bonnie J., and Carolyn F. Creed. 1990. 'Social Structure and Debates on Fisheries Management in the Mid-Atlantic Surf Clam Fishery.' *Ocean and Shoreline Management* 13: 199–229.

– 1994. 'Individual Transferable Quotas in Clams and Fish: A Comparative Analysis.'
Paper presented at Theme (T), 82nd Statutory Meeting, International Council for the
Exploration of the Seas, St John's, 22–30 Sept. 1994.

McCay, Bonnie J., John B. Gatewood, and Carolyn F. Creed. 1990. 'Labor and the Labor
Process in a Limited Entry Fishery.' *Marine Resource Economics* 6: 311–30.

McCay, Bonnie J., and Svein Jentoft. 1996. 'Unvertrautes Geläde: Gemeineigentum Unnnter Der Sozialwissenschaftlichen Lupe' ('Uncommon Ground: Critical Perspectives on Common Property Theory'). *Kölner Zeitschrift für Soziologie und Sozialpsychologie*, Nr. 36: 272–91.

McIlroy, Anne. 1995. 'Newfoundland Crabbers Make Millions while Neighbours Hurt ...' *Toronto Star*, 28 Dec.; reprinted in *Fogo Island Flyer* 2, no. 8, Jan. 1996: 6–7, with editorial comment by Cheryl Penton.

Neher, Philip A., Ragnar Arnason, and Nina Mollett, eds. 1989. *Rights Based Fishing. Proceedings of the NATA Advanced Research Workshop on Scientific Foundations for Rights Based Fishing, Reykjavik, Iceland, June 27–July 1, 1988*. Dordrecht: Kluwer Academic Publishing.

Pálsson, Gisli, and Agnar Helgason. 1995. 'Figuring Fish and Measuring Men: The Quota System in the Icelandic Cod Fishery.' *Ocean and Coastal Management* 28, nos. 1–3: 117–46.

Squires, Dale, James Kirkley, and Clement A. Tisdell. 1995. 'Individual Transferable Quotas as a Fisheries Management Tool.' *Reviews in Fisheries Science* 3, no. 2: 141–69.

Young, Michael D., and Bonnie McCay. 1995. 'Building Equity, Stewardship, and Resilience into Market-Based Property Right Systems.' In Susan Hanna and Mohan Munasinghe, eds., *Property Rights in Social and Ecological Context: Concepts and Case Studies*, 87–102. Washington, DC, World Bank.

16

A Future without Fish? Constructing Social Life on Newfoundland's Bonavista Peninsula after the Cod Moratorium

PETER R. SINCLAIR, WITH HEATHER SQUIRES AND LYNN DOWNTON

> It seems like there's not enough people preparing ... for what is about to happen. They're just ... you know, taking advantage of what is happening. Some people are ... I mean, some people are and some people ... maybe they don't realize, or, you know, they just can't grasp, you know, you just can't grasp what ... what is really happening.
> BONAVISTA PENINSULA INTERVIEW, date on file with author

What happens when an ocean fish population is exploited to the point of collapse? Obviously, the ecological system in the area once inhabited by that fish will be altered, and in an especially significant way should the fish constitute a relatively large portion of the living organisms. The economic and social impact on people who had depended on this fishery may also be great, particularly in areas with few alternative fisheries and few other occupations. Whereas Barbara Neis in her chapter considers changes in the Newfoundland fishery over a longer period and focuses on their effect on women, here we explore the impact of unsustainable fishing practices on the people of the Bonavista Peninsula in eastern Newfoundland (see Figure 11.1, Neis et al., this volume), where the cod fishery dominated the local economy for over two hundred years. We consider the responses of local people to ecological and economic disaster after the cod fishery collapsed and a moratorium was implemented in 1992.

The ban on cod fishing wrenched away the economic underpinning of the local social structure. In these circumstances, we might expect a variety of reactions at both individual and collective levels. Some people might leave in the hope of finding employment elsewhere. Some might hang on in anticipation of a revived fishing economy in the next few years. A few might attempt to establish their own businesses. Unless an alternative source of local employment emerges, the only other possibility would be long-term dependence on state income support at a level even more extensive than in the past, but this appears unlikely if current ideological and financial circumstances persist. Some form of collective action might also be expected from one or more of the area's voluntary service organizations, unions, development associations, and town councils. Some new organization might also be formed to deal with one or more of the problems faced by local people. However, no sustained mass protest action has been evident. We try to explain both the range and predominance of certain types of individual response and the weakness of local collective action.

Embedded Relations

A sudden economic disaster, such as the closure of the cod fishery, brings with it social change that may best be captured in the label 'disembeddedness.' This is a process that implies breaking up of an earlier state of social embeddedness, though we caution against any static image of people and social structures in the past. Mark Granovetter (1990: 98) refers to economic activity as being 'embedded' in the sense that 'economic action, outcomes, and institutions are affected by actors' personal relations, and by the structure of the overall network of relations.' As Bonnie McCay also discusses in her chapter in this volume, economic activity is explained not, as in classical economics, by reference to the rational decisions of human actors, but in terms of partial mediation by social context. Granovetter, like other recent economic sociologists (for example, Etzioni 1988) treads carefully between an undersocialized and an oversocialized conception of the human actor – i.e., between viewing behaviour as a rational response to individual preferences and seeing it as the product of social learning and external stimuli. Not only economic activity, but all action and institutions, might be considered as to some degree embedded or nested in other structures.

Applying this approach to fishing villages in the North Atlantic, Barrett, Jentoft, and McCay et al. (1995) describe how global or at least external processes of excessive resource exploitation led to local crises in which hitherto-embedded communities became disembedded – individuals were separated from established institutional patterns. This process seems akin to social disintegration in more traditional language. At the same time, Barrett, Jentoft, and McCay (1995) point to some possibilities (with examples from northern Norway and Atlantic Canada) of how such local fishing communities may become re-embedded when appropriate collective action occurs. Their focus is on how fisheries may again become embedded in the community. Bonnie McCay (this volume) speculates on embeddedness and possibilities for its reconstruction in the contemporary crab fishery of Fogo Island, Newfoundland. Referring to the influence of social networks and values on individual choice and action, Barrett, Jentoft, and McCay (1995) argue that 'resource crises and globalization are forces that may well dis-articulate these relations unless they are effectively addressed by the community cooperatively.'

This perspective explicitly avoids social or structural determinism. We believe that it fits well with Anthony Giddens's structuration theory (Giddens 1984), in which what people see as specious and what they can actually do are conditioned by the resources that enable their actions in certain respects while constraining them in others. These resources and constraints are themselves the products of past action and are continually reconstructed. The process of reconstruction makes possible social change, including actions that are part of the process of re-embedding. Individuals develop coping strategies or at least attitudes towards their personal circumstances that point to certain courses of action, such as to migrate or remain, to retrain or hope for the fishery to return. For the Bonavista Peninsula, in particular, we ask how it is reasonable or even possible for so many people to imagine that they will still be living there in five years.

Embeddedness and its related concepts point us in the direction of understanding people's experience of crisis by identifying how that crisis may have altered a previous level of social connectedness or integration. (This is not meant to imply a reified state of harmony or common perspectives in the past.) In contrast with the examples in Barrett, Jentoft, and McCay (1995), but like McCay, in her chapter in this volume, we see a preponderance of individual rather than collective responses to the fisheries crisis. Our explanation for the weakness of collective action focuses on the cultural and structural features of the area in the period prior to the fisheries crisis, which, we claim, made anything more than individual or household responses unlikely. In this regard, the existence and nature of the government's income support program for fishery workers was critical in dampening the potential for protest. We also seek to explain the variation among individuals in how they perceive and respond to this crisis.

We show that those who are younger, and who have thus developed fewer commitments to remaining in the area than older residents, are more likely to leave. Furthermore, people with the educational levels that might allow them to succeed elsewhere are more likely to leave in the future. State income support and a vibrant informal sector of unpaid and officially unrecognized work make it reasonable for others to remain, many even with the expectation that they will be more prosperous in the future. In this analysis, we do not attempt to investigate the psychological state of the unemployed or the impact of unemployment on family relations and larger-scale local social structures, on which the evidence is mixed (see Redburn, with Waldron; Allen et al. 1986; Burnham 1988; Buss 1983; Perucciet al. 1988). Instead, we investigate people's attitudes to life in the area, their plans for the future, and their current use of resources in their coping strategies. We highlight, where appropriate, social structural differences, especially based on age and gender.

The Bonavista Peninsula of Newfoundland

In 1991, there were 10,979 people in the area covered by this research. They lived in eleven towns and unincorporated areas stretching from Keels, about half-way inside Bonavista Bay on the north side, to Bonaventure, Trinity Bay, on the south side of the Bonavista Peninsula. With 4,597 residents in 1991, Bonavista, close to the headland, is the largest town and key local service centre. However, many people, especially those closer to the head of the bay, routinely travel to Clarenville (about 1.5 hours' drive from Bonavista) for major purchases. In Bonavista we find an old fish plant that remains open, even today, during the crab-processing season and also a marine service centre. Just around Cape Bonavista on the Trinity Bay coast is a cluster of small towns that merge into each other (Catalina, Little Catalina, Port Union, and Melrose), but which are usually known collectively as Catalina. Here, Newfoundland's second largest deep-sea fish plant, once operated by Fishery Products International (FPI), but now shut down completely, is located. Further along the southern shore of the peninsula is the Trinity area, less active in recent years as a fishing base, but increasingly significant as a centre of small-scale tourism. However, crab are important, as there is an active plant in Trouty. Throughout the study area, the key to employment until 1992 was the fishing industry, both harvesting and processing. There is no other industry of any size apart from seasonal tourism.

Cod Fishery in Crisis

Though this chapter is not about the sociological or biological sources of the cod crisis (for that, see Villagarcia et al., Hutchings, this volume), a brief account of the recent history of the Bonavista Peninsula area's cod fishery provides necessary context. Cod, redfish, and various flatfish species comprised the groundfish on which fishing in Newfoundland has concentrated in the twentieth century. Apart from the probability that some inshore fishers caught local bay stock, 'northern' cod, which ranged from southern Labrador to the northern half of the Grand Banks off eastern Newfoundland, was the mainstay of the peninsula's industry. Canada placed a temporary moratorium on northern cod in July 1992 and, by the spring of 1995, had closed indefinitely all commercial groundfish stocks around Newfoundland, except redfish. These events devastated a region that had lived on what many believed to be a permanently productive fishery.

As a result of the moritorium, fish plants closed down and boats were tied up. Fishing companies sold off their trawlers and tried to restructure their operations. In the case of the largest, FPI, this process made the company more international and more profitable. Though annual sales were at record highs in 1994, this level was achieved at the cost of closing ten of its fish plants, eliminating thirty-eight of its trawlers, and cutting its workforce from about eight thousand to two thousand in Newfoundland. Its only growth point in Newfoundland was the modernization of its shrimp plant in Port au Choix on the west coast (*Evening Telegram* 1994). In recent years, FPI has been more profitable than National Sea Products, the other major Atlantic coast fish company, because its recovery strategy proved sound. FPI might eventually have reorganized itself in any case but appears to have been encouraged to do so by the moratorium.[1] On the Bonavista Peninsula, it closed its deep-sea plant at Catalina, which employed up to one thousand people, along with a small inshore plant at Charleston, just outside our study area. FPI's plant in the town of Bonavista remained open only to process crab, one of the few remaining fisheries that did provide some weeks of plant employment and prosperity to a small group of fishers. In addition, a smaller crab plant operates at Trouty, Trinity Bay.

Ordinary fishers and plant workers now found themselves unemployed, though partially protected by massive government compensation. The initial emergency program was replaced after two years by The Atlantic Groundfish Strategy, or TAGS (see Neis, this volume), which was scheduled to run only until 1998. TAGS provides up to $382 per week, depending on recipients' previous experience, but recipients must enroll in counselling and retraining programs as part of the official plan to cut employment in any future fishery by 50 per cent. The numbers affected in Newfoundland are estimated at roughly thirty thousand, or about 12 per cent of the total labour force (*Globe and Mail*, 28 Jan. 1995; Schrank 1995).

Disembedded People

Without intending to create an image of perfectly integrated institutions, we claim that life in the Bonavista area for the twenty or so years prior to 1992 was maintained because the inhabitants were adequately embedded in local institutions and to some degree in wider national structures. People reproduced or maintained their lives in a setting in which they derived the means to do so from a combination of personal subsistence labour, paid work, help from others, and state aid (see also Ommer, Neis, McCay, this volume). Inevitably, there was variation among people and their households, but most were embedded in a local economy that included both formal and informal sectors, while also drawing on unemployment insurance (UI) payments and cash transfers from provincial and national governments. There was no perfect formula, as many people in the past moved away, thus reducing pressure on the local economy, though return migration was also significant.

In so far as it worked, the local economy required for most people and households that formal and informal relationships should complement each other; both were necessary. Unfortunately, practices in the formal economy were largely outside the control of local people or even of FPI itself. We have seen that a crisis in the fishery pervaded the whole northwest Atlantic groundfish industry, particularly northern cod. At the local level, the crisis disembedded or disarticulated individuals by eliminating their connections to the formal economy. As they had no obvious alternative to the fishing industry, it was unclear what the future would hold once state income support was reduced significantly. What did people think about their lives at this time? What would and could they do individually and collectively? We proceed by examining individual reactions prior to considering the limitations on collective action.

Research Data

Information on the experience of crisis on the part of residents of the peninsula comes from two sources. First, from September 1994 to January 1995, we attempted to interview all persons sixteen years and older in a randomly selected sample of 320 households, based on the most up-to-date, computerized list of residential numbers as supplied by Newfoundland Telephone. This process generated 619 completed interviews, from which we selected for analysis those in the normal labour-force age range (sixteen to sixty-five). Of the 473 persons who reported a current or previous occupation, 9.9 per cent were fishers, 26.2 per cent worked in fish plants, and 63.9 per cent held other occupations. In this analysis, we combined the categories 'fishers' and 'plant workers' as 'fishery workers.' Second, field researchers worked as participant observers in Bonavista and Catalina–Port Union for one year, beginning in September 1994. During this time, they kept field journals and conducted interviews for forty detailed life histories. The field work was designed to collect information complementary to that of the survey; we draw on both sources in this chapter.

Attitude to Life in the Area: Leaving or Staying?

What then does the moratorium mean to the people who most depend on the fishery? When the economic base of an area erodes or is suddenly erased, unemployment and out-migration might be expected. We seek to demonstrate that the experience of unemployment is widespread and that out-migration, driven by economic or educational factors, is likely, especially among young men and women (and see Neis, this volume).

Almost every worker in the fishing industry had been unemployed in the previous year and was still unemployed at the date of the interview. Indeed, 97.7 per cent (168 of 171) were unemployed in the week prior to being interviewed. Those with other occupations were less likely to have been out of work, but the comparable proportion is still dramatically high (63.6 per cent without employment). Many people accustomed to comfortable incomes experienced the moratorium as a shock, even if they had been aware of serious problems in the fishery.

An example from Catalina, once a vibrant deep-sea fishing port, can convey the nature of this experience. We take this passage from an interview with a married man, who had worked on a trawler for thirteen years. His income had dropped from over $40,000 to about $15,000 per annum.

Interviewer: So how have you been affected?
Respondent: Only way I was affected now is money-wise ... just like that! Snapped from you, just like that!
Interviewer: How did you or have you adjusted to that?
Respondent: I can't say it was easy ... I don't worry. It's pointless to worry.
Interviewer: Are families feeling the strain?
Respondent: Oh, God, yes, definitely ... I think in a lot of cases it's tearing families apart. There's a strain on my family ... All of a sudden I'm home all the time. Even [our daughter] is having a hard time coping with me being around, and I'm having an even harder time ...

Will people leave the area? Respondents were asked if they expected to be living elsewhere within one and within five years. On this basis, we divided the sample into leavers and stayers. Even if some leavers choose to remain and some who expect to remain end up moving away, the question does capture the plans of the population at a critical time. Migration plans do not differ significantly between fishery workers and those with other occupations; 31.0 per cent of the former and 34.4 per cent of the latter expect to leave. Of those who have recently considered leaving, two-thirds (66.5 per cent) would go mainly to find a job, and more than half (54.3 per cent) would move away from Newfoundland. That so many are thinking of leaving is a symptom of the stress created by the area's economic limitations and a cause for concern about the viability of

Table 16.1 Migration intention by attitude to living in the area (percentage)

Level of satisfaction	Leavers	Stayers	Total
Very satisfied	17.4	32.1	27.5
Satisfied	51.3	59.2	56.7
Neither satisfied nor dissatisfied	16.4	6.7	9.8
Dissatisfied	12.3	1.9	5.2
Very dissatisfied	2.6	0.0	0.8
Total	31.9	68.1	100.0

$N = 612$; chi-square $p. < 0001$.
Cramer's $V = .32$; lambda with migration intention dependent $= .12$.

local social life, especially since more recent fieldwork points to growing despair and heavier migration.

We have claimed that people's perceptions or interpretations of their circumstances must form part of an adequate explanation of their actions. What is driving some away is certainly not overall dissatisfaction with living on the Bonavista Peninsula. Indeed, leavers and stayers differ only to a modest degree in various measures of attachment or quality of life there. In response to our general question about satisfaction with living in the area, it is evident from Table 16.1 that leavers are significantly more likely than stayers to have a negative outlook. Yet the powerful impression conveyed by this table is the high proportion, even of leavers, who are satisfied or very satisfied. Most are not being driven away by a distaste for life there.

When respondents were asked to compare their area with others in nine aspects broadly connected with quality of life, leavers were somewhat more likely to rate Bonavista worse than others with respect to its being a suitable place to raise children, availability of jobs, cost of living, and having relatives at hand, as indicated in Table 16.2. Still, the degree of association was weak (as measured by Cramer's V), and our general impression is that both groups regarded the area well in all aspects except for availability of jobs and recreational opportunities.

Attitude to the area is thus a poor way of separating those who will leave from those who think that they will remain. The vast majority perceive the Bonavista area as relatively poor in job prospects, and most people do realize that they need employment, at least for part of the year, to maintain an adequate standard of living. Even those attached to the area in other respects might be pushed out on the grounds of employment. Yet people are also aware that the range of jobs open to them elsewhere is likely to be restricted if they lack the resources or characteristics that employers expect. Increasingly, low education levels and

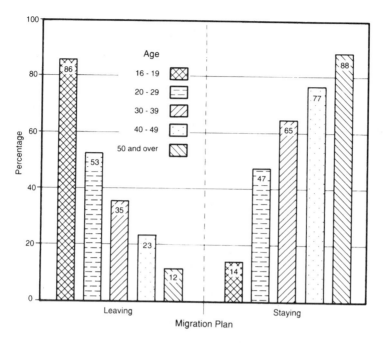

Figure 16.1   Migration plans by age, Bonavista

middle age are recognized as barriers to employment. In conjunction with this, people know that the cost of living is much higher in urban areas, where they would not own a home and where subsistence production is likely to be more restricted. Thus, provided that there is some hope of maintaining life locally, we should not expect many older and less-educated people to plan to leave.

These observations are hardly novel, but they are relevant to this analysis, and they are supported by our data. Figure 16.1 shows a dramatic, if anticipated decline in the proportion of people who expect to be living elsewhere as age increases. As many as 85.7 per cent of persons under age twenty expect to leave, but only 11.1 per cent of those over fifty. As to education, Table 16.3 demonstrates a strong link with propensity to migrate. Of those with less than grade 9 education, only 18.2 per cent expect to leave, in contrast with 50.4 per cent of people educated beyond high school. It is the latter group that is best equipped to find satisfying employment elsewhere.

Another factor that may be connected to keeping people in the area is home-ownership. We might expect homeowners to be more inclined to remain in the Bonavista area, because they would almost certainly be unable to live in equiv-

Table 16.2 Migration intention by perception of area characteristics (residents 16–65 years old)

| Characteristic | Migration intention | Percentage who perceive area compared with elsewhere as | | | Chi-square probability; Cramer's V |
		better/much better	about the same	worse/much worse	
Having relatives close by	Leavers	71.0	15.1	14.0	$p. < .01$
	Stayers	78.5	15.3	6.2	$V = .13$
Having many friends around	Leavers	78.5	15.6	5.9	NS
	Stayers	82.5	12.8	4.7	$V = .05$
Place to raise children	Leavers	69.9	18.3	11.8	$p. < .001$
	Stayers	83.6	11.7	4.7	$V = .17$
Availability of jobs	Leavers	2.2	3.8	94.1	$p. < .05$
	Stayers	1.8	10.8	87.4	$V = .12$
Cost of living	Leavers	38.7	49.5	11.8	$p. = .05$
	Stayers	40.5	53.6	5.8	$V = .11$
Crime level	Leavers	79.6	14.0	6.5	NS
	Stayers	83.6	13.8	2.6	$V = .09$
Weather	Leavers	14.5	57.0	28.5	NS
	Stayers	17.5	58.8	23.7	$V = .06$
Recreational opportunities	Leavers	17.3	19.5	63.2	NS
	Stayers	15.6	25.3	59.1	$V = .07$
Affordable housing	Leavers	60.8	33.9	5.4	NS
	Stayers	64.2	32.8	2.9	$V = .06$

Table 16.3 Migration intention by education ($N = 535$)

Education	% Leaving	% Staying	Total
Less than grade 9	18.2	81.9	31.8
Some high school	48.6	51.4	20.0
High school	34.0	66.0	26.9
Beyond high school	50.4	49.6	21.9
Total	35.5	64.5	100.0

Cramer's $V = .28$; chi-square $p. = < .0001$.

alent housing elsewhere in the country for some considerable time after their move (if ever). The relationship is highly significant for the whole sample, but when we introduce a control for age by dividing the sample into four age categories, with thirty as the cutoff point for the youngest category, we find that the relationship between homeownership and plans to stay or leave exists only for persons under thirty. The strong majority of this group of young adults (83.5 per cent) do not own their homes, and most of the non-owners expect to leave (65.9 per cent). In contrast, only 38.5 per cent of young homeowners expect to go elsewhere. In the other age groups, movers form a minority, and their ownership of a home does not tie them to Bonavista. Other indicators of social ties to the area, such as whether or not the respondent was born there, or was married, or had children, failed to distinguish leavers from stayers after controls were introduced for age. Exactly what is leading some to think of leaving and others not remains unclear.

Those Who Stay

Are the people who expect to stay in the Bonavista area irrational dreamers or realistic assessors of their situation? We noted above that people are positive about living in the area to a remarkably high degree. What in addition keeps them there is probably a combination of factors, among which recognition of the lack of opportunity elsewhere, availability of state income support, and the informal economic sector are most important.

Our fieldwork, as reflected in the opening quote, told us that local people were divided on the question of whether a reasonable future was likely in this area. Some felt that the fishery had little or no future. They saw no other industry to replace it and believed that the federal government would like to get rid of the inshore fishery. On many occasions, we heard about the young people who had left for St John's or the mainland, leaving villages populated by almost nobody but older adults. Even if these perceptions are exaggerated, as we believe them to be, they bring out the depressed mood of many of our respondents.

Others, however, expected that the fishery would eventually come back and that their jobs would be restored. Failing that, they did not expect the state to leave them to starve; some kind of ongoing income support, perhaps less generous than TAGS, could be anticipated, especially for those who lacked the characteristics that would make them easily employable elsewhere. For them, sticking out the tough times, while continuing to enjoy the advantages that the area offers, makes sense – with adjusting to changing conditions being perhaps less risky than leaving or protesting (Hirschmann 1970).

Staying is a serious option in part because people are accustomed to making use of their local environment to satisfy many of their food and shelter needs in an informal sector that integrates many different households through the provision of various unpaid services across households.[2] Informal economic activity generates resources that enable people to survive in this economically depressed area at a level of living that they find tolerable. However, this type of informal sector is certainly not exclusive to rural areas or to Newfoundland, but it is too little recognized and supported in official policy, despite being so central to sustainable social life.

Unofficial or unrecorded work for pay is thought to be widespread, especially since the moratorium. Several informants told us that there is a great deal of 'under the table' employment, especially by TAGS recipients, who are free to offer their skills now that the fishery is closed, but we cannot provide a firm quantitative estimate. The extent of such occasional odd jobs is surely underestimated in our survey, in which 12.2 per cent of respondents reported such employment. The most common activities were babysitting (30.1 per cent), which Barbara Neis has examined (this volume), carpentry (16.4 per cent), and mechanical work (9.6 per cent). In addition, another 4.8 per cent made some item for sale, with knitwear being the most common.

We suspect that a much greater proportion of the informal sector is found in unpaid work activities. With few exceptions, all households must organize their own domestic labour and, where necessary, child care. This work, mostly done by the women of the houselhold, if there are any present, is not discussed here (but see Neis, this volume), event though it is essential to the maintenance of social life. A large proportion of people helped build the homes in which they live (24.3 per cent of the whole sample and 37.1 per cent of the men), and many respondents live in households in which at least one person draws on local natural resources to meet his or her subsistence needs. In all probability this emphasis on home building and repair is much more widespread here than in urban areas and offers a powerful incentive to stay. Since 1992, former fishery workers have had more time to devote to these activities, and many have improved their homes.

Table 16.4 Reports of unpaid work provided for and received from other households.

Activities done without expecting pay	% reporting tasks typically done in the area (rank order)	% reporting having done this work for others (rank order)	% reporting having received this assistance from others (rank order)
House cleaning	21.8 (1)	9.0 (4)	5.2 (7)
Babysitting	16.3 (2)	10.3 (2)	5.3 (6)
Painting	14.8 (3)	6.8 (5)	7.3 (2)
Carpentry	10.5 (4)	3.9 (9=)	5.6 (5)
Helping elderly	10.4 (5)	3.8 (11)	1.9 (10)
Clearing snow	8.9 (6)	5.4 (7)	3.6 (9)
Cutting wood	8.4 (7)	5.3 (8)	6.0 (4)
House repairs	8.6 (8)	12.0 (1)	13.7 (1)
Cooking meals	8.1 (9)	9.3 (3)	6.5 (3)
Giving lifts	6.4 (10)	5.5 (6)	4.1 (8)
Cutting grass/yard	5.3 (11)	3.9 (9=)	1.5 (11)

Food is often obtained from the local environment. This area has abundant blueberries and partridge berries, which have been picked in fall as long as people have lived there. In our study, 81.2 per cent of respondents reported berry picking as a subsistence activity; it was the most commonly reported. Growing of vegetables appears to have revived since the crisis, though not everyone has access to suitable land; 38.2 per cent did cultivate some vegetables, according to our study. Many people expressed concerns about the declining supply of accessible, high-quality wood. Nearly half (47.6 per cent) of our sample lived in households where wood was obtained from the local forest. Cod (in years prior to the moratorium) or trout were taken by 70.0 per cent, while 59.2 per cent lived in homes where someone hunted such species as rabbit, moose, and turr (the local name for sea-birds of the murre family).

Beyond what people do for themselves are the resources they provide for and receive from members of other households through exchange. Though our informal interviews often produced statements to the effect that people were much more individualistic now than in the past and that informal help was less common, most respondents (79.2 per cent) could list at least one thing that would commonly be done without the expectation of pay. More than half also stated that they had helped other households in some way (64.5 per cent) or had received such help (50.5 per cent) in the past year.

Table 16.4 shows the tasks that respondents most commonly believe people in the area generally do for each other without expecting to be paid. There is a considerable difference in the rank order of the unpaid tasks that respondents believe are typically performed for and by people in the area and the most commonly mentioned unpaid task that the respondents themselves undertake for people in other households or have people in other households provide for them. The majority of these eleven most frequently mentioned tasks are gender-specific. Almost no man reported cleaning house, babysitting, or cooking meals, for example (data not reported here). Only two tasks – giving rides and, to a lesser degree, painting – were gender-neutral. Relatives were assisted on almost half the occasions reported, which again emphasizes the practical kinship links that integrate many people in the area. Other assistance was directed to friends, neighbours, and voluntary associations.

Collective Action?

When embedded relations are disarticulated, individuals may try to reconstruct an adequate institutional environment or re-embed themselves in some way. They may act as individuals but also as participants in wider social groups, ranging from households to regional associations. So far we have considered only individual responses. Now we turn to consider the limits to collective action. Though mayors, union officials, and leaders of voluntary organizations have certainly expressed concern and demanded remedial action, the people of the area have rarely protested what is happening to them or promoted new strategies. For example, in order to discuss the provincial government's plan to restructure funding and organization of regional development, the Bonavista area development association held a series of public meetings throughout its territory, at which the highest turnout was twenty-five (in the town of Bonavista). An exception is the large demonstration (about two thousand people) in Bonavista that was organized by the Fisherman, Food and Allied Workers as a protest against a reduction in TAGS payments.

Collective action does not arise simply because large numbers of people are dissatisfied with some aspect of their lives. The experience of deprivation is not by itself enough to propel people to action. They need resources and a cohesive social network that will allow or even encourage them to mobilize these resources (Ferree and Miller 1985; Jenkins 1983; Klandermans 1984; Tilly 1978). Of course, people do interact and know each other, but our respondents repeatedly told us that residents are more individualistic than they used to be, especially in Bonavista. Community feeling or spirit was said to be lower, with people turned more towards themselves or their immediate families. One woman, who had lived all her life in the area except for a few years in St John's, proclaimed: 'There's absolutely no sense of community here.' She thinks that people prefer to spend their time backbiting. Another person, a fairly recent newcomer to Bonavista, described people in atomistic terms: 'I don't know if it's ... just the way I'm thinking, but ... people seem to be in their own little world a lot. You say hello or good day to someone and sometimes you find you don't get much of a response, if any. And whether they're wrapped up in their own little world ... and ya see lots of times people don't seem, don't seem to carry a lot of smiles.'

In similar fashion, a local skipper described Bonavista people: 'Bonavista is such a different group of people here ... People wants their own way and ... I've always found – and I'm from here – and I've always found that you can't get people to agree on anything. You can't get people to pitch in on anything. There's a bit of pride ... community pride from ... being from here, but people don't really work together well ... here, right. And it comes from ... from the sea, same thing. I mean, you got to take t'ings as they are; you can't paint it prettier than what it is, you know. And that's the way it is. I find the people of Bonavista contrary.'

If they are to act collectively, people must have some hope that they can make life different. This middle-aged woman clearly saw local people as having little to offer in recent years: 'Don't seem like there's much community spirit here. The moratorium have changed a lot of people – not changed, but it gave them a different outlook; like, they feel that they're not as useful as they were. I guess after a while you'd get to a point where you got nothing to do and nowhere to go, and you know, there's nothing else to go to work at here. So what do you do? It's either move or stay and stay on the moratorium [i.e., TAGS income support].'

Bonavista may have been different from Port Union–Catalina, more controlled by merchant families, so that ordinary people lacked a tradition of involvement in public issues. The same respondent noted the inability to co-operate around the proposed John Cabot celebration plans for 1997 and felt that it might be:

because of the background, the upbringing. Down through the years there was two main merchants in the town.
Interviewer: The Swyers and the Ryans?
Respondent: Yeah, the Swyers and the Ryans. So what I've learned is that down through the years they were more or less controlling the town, and they called the shots, right. I mean, you had fishermen working for them so they were pretty well controlled by these two merchants, and I suppose the fishermen worked for them. And I suppose up through the years it probably got embedded in their minds that ... well, if it's OK for their grandfathers it's OK for them. And then probably, you know, where it's located so far away

from any other major centre there's no ... influence from other major centres. And even council, I mean, it seems to be so difficult for even council to go along with some of the changes to make things better for the future. There's no movement at all.

Many people appeared despondent (and thus unlikely to have the motivation to act collectively): 'I would say that people are to the point where ... well, they just ... they just don't seem to ... care any more. A lot of people are depressed from the closure of the fishery. There's really not a whole lot to look forward to from day to day. And I think when you get into that state of mind that you just don't ... you just don't seem to have, well, a whole lot of care. You don't ... there's really ... no energy generated. You don't get excited about a whole lot, and ... ah, there seems to be so much negativity around here. And if you dwell on negativity you sort of have a negative effect; it rubs off on someone else.'

Another source of division that militates against collective action is that many people feel that fishery workers have received more favourable treatment than others through the TAGS program. For fishery workers, the TAGS plan, which resulted in part from union pressure, probably reduced significantly the possibility of local unrest and collective action. In addition, it has been divisive and has inhibited the possibility of other collective action. TAGS recipients are sometimes said to be 'milking the system' by using this income to buy new cars and build homes without planning seriously for the future. For example, this interview brings out the negative attitude:

Interviewer: Do you think there's been much conflict between people that have been receiving TAGS and those that haven't?
Respondent: You got a lot of rumblings, you know, people looking for a job – and you get all these qualifications ... Most people go away and go to school and come back. The only jobs now are – there's all the benefits now for an employer if they hires a TAGS person. It's not really a level playing field. There's some resentment there.
Interviewer: Yeah? Ive heard some people say there's been some undercutting for a while there, you know, people getting TAGS and working under the table?
Respondent: Yeah, there's been some of that. Contractors – they're bitter, I don't think that's too harsh a word. They're bitter, because that's their livelihood and the TAGS people are getting their income plus they're cutting in on the people doing construction. They don't like it. They don't like it.

Divisions based on different personal circumstances occurred no doubt in each town and village, but others cut across the region. For example, there has long been rivalry between Catalina and Bonavista, which was sometimes evident in the past in fights that would break out between groups of young people.

As a well-off Bonavista skipper remembered it:

> There was always a rivalry between Bonavista and Catalina, you know. You wouldn't go over there takin' ... lookin' for their women, you know! And they was comin'over here ... so you was beatin' the shit out of them and they were beatin' the shit out of you, you know!
> *Interviewer:* Oh yeah?!
> *Respondent:* That was pretty well gone time ... I saw it, but time I came along that was pretty well gone, but ... But it's still there.

A critical recent instance is found in the conflict over FPI's desire to move crab production from the older plant in Bonavista to the newer one in Catalina. The workers in Bonavista feared that they would eventually lose out to Catalina workers and that their town might die without its own plant. The Catalina plant workers (who included many people from Bonavista as well) were willing to accept the transfer of crab workers from Bonavista to 'their' plant in the hope that the Catalina plant, once it was open again, might expand and process some groundfish (perhaps previously underused species or imported cod, as in National Sea's plant at Arnold's Cove, in Placentia Bay).

With so many people discouraged about the future, accustomed to relying on state income security, and divided within and across communities, it is not surprising that attempts at collective responses to the crisis have been rare and that public support has been luke-warm for those who have tried to provide some leadership.

Conclusion

We have claimed that the cod crisis disembedded large numbers of people, who could no longer function as workers in the formal economy. It was as if Bonavista society as a whole was now drifting rather than being embedded in its local economy. In their examination of roughly similar situations, Barrett et al. (1995) describe the positive collective action taken by those affected to re-embed their communities. In their examples, this was most evident in Bugøynes, in northern Norway, and in Sambro, Nova Scotia, but less clearly so in Fogo Island, Newfoundland (see McCay, this volume).

Our study of the Bonavista Peninsula provides a much less positive image of rural reconstruction. We found individual action predominant. People worked out strategies for the future, whether staying or leaving, that depended on themselves or their immediate families and on state support rather than on the reality or possibility of collective action. We examined some social structural and attitudinal factors that might be expected to help us distinguish between the critical individual responses of leaving and staying. We also considered the dimensions of local social structure that made this area a weak organizational base for collective action.

The concept of embedded relations is useful because it points to the socially disruptive impact of locally uncontrollable events such as the cod moratorium and also to the kind of integration of economic and other relationships that must be constructed if social life in the area (and others like it) is to be maintained. In the Bonavista Peninsula, we cannot expect the kind of concerted political action that took place in Bugøynes. Rather than waiting for some external, paternalistic state program, local residents should themselves overcome their divisions and take steps within their means to solve their own problems. Working out an economically rational and socially responsible use of the area's fish-processing capacity would be a starting-point, to which local organizations, FPI, and the various levels of government should accord maximum attention. Failure will mean loss of so many young people that the area will lack vitality and become seriously depopulated, if not altogether dead.

NOTES

1 Thus Ernest Bishop, vice-president of marketing, reflected on the announcement of the moratorium: 'Everybody looked forward and said, "We've got no resource, so where do we go from here?" That's when we had to make some changes' (*Globe and Mail* 1994).
2 For a more extensive analysis of the informal sector elsewhere in Newfoundland, see Felt and Sinclair (1992) and Felt, Murphy, and Sinclair (1995).

WORKS CITED

Allen, Sheila, Alan Walton, Kate Prucell, and Stephen Wood, eds. 1986. *The Experience of Unemployment*. London: Macmillan.
Barrett, Gene, Svein Jentoft, and Bonnie J. McCay. 1995. 'Global Fishing Villages in Crisis: Embeddedness and Community Development on the North Atlantic Rim.' Paper presented to the European Society for Rural Sociology, Prague, 31 July–4 Aug.
Burnham, Patrick. 1988. *Killing Time, Losing Ground: Experiences of Unemployment*. Toronto: Wall and Thompson.
Buss, Terry F., and F. Stevens Redburn, with Joseph Waldron. 1983. *Mass Unemployment: Plant Closings and Community Mental Health*. Beverly Hills, Calif.: Sage.
Etzioni, Amitai. 1988. *The Moral Dimension: Toward a New Economics*. New York: Free Press.

Evening Telegraph. 1994. 24 Aug. 1994. St John's.

Felt, Lawrence F., and Peter R. Sinclair. 1992. '"Everyone Does It": Unpaid Work in a Rural Peripheral Region.' *Work, Employment and Society* 6, no. 1: 43–64.

Felt, Lawrence F., Kathleen Murphy, and Peter R. Sinclair. 1995. 'Everyone Does It: Unpaid Work and Household Reproduction.' In Lawrence F. Felt and Peter R. Sinclair, eds., *Living on the Edge: The Great Northern Peninsula of Newfoundland,* 77–102. St John's: ISER, Memorial University.

Ferree, Mary Marx, and Frederick D. Miller. 1985. 'Mobilization and Meaning: Toward an Integration of Social Psychological and Resource Mobilization Perspectives on Social Movements.' *Sociological Inquiry* 55: 38–62.

Giddens, Anthony. 1984. *The Constitution of Society.* London: Oxford University Press.

Globe and Mail. 1994. 17 May. Toronto.

– 1995. 28 Jan. Toronto.

Granovetter, Mark. 1990. 'The Old and the New Economic Sociology: A History and an Agenda.' In Roger Friedland and A.F. Robertson, eds., *Beyond the Marketplace,* 89–112. Hawthorne, NY: Aldie de Gruyter.

Hirschmann, Albert O. 1970. *Exit, Voice, and Loyalty.* Cambridge, Mass.: Harvard University Press.

Jenkins, J. Craig. 1983. 'Resource Mobilization Theory and the Study of Social Movements.' *Annual Review of Sociology* 9: 527–53.

Klandermans, Bert. 1984. 'Mobilization and Participation: Social-Psychological Expansions of Resource Mobilization.' *American Sociological Review* 49: 583–600.

Perucci, Carolyn C., Robert Perucci, Dena B. Targ, and Harry R. Targ. 1988. *Plant Closings: International Context and Social Costs.* Chicago: Aldine de Gruyter.

Schrank, William E., 1995. 'Extended Fisheries Jurisdiction and the Current Crisis in Atlantic Canada's Fisheries.' *Marine Policy* 19: 285–99.

Tilly, Charles. 1978. *From Mobilization to Revolution.* Reading, Mass.: Addison-Wesley.

17

Directions, Principles, and Practice in the Shared Governance of Canadian Marine Fisheries

EVELYN PINKERTON

Political ecologists have always been particularly interested in cases of successful resource-management problem-solving by self-organizing local and regional bodies and have generated middle-range propositions or hypotheses about the conditions that permit such success. In the case of fishing communities and amalgamations thereof, they define 'success' in terms of the ability of such bodies to take on the powers and responsibilities of resource management, and to deal with resource management problems, both directly and indirectly.

Local or regional bodies, of course, may approach resource problems quite differently from bureaucratic line agencies. While the latter have the legal mandate to manage and conserve fisheries, the former often mediate or contain conflicts over allocation, gear regulation, and harvest timing and location and thus control fishing effort both directly and indirectly. There are over fifty well-documented cases of such problem-solving by local bodies in pre-industrial and post-industrial settings (Cordell 1989, Dyer and McGoodwin 1994; Schlager and Ostrom 1993; Wilson at al. 1994), and many more that are incompletely or poorly documented.

Basic Assumptions

Cultural ecology (now called political ecology) is a subarea of anthropology that originally analysed the way in which small-scale societies adapted to their local environments over long periods, particularly by developing customs and rules governing access, use, and distribution of resources. These societies, it is now known, learned to do so by trial and error, adapting over generations to the long-term carrying capacity of the area and resources they used. In some cases they developed mechanisms to limit their own population growth; in others they simply controlled access to their resources so tightly that they could not be overharvested.

While some cultural ecologists saw the adaptive process as purely material in nature (Harris 1979), others examined the way in which such learning became encoded – not only in rules, but also in rituals, beliefs, norms, and values – in such a way that the rules of resource use became deeply embedded in cultural assumptions about how the world works, what constitutes proper human behaviour, and who has the authority to enforce the rules. Thus Swezey and Heizer (1977) examined how shamans on the Klamath River in California regulated the salmon fishery in former times by overseeing group construction of a weir and then setting the timing of the ritual opening and closing of the weir in accordance with their understanding of salmon escapement needs.

In the present day, knowledge and the ability of local authorities to make effective rules on resource use reinforce the behavioural norms that produced sound management practices. Social controls of this kind are strong because they operate both at the level of political rules (secular authority) and at the level of beliefs and values (religious authority). An individual who breaks the rules is labelled as morally or spiritually bankrupt, in addition to being a rule-breaker. Whether material, political, or ideological factors are seen as primary inspirations for such rules, cultural ecologists and others agree that pre-industrial societies were often able to live within the natural limits of resource abundance in the fishing and hunting territories they occupied (McCay and Acheson 1987; Williams and Hunn 1982; Usher and Tough, Newell, this volume). Success in not overfishing is understood to derive from socio-political and ideological controls, because the technology and incentives existed then as now to overharvest many fish populations (Pinkerton 1994; Cadigan, this volume), and because levels of harvest were stable over long periods – fur, fish, and other resources were reported by Europeans as being in abundance at the time of first contact.

Culture and Ecology in Industrialized Settings

The types of mechanisms for encoding resource knowledge into enforcable rules that worked in small-scale pre-industrial settings have also been found in fishing subgroups within industrialized societies. James Acheson's (1975) ground-breaking study of harbour gangs in Maine outports showed that fishing communities there had organized themselves to enforce informal community rules that protected the territory used by each individual lobsterer. The local community helped the fisher exclude outsiders from the territory that he or she worked, which fronted on his on her property. This exclusion in effect granted a right of access to each individual fisher, who could then regulate his or her effort in the most cost-effective manner, without fear of a competitive race for first capture. Acheson found a significantly higher catch per unit of effort (CPUE) in the outports where harbour gangs were able to defend individual territories, than in the more urbanized areas, where social controls were weaker and territories more difficult to defend. A similar system, in operation since the late nineteenth century, was later documented in Little River, Nova Scotia (Brownstein and Tremblay 1994, and see Manore and Van West, this volume).

The boundary mechanisms for local-group definition and exclusion are often complex and culturally specific (Douglas 1992), and this is reflected in a rich and varied scholarship. Berkes (1977, 1981), Feit (1973, 1988), Langdon (1984), Usher (1987), and Wenzel (1986) have all studied contemporary hunters and fishers in the Canadian and Alaskan north, who regulated their use of multiple species in complex ways, depending on the presence or absence of 'outsiders.' Marine biologist Milton Freeman has also found parallels in the community-based management of marine mammals in the Canadian Arctic and in Japan (Freeman and Carbyn 1988, Freeman et al. 1988). Mechanisms for achieving de facto or de jure exclusion from local fishing territories have been key to the success of all these systems, because state regulatory systems and the unimpeded operation of industrial markets would probably have led to over-exploitation of local resources by outsiders if locals had not had the ability to exclude them.

McCay's early work on informal self-regulation within industrial fisheries examined issues such as the indirect effects of informal supply management by New Jersey whiting fishers (McCay 1980, 1987). She found, for example, that a local marketing cooperative's efforts to control the flow of raw fish into the market (in order to increase prices) had the indirect effect of limiting fishing effort. McCay (1984, 1987) called attention to the informal poaching traditions of local populations formally excluded from the commons by enclosures and allocations of rights based on class and power and to the importance of legitimacy in effective management systems. Thus local populations have often developed informal mechanisms for their own inclusion (illegal fishing) even when they were formally excluded, because their exclusion violated local traditions and was not considered legitimate. In these situation, state enforcement of exclusion has proven exceptionally difficult or virtually impossible to maintain (see also McCay, Thoms, this volume).

Reconceptualization of Local Self-Management as Nested in Co-management Agreements

The earlier work of these scholars analysed local systems as regulating themselves independently of, and often in spite of, governmental authority. Though government usually played some role, and sometimes even adopted certain local rules into law or regulation, the creative potential of this relationship was little analysed. On the west coast of North America, however, Washington state presented a living model of how treaty Indian tribes (the U.S. term), as informally self-regulating fishing communities, could enter into agreements with government to formalize and protect their self-management rights. *U.S. vs. Washington* (1974), known as the *Boldt* decision, recognized a Washington treaty Indian tribe's right of access to a fair share of the fish, but Judge Boldt concluded that the tribes would be able to exercise this right only through co-management of all aspects of harvest planning and implementation with the state department of fisheries. He also specified a number of tests that tribes had to meet before qualifying as self-regulating bodies with co-management rights – they had to possess, for example, clearly defined membership criteria and enrolment procedures, as well as to work with professional staff biologists.

This model was in effect a self-managing body nested inside a co-management relationship with government. Certain rights to internal self-regulation were retained, with the Indian tribes making their own rules for internal allocation, licensing, transfer of licences, and running their own monitoring and enforcement systems. Other rights they shared with the state after a co-management agreement, called the Puget Sound Salmon Management Plan, was finally negotiated in 1984. The plan consisted of a schedule for joint tribal–state decision-making in all stages of stock assessment, pre-season harvest planning, and in-season harvest management, as well as coordination of tribal and state monitoring and enforcement (Pavel 1989; Pinkerton and Keitlah 1990).

Canadian fisheries policy-makers viewed the successful implementation of this model as a milestone, for three reasons. First, it reinforced an existing positive impression of the benefits of co-management, because the Washington case parallelled the relationship of the Japanese government with the highly successful Japanese inshore fishing cooperatives. Under their system, Japan's inshore fisheries exceeded their offshore sector in productivity and economic value (Pinkerton and Weinstein 1995). Second, reaching this state-level agreement enabled negotiators to break a twenty-five year impasse by achieving cooperation at the international level: Indian tribal–state cooperation in Washington state allowed completion of the Pacific Salmon Treaty between the United States and Canada in 1985 (Cohen 1986, 1989). Clarification and implementation of tribal rights allowed the Indian tribes along the Columbia to join Washington state officials in bringing Alaska into the treaty. Otherwise, the tribes could have implemented *U.S. vs. Washington* on the Alaska catch of Washington-bound salmon. Third, the model suggested a way for dealing internally with Canada's First Nations and was used to make peace with the Quebec

Micmac in the Restigouche Accords of 1982. Canadian authorities also saw the model as applicable far beyond relations with First Nations. Indeed, fisheries minister Roméo LeBlanc began using the term 'co-management' in 1982 with respect to preliminary experiments that he conducted in the Maritimes (Lamson and Hansen 1984).

Pinkerton (1988, 1989) conceptualized the potentially broad scope, and different levels of operation, of fisheries co-management systems by analysing the example of the Washington Indian tribes and through a comparison of a range of other North American cases. Taken together, these examples illustrated the full potential of the model as a power-sharing arrangement and as a tool to deal with resource management problems, capable of including policy-making, data gathering and analysis (monitoring, stock assessment), harvest planning, allocation, enforcement, enhancement, coordination of functions locally and regionally, and habitat protection. Twenty propositions (testable hypotheses) were generated from these cases about the logistical conditions favouring the genesis of co-management and about the resource problems that could be addressed through co-management arrangements. Meanwhile Norwegian sociologist Jentoft (1989) and his students (Jentoft and Kristoffersen 1989) documented similar historical arrangements in Europe, particularly in Norway, where fishers elected an inspector in the northern cod fishery to regulate the fishing effort by time closures.

Appropriate Scale, Scope, Parties, and Levels of Collaboration

A second generation of co-management studies examined more complex arrangements involving multiple parties instead of just one government agency and one local group. These studies also explored in more depth the implications of agreements at different geographical scales (small, medium, large), at different governmental levels (local, regional, national), and with different scopes (broad and multiple issues, few and narrow issues). They developed further hypotheses concerning the appropriate conditions for different types of co-management and then analysed the ideal conditions for any collaborative arrangement.

Sometimes it is the nature of the problem that dictates the necessary parties and level of governance. Many resource problems, such as habitat protection, cannot be solved through an agreement between one community and one government agency. Even if a fishing community holds formal rights to protect fish habitat, as the Washington Indian tribes did after 1981, this right cannot be implemented without some institutional arrangement to deal with the conflicting rights and political power of other parties. Complex policy and implementation problems therefore involve many parties and often multiple levels of governance.

For example, the state of Washington attempted to implement the tribal right to protect fish habitat by working with a policy-making body of four state agencies (Ecology, Fisheries, Natural Resources, Wildlife), local Indian tribes, environmentalists, and industrial watershed users (mainly logging companies, hydro-power companies, and irrigators). This body of cooperators in the Timber, Fish, Wildlife Agreement (TFW) hammered out policy-level agreements on general principles and directions, including what kinds of data were necessary for them to agree on further policies. They co-managed the policy-making function and operated on a state-wide level only. The TFW cooperators' policy-level decisions were implemented on the local level by interdisciplinary teams representing the perspectives of the cooperators. The teams reviewed timber-harvesting plans, and their consensus decisions provided the implementing state agency (Department of Natural Resources) with information about the degree of habitat protection required for each logging permit. This meant that the teams co-managed habitat protection at the local level.

The TFW Agreement was successful as a political exercise in getting the appropriate parties to work together, recognizing common interests and political necessity in solving the problem in the least disruptive way – the diversity of the parties reflected the complexity or scope of the problem. It was less successful at getting these actors together at the appropriate level and scale of operation, a problem explored in greater depth by Wilson and Dickie (1995) and Ostrom (1995). However this state-level and local-level co-management agreement on policy-making, data collection, data analysis, and monitoring of habitat protection did precipitate further arrangements that came closer to the appropriate scale of attack – now agreed to be that of the watershed (Pinkerton 1992). The later addition of the regional (watershed) level of analysis and planning filled this gap and allowed the process to follow the general systems-theory 'law of requisite variety' (Ashby 1968) – the 'law of requisite adequacy' (Ramos 1981) or the 'law of the situation' (Follet 1940). All these scholars agreed that any regulatory system needs as much variety in the actions that it can take as exists in the system that it is regulating and that managers need to work within the norms existing in specific situations. Destruction of fish habitat occurs on both a local and a regional scale through cumulative effects in different parts of a watershed that ultimately affect the entire basin. The original TFW Agreement was too restricted in its terms of reference, because the balance of power at the time did not allow a more scale-appropriate agreement. The balance of power also changed as the problems – and the solutions – became apparent, at the same time as public opinion was sufficiently mobilized to support these changes.

The Skeena Watershed Committee

This type of multi-party model was sufficiently successful at involving the appropriate parties at appropriate governance levels to persuade others to apply it, with key alternations in scope and scale where required. British Columbia's fisheries managers learned from the successes, and the limitations, of the Washington state examples and launched a home-grown, regional version of multi-party co-management. This scheme was far more comprehensive in scope but had enough management functions included to address the problem, and it operated at an appropriate intermediate (watershed) scale, making its success more likely.

The Skeena Watershed Committee brought together sport, commercial, and Aboriginal fishers on the Skeena River watershed, along with federal and provincial regulators, to solve conflicts over harvesting levels for salmon species of differing abundance. All partners in the system agreed on a three-year pre-season harvest-planning process, and then continued to oversee stock-assessment studies, monitor harvest, oversee habitat restoration, and make policy on enhancement and other long-term planning issues (Pinkerton 1996). This multi-party watershedwide process had a good chance of success because it dealt with resource problems at a governmental level where conflicts among differing parties can be addressed through a common interest in the productivity of the watershed. The management functions addressed (stock assessment, harvest planning, habitat protection, enhancement) have a broad enough scope to reflect the complexity of the problem, and the regional scale is large enough for people to have a common interest in a meaningful geographic unit but small enough to allow the building of long-term cooperative relationships based on increased shared knowledge about the resource, enlightened self-interest, and even stewardship.

Under these kinds of favourable conditions, new and unforeseen forms of cooperation can emerge out of the most difficult conflicts. Discussion of entirely new cooperative forms was occurring on the Skeena River by 1996, the third year of the agreement. For example, the Aboriginal sector (made up of three major groupings located at the river mouth, mid-river, and the top of the watershed) proposed a system of harvest sharing of the surplus salmon taken at the top of the system among all Aboriginal groups and the watershed committee as a whole. Further discussion has included forms of risk-sharing and benefit-sharing, whereby some of the salmon would be shared wherever it was caught. This solution, if accepted, would spread risks and benefits among those Aboriginal fishers who sometimes cannot fish at the river mouth for conservation reasons, or in river because of flow regimes, or at the top of the system because of marketing and quality issues. If harvest regimes can meet the test of adequate selectiveness at key times, and if sharing can be equitable, all watershed participants will benefit from being able to adapt to the physical and economic conditions of any particular season. Innovative discussion of this kind can emerge when the appropriate parties, governance level(s), scale, and scope are taken into account in the management process.

Of course, even when the scale, scope, parties, and level of governance are appropriate, success is not guaranteed, especially where the overlap of interests is not great. The Skeena Watershed Committee was highly vulnerable to losing the commercial sector if allocation or conservation pressures on it overshadowed too greatly the future hopes held out to it through potential improvements in data collection and analysis, enhancement, enforcement, and real power-sharing. Not only does it take time for confidence to build so that future benefits will be realized, but all sectors must be convinced that the other sectors genuinely accept each other's right to exist. The Skeena process ultimately failed on these two counts. For one thing, the small-gear (gillnet) commercial sector eventually got the impression that the sport sector really did want to eliminate its fishery in the river mouth. For another, the sport sector also made the error of pushing government too far too fast to extend the conservation plan into adjacent areas, thus affecting large gear (seine) and the major processors – and incurring the opposition of these major parties. With the majors strongly opposing the commitee, the small gear alienated, and discontent brewing over perceived unilateral government action (Pinkerton 1996), the commercial sector withdrew from the Skeena Watershed Committee in March 1997. The fact that several commercial representatives expressed strong regrets and mixed emotions to government indicates that they perceived 'how it could have been,' had conditions been a bit different. This example illustrates the overriding importance of primary bonds of trust between parties as the final determinants of whether institutions succeed or fail.

Parallel Findings in Political Science and Institutional Economics

While political ecologists examined institutions for their appropriateness and adaptiveness, political scientists and economists framed parallel findings largely in terms of economic rationality. Garrett Hardin's classic discussion of the so-called tragic problems associated with resource commons acknowledged that mutual coercion, mutually agreed to, would work – though he did not seem to know instances of it. Now we have many examples. Summarizing much of her own and other scholars' earlier work, Ostrom (1990) examined institutional design principles for 'nested quasi-autonomous units operating from very small up to very large scales.'

These principles were based on empirical studies of long-surviving, self-governing institutions. Some are concerned with the logistics of design and parallel seven of Pinkerton's propositions about the elements of successful co-managing units: 1) clearly defined boundaries of the resource itself and of authorized users; 2) harvesting regulations (about how, when, where, and how much of the resource may be taken and what kinds of 'provision rules' or fees) that are congruent with local conditions (and are often designed by locals). Others are about power-sharing aspects of design, but mostly power-sharing among harvesters, not between government and harvesters; 3) the necessity for individuals affected by operational rules (such as harvest regulations) to be able to participate in making them; 4) the accountability to harvesters of monitors who audit resource condition and harvesting; 5) graduated sanctions imposed on violators by other harvesters or their representatives; 6) rapid access by harvesters to low-cost local forums to resolve conflicts; and 7) the (formal or informal) right of harvesters to devise their own institutions without interference by external governmental authority.

While Pinkerton's conditions for co-management also include formal cost-sharing arrangements between government and communities, Ostrom focuses on the automatic cost-effectiveness of self-managing systems at the local level, in which harvesters can monitor and enforce regulations on fellow harvesters at the same time as they are fishing.

Ostrom's interest in small-scale systems may have been influenced by collective-choice theorists such as Mancur Olson, whose *Logic of Collective Action* (1965) predicted that incentives for collective action and self-organization would be far weaker in medium and larger-scale institutions, in which 'free riding' would be more difficult to monitor and police (i.e., costs of self-organizing would not be equitably shared). However, Ostrom also studied larger-scale settings in which collective action by many individuals could be organized through joint hiring of a monitor. She noted that the same design principles applied to self-organized groundwater management boards in California, with fairly dense urban populations and hetereogeneous actors (water companies, domestic users, irrigators).

Whatever the scale, this approach conceptualizes individual actors as making a rational economic calculation that the benefits of cooperation exceed the costs of not cooperating. For example, if the water users do not figure out how to regulate themselves, they face higher costs by having to import water and fight about allocation in court. Game theory, especially variations on the 'prisoners' dilemma,' is often used as a modelling tool to predict how individuals perceive the costs and benefits of various scenarios in multiple encounters of individual actors (Axelrod 1984; Ostrom 1990; Taylor 1987). In other words, it is economically rational at the individual level to act collectively.

This approach thus differs from political ecological approaches, which emphasize not only the existence or discovery of overlapping interests but also the existence or forging of common values and the existence or development of stewardship (values about how people can and should cooperate as well as values about the importance of maintaining a stable relationship with the resource). The 'social learning' aspect of political ecology also parallels the work of political scientists such as Robert Reich (1988) on the 'civic discovery' of common goals through the public discussion of policy. Reich, of course, envisages a larger scale and shorter public process than the medium-scale watershed planning model. The study of collaboration among multiple parties is now greatly assisted by an inter-organizational sociological literature (Gray 1991) and a corpus of trained mediators teaching principled negotiation and helping new self-regulating institutions develop their own operational rules. These theoretical and practical tools will only grow in utility as conflicts among sport, commercial, and Aboriginal fishing communities increase, at least on Canada's west coast.

Property Rights and Economic Problems in Fisheries

Many of these scholars also showed that self-governing institutions can solve economic problems – such as overcapitalization of fishing units – that many conventional economists considered could be dealt with only through private property rights or through state property and regulation structures. Schlager and Ostrom (1993), however, demonstrated empirically how small-scale, self-regulating fisheries institutions solved three resource-management problems that were considered crucial by fisheries economists. Their study compared forty-four self-organized local fisheries around the world in terms of the degree of property rights they possessed over their fisheries and the extent to which they dealt successfully with technical externalities (gear conflicts), appropriation externalities (overcapitalization), and assignment problems (allocation conflicts).

Schlager and Ostrom chose only cases in which groups held at least some rights over access and withdrawal (licence to fish); internal use patterns (how, when, and where to fish); exclusion and membership (who has an access right and how that right can be transferred); and alienation (sale or lease of other rights). Significantly, they found that the larger the bundle of rights held by local self-regulators, the more effort they invested in the creation of rules that solved the three economic problems. Pinkerton (1995) offered a friendly amendment to this formulation, noting that in many contexts rights should be more broadly conceptualized to include rights to protect habitat, to coordinate their activities with those of other users, to obtain government data, and to make policy. From a political-ecology perspective, management (instead of property) rights emphasize the responsibilities of rights-holders for multi-generational resource stewardship, in addition to purely extractive rights.

The concept of property rights suggests private access or ownership, whereas – as discussed above, and in Pauly (this volume) – many fisheries problems cannot be handled successfully without involving multiple parties, often with very different formal rights. Fisheries are not even the property of fishers at the local level, or of the communities in which they live. Some scholars (Pinkerton 1995, Pinkerton and Weinstein 1995) argue that it is more appropriate to think of these communities as ideally situated to exercise management rights. Viewed more broadly, management problems are best solved when rights accrue to the parties best situated to be accountable to sustainable management practices and to resolve inter-sectoral conflicts. These usually turn out to be those parties with long-term relationships to, and dependence on, local fishing territories, as Francis Christy noted in a classic (1982) discussion of 'territorial use rights fisheries' (TURFS).

Conclusion

Some scholars have viewed the concept of shared governance discussed above as applicable to 'virtual communities' – that is, those not geographically attached or bounded. Some assert, for example, that groups of holders of individual transferable quotas (ITQs) – private rights to a fixed amount of fish (see McCay, Pauly this volume) – can take over management functions in the same way as geographical communities do. There is a fundamental difference, however, with geographical communities. The exclusion of outsiders from geographical territory reinforces the long-term relationship of people to a healthy, sustainably managed territory where all the externalities are evident, such as healthy habitat, age classes of the stocks, and species balance in the ecosystem. It is difficult to overexploit one part of the system without this being detected in another part when local institutions have the appropriate scope, scale, parties, and level of operation to deal with the nature and complexity of the kinds of resource problems raised here. In contrast, if individual quota-holders may transfer their rights by simple sale, it is difficult to see how virtual communities (unlike real geographical ones) would be forced to bear the costs of non-sustainable use. In other words, the institutional arrangements allow them to avoid the costs of non-sustainable use by selling out when the discount rate is high (Clark 1981). In many fisheries, overexploitation and its effects may not be detectable for years. This situation leaves fish stocks highly vulnerable to individual 'fish-and-get-out,' highgrading, quota busting, price-dumping, quota-ratcheting, and data-fouling strategies (Copes 1986, 1996).

In summary, as many others in this volume demonstrate, growing problems in fisheries management have outstripped the ability of our current institutions to manage matters effectively. New theoretical directions from political ecology, political science, institutional economics, and organizational sociology have been brought together in the work of many interdisciplinary scholars, often known collectively as common-property theorists. As pressure increases for shrinking governments to transfer the costs of management to 'stakeholders,' and as conflict increases among traditional stakeholders and between traditional and new stakeholders over their rights to influence policy goals of the management system – for example, maximum sustainable yield (MSY) and maximum economic yield (MEY), ecosystem management, or socio-political goals such as meeting constitutional obligations to First Nations and optimizing employment in coastal communities (see Gallaugher and Vodden, this volume) – these new theoretical directions, tested principles, and practical examples are being more intensely explored by fishing communities and governments for their potential to address present-day pressures on both the resource and the people who are sustained by it.

WORKS CITED

Acheson, J.M. 1975. 'The Lobster Fiefs: Economic and Ecological Effects of Territoriality in the Maine Lobster Industry.' *Human Ecology* 3, no. 3: 183–207.

Ashby, W.R. 1968. 'Variety, Constraint and the Law of Requisite Variety.' In W. Buckley, ed., *Modern Systems Research for the Behavioral Scientist: A Sourcebook*, 129–36. Chicago: Aldine.

Axelrod, Robert. 1984. *The Evolution of Cooperation*. New York: Basic Books.

Berkes, F. 1977. 'Fishery Resource Management in a Subarctic Indian Community.' *Human Ecology* 5: 289–307.

– 1981. 'The Role of Self-regulation in Living Resource Management in the North.' In M.M.R. Freeman, ed., *Renewable Resources and the Economy of the North*, 166–77. Ottawa: ACUNS/MAB.

Brownstein, J., and J. Tremblay, 1994. 'Traditional Property Rights and Cooperative Management in the Canadian Lobster Fishery.' *Lobster Newsletter* (DFO, Halifax) 7, no. 1: 5.

Christy, F. 1982. 'Territorial Use Rights in Marine Fisheries: Definitions and Conditions.' *FAO Fisheries Technical Paper no.* 227. Rome: FAO.

Clark, C.W. 1981. 'Bioeconomics of the Ocean.' *BioScience* 31 no. 3: 231–7.

Cohen, F. 1986. *Treaties on Trial: The Continuing Controversy over Northwest Indian*

Fishing Rights. With contributions by Joan La France and Vivian Bowden. Seattle: University of Washington Press.

– 1989. 'Treaty Indian Tribes and Washington State: The Evolution of Tribal Involvement in Fisheries Management in the U.S. Pacific Northwest.' In E. Pinkerton, ed., *Co-operative Management of Local Fisheries: New Directions for Improved Management and Community Development*, 37–48. Vancouver: University of British Columbia Press.

Copes, P. 1986. 'A Critical Review of the Individual Quota as a Device in Fisheries Management.' *Land Economics* 62, no. 3: 278–91.

– 1996. 'Adverse Impacts of Individual Quota Systems on Conservation and Fish Harvest Productivity.' Discussion Paper 96-1. Institute of Fisheries Analysis. Burnaby, BC: Simon Fraser University.

Cordell, J., ed. 1989. *A Sea of Small Boats.* Cambridge, Mass.: Cultural Survival Inc.

Douglas, Mary. 1992. *Risk and Blame: Essays in Cultural Theory.* London: Routledge.

Dyer, C., and J.R. McGoodwin, eds. 1994. *Folk Management in the World's Fisheries.* Niwot: University Press of Colorado.

Feit, H. 1973. 'The Ethno-ecology of the Waswanipi Cree: Or How Hunters Can Manage Their Resources.' In B. Cox, ed., *Cultural Ecology: Readings on the Canadian Indians and Eskimos*, 115–25. Toronto: McClelland and Stewart.

– 1988. 'The Power and the Responsibility: Implementation of the Wildlife and Hunting Provisions of the James Bay and Northern Quebec Agreement.' In S. Vincent and G. Bowers, eds., *James Bay and Northern Quebec: Ten Years After*, 74–88. Montreal: Recherches amerindiennes au Québec.

Follett, M.P. 1940. *Dynamic Administration: The Collected Papers of Mary Parker Follett.* Ed. Henry C. Metcalf and L. Urwick. London: Harper and Brothers.

Freeman, M.R.R., and L. Carbyn, eds. 1988. *Traditional Knowledge and Renewable Resource Management in Northern Regions.* Edmonton: Boreal Institute for Northern Studies.

Freeman, M.R.R., et al. 1988. *Small-type Coastal Whaling in Japan.* Edmonton: Boreal Institute for Northern Studies.

Gray, B. 1991. *Collaborating: Finding Common Ground for Multi-party Problems.* San Francisco: Jossey-Bass.

Harris, M. 1979. *Cultural Materialism. The Struggle for a Science of Culture.* New York: Random House.

Jentoft, S. 1989. 'Fisheries Co-management.' *Marine Policy* 13: 137–54.

Jentoft, S., and T. Kristoffersen. 1989. 'Fishermen's Co-management: The Case of the Lofoten Fishery.' *Human Organization* 48, no. 4: 355–65.

Johannes, R.E. 1981. *Words of the Lagoon.* Berkeley: University of California Press.

Lamson, C., and A. Hansen, eds. 1984. *Atlantic Fisheries and Coastal Communities: Fisheries Decision-making Case Studies.* Halifax: Dalhousie Ocean Studies Program.

Langdon, S. 1984. *Alaska Native Subsistence: Current Regulatory Regimes and Issues. Alaska Native Review Commission.* Vol. 19. Anchorage, Alaska.

McCay, Bonnie. 1980. 'A Fishermen's Cooperative, Limited: Indigenous Resource Management in a Complex Society.' *Anthropological Quarterly* 53: 29–38.

– 1984. 'The Pirates of Piscary: Ethnohistory of Illegal Fishing in New Jersey.' *Ethnohistory* 31, no. 1: 17–37.

– 1987. 'The Culture of the Commoners: Historical Observations on Old and New World Fisheries.' In Bonnie McCay and J. Acheson, eds., *The Question of the Commons: The Culture and Ecology of Communal Resources*, 195–216. Tuscon: University of Arizona Press.

McCay, Bonnie, and J. Acheson, eds. 1987. *The Question of the Commons: The Culture and Ecology of Communal Resources.* Tuscon: University of Arizona Press.

Olson, M. 1965. *The Logic of Collective Action: Public Goods and the Theory of Groups.* Cambridge, Mass.: Harvard University Press.

Ostrom, E. 1990. *Governing the Commons: The Evolution of Institutions for Collective Action.* New York: Cambridge University Press.

– 1995. 'Designed Complexity to Govern Complexity.' In S. Hanna and M. Munasinghe, eds., *Property Rights and the Environment, Social and Ecological Issues*, 33–46. Washington, DC: Beijer Institute of Ecological Economics and the World Bank.

Pavel, J. 1989. 'The Puget Sound Salmon Management Plan: A Co-operative Management Tool.' In B.L. Smith and P. Hurt, eds., *Fisheries Co-management: a Response to Legal, Social, and Fiscal Imperatives*, 22–9. Odanah, Wis.: Great Lakes Indian Fish and Wildlife Commission.

Pinkerton, E. 1988. 'Co-operative Management of Local Fisheries: A Route to Development.' In J. Bennett and J. Bowen, eds., *Production and Autonomy: Anthropological Studies and Critiques of Development*, 257–73. Lanham, Md: University Press of America.

– 1989. 'Attaining Better Fisheries Management through Co-management: Prospects, Problems, and Propositions.' In E. Pinkerton, ed., *Co-operative Management of Local Fisheries: New Directions for Improved Management and Community Development*, 3–33. Vancouver: University of British Columbia Press.

– 1992. 'Translating Legal Rights into Management Practice.' *Human Organization* 51 no. 4: 330–41.

– 1994. 'Northwest Coast Indian Conservation of Salmon Fisheries: Why Here?' Paper presented at the meeting of the Canadian Anthropology Association, University of British Columbia, Vancouver, B.C., 7 May.

– 1995. 'Alternatives to the 'Property' Dilemma: What Is the Most Useful Way to Conceptualize Co-management Rights?' Paper presented at the Conference of the International Association for the Study of Common Property, Bodo, Norway, 28 May.

– 1996. 'The Contribution of Watershed-based Multi-party Co-management Agreements to Dispute Resolution: The Skeena Watershed Committee.' *Environments* 23, no. 2: 51–68.

Pinkerton, E., and N. Keitlah. 1990. 'The Point No Point Treaty Council: Innovations by an Inter-Tribal Fisheries Management Co-operative.' U.B.C. Planning Paper DP No. 26. School of Community and Regional Planning, University of British Columbia, and the Nuu-chah-nulth Tribal Council.

Pinkerton, E., and M. Weinstein. 1995. *Fisheries That Work. Sustainability through Community-Based Management.* Vancouver: David Suzuki Foundation.

Ramos, A.G. 1981. *The New Science of Organizations: A Reconceptualization of the Wealth of Nations.* Toronto: University of Toronto Press.

Reich, Robert. 1988. 'Policy-Making in a Democracy.' In R. Reich, ed. *The Power of Public Ideas*, 123–56. Cambridge, Mass.: Ballinger.

Schlager, Edella, and Elinor Ostrom. 1993. 'Property Rights Regimes and Coastal Fisheries: An Empirical Analysis.' In Terry L. Anderson and Randy T. Simmons, eds., *The Political Economy of Customs and Culture: Informal Solutions to the Commons Problem*, 13–41. Lanham, Md.: Rowman and Littlefield.

Swezey, A., and R. Heizer. 1977. 'Ritual Management of Salmonid Fish Resources in California.' *Journal of California Anthropology* 4, no. 1: 6–29.

Taylor, M. 1987. *The Possibility of Cooperation.* Cambridge: Cambridge University Press.

United States vs. State of Washington (Boldt Decision), 384 F. Supp. 312 (1974).

Usher, P. 1987. 'Indigenous Management Systems and the Conservation of Wildlife in the Canadian North.' *Alternatives* 14, no. 1: 3–9.

Wenzel, G. 1986. 'Resource Harvesting and the Social Structure of Native Communities.' In *Native People and Renewable Resource Management: Proceedings of the 1986 Symposium of the Alberta Society of Professional Biologists, Edmonton, Alta.*, 10–22.

Williams, N., and E. Hunn, eds. 1982. *Resource Managers: North American and Australian Hunter-gatherers.* Canberra: Australian Institute of Aboriginal Studies.

Wilson, J.A., Acheson, J., Metcalfe, M., and Kleban, P. 1994. 'Chaos, Complexity and Community Management of Fisheries.' *Marine Policy* 18, no. 4: 291–305.

Wilson, J.A., and L.M. Dickie. 1995. 'Parametric Management of Fisheries: An Ecosystem-social Approach.' In S. Hanna and M. Munasinghe, eds., *Property Rights in a Social and Ecological Context: Case Studies and Design Applications*, 153–65. Washington, DC: Beijer Institute of Ecological Economics and the World Bank.

18

Fisheries Management:
Putting Our Future in Places

DANIEL J. PAULY

The crisis of fisheries is real, and global.[1] It is an ecological crisis in that fisheries production systems worldwide are losing their productive capacity. It is a socioeconomic crisis in that industrial fisheries now rely globally on U.S.$50 billion worth of subsidy per year (Garcia and Newton 1997), while simultaneously undermining the livelihood of millions of small-scale fishers (Pauly 1997).

It is also an intellectual crisis in that fisheries science has recently lost much of its hard-won credibility, partly because of its continued perception, in spite of several fisheries-induced collapses, of narrow industry interests as the only legitimate 'clients' for its services (see, for example, Charles 1995), and of the apparent unwillingness of many fisheries scientists to rely on the broad ecological knowledge so far accumulated to support management approaches based on precautionary principles (see McGuire 1991, for a tropical, and Hutchings and Myers 1994, for a cold-water example; see also Hutchings, Neis et al., this volume).

It is finally an ethical crisis, and certainly one of alienation as well, as much of the fisheries sector has discarded, along with an estimated 27 million tonnes of by-catch per year (Alverson et al. 1994), all notions of guardianship of the resources on which its survival depends. In effect, the fisheries sector has abdicated this role to conservationist organizations, which industry representatives, from the CEOs of fishing corporations to lobbyists for sport fishing, too often criticize for their lack of 'realism.'

Canada's fisheries provide examples of all the above-mentioned ills. Indeed, Canada offers one of the few instances of a major stock's being driven into collapse by a fishery that largely followed mainstream scientific advice (Finlayson 1994; Hutchings, this volume). Also, Canadian fisheries reproduce numerous aspects of the global fisheries crisis, as they pit a corporate fishing fleet against owner-operated craft, both against sport fishers, all against the First Nations, all of these users against the Canadian Department of Fisheries and Oceans (DFO), and the entire sector against the taxpaying public at large – the ultimate sovereign. Walters (1994) provides a detailed analysis of this buck-passing exercise on the Pacific coast of Canada. Thus Canada can serve as a microcosm of the world's fisheries, and in this chaper[1] I alternate between Canadian and other examples, as when discussing solutions proposed to address these problems.

Dealing with Real People

The main theme of this chapter, written from the perspective of a fisheries biologist, is my contention that to be successful future management schemes, whether based on market incentives, on co-management, or on governance arrangements, must involve local communities living in real places and exploiting stocks that have places as well. However, before we look at the places of fish, we should discuss at least one of the half-truths that pollute such discussion.

The cliché often used by fisheries scientists to counter conservationist arguments is that 'management is not about fish, but about people.' This is a statement that may resonate nicely in empty heads but which implies that we, fisheries scientists, are or should be equipped to deal with 'people issues.' Our collective inability, nay unwillingness, to engage in collaborative work with social scientists (sociologists, anthropologists, historians) and our tendency to reduce what they commonly use as fact or evidence to the status of anecdote (Pauly 1995) show our discipline to be conceptually ill-equipped to deal with people issues. Moreover, and perhaps more important, applied sciences dealing with real people (as opposed to our nominal fishers and hypothetical managers), such as medicine or psychology, have by necessity developed strong codes of conducts to regulate the ethical aspects of their interactions with people (see, for example, contributions in Roy, Wynne, and Old 1991). Thus, psychologists must obtain informed consent, usually in writing, before performing even seemingly harmless teaching experiments on graduate students, while complex protocols regulate the conduct of medical experiments, with both animals and people. When human beings are involved, experiments on the efficacy of a potentially life-saving drug must be interrupted when the subjects receiving placebos can be shown to suffer from not being included in the treatment group. This and similar ethical issues, in many medical schools, are part of the curriculum and taught by faculty specialized in medical ethics.

How about fisheries scientists? Do we really want to be dealing with real people and become personally responsible (as medical doctors are) for what we do, for the advice that we give, and for the consequences of such advice?

Types of Solutions

Three classes of approaches have been proposed to address the global fisheries crisis and/or its local manifestations, usually as a complement or as an alternative to national and/or international 'top–down' regulatory approaches. These are market-based approaches, community-based approaches, and ecology-based approaches.

Market-Based Approaches

The open-access nature of most fishing grounds, in Canada and elsewhere, has long been identified as the major cause of the 'race for fish' and its attendant ills, such as overcapitalization, rent dissipation, and stock collapses. Consequently, there are numerous account by economists attributing these problems to 'market failures' – i.e., to the market's being unable to properly account for, or 'internalize,' the social and environmental costs of fishing (review in Clark 1990). There are, in contrast, far fewer accounts by economists of the active collusion (what other word is there?) between governments and large fishing enterprises, as made manifest in the boat-building and other subsidies that have enabled large fishing enterprises to exploit coastal resources far from their home ports (Garcia and Newton 1997) and, in the process, to marginalize otherwise efficient, localized, small-scale or artisanal fisheries (Pauly 1997). Thus it is not surprising that the market-based mechanism recently proposed to overcome market failures – individual transferable quotas (ITQs) – is often perceived to be not a tool for resource management per se but a ploy for transferring more public assets from public to corporate ownership (see, for example, Davis 1996). When firmly implemented, however, ITQs seem to achieve much of what is expected of them, reintroducing rationality – albeit in its narrowest, economic sense – into an industry that had gone irrational several decades ago (Arneson 1996; McCay, this volume).

Community-Based Approaches

To emphasize community-based approaches is to imply that they are something new – yet we know that in earlier times, and for obvious reasons, local communities were the only entities required for and capable of managing fisheries (see, for example, contributions in McCay and Acheson 1987; Ruddle and Johannes 1985; and Gallaugher and Vodden, Newell, Pinkerton, Thoms, this volume). The reason why community-based management is not a trivially obvious thing to do is that, in parallel with the development of large fishing boats (typically bottom trawlers, but also purse seiners and other type of industrial craft), national governments in recent decades have created centralized agencies to regulate fisheries, with an internal culture, and often an explicit mandate, that favoured industrial (and distant-water) fleets over smaller, locally based artisanal fleets (see Pauly 1988 for examples pertaining to tropical fisheries, and Finlayson 1994 and Charles 1995, for the case of northern cod).

The resulting inequities and alienation, in both developed and developing countries, are well documented in the literature and have spawned the concept of co-management, a sharing of responsibility between national governments (usually represented through a regulatory agency) and fisher communities (see contributions in Pinkerton 1989; Pinkerton and Weinstein 1995; and Pinkerton, this volume). Co-management, as currently conceived, may range from the right of communities to be consulted during a decision-forming process to their nearly full autonomy (with respect to fishing). Usually these arrangements assume that owners of fishing vessels have legitimate interest in co-managing a fishery, but not their salaried crew or their wives, who are often involved in processing (Neis, this volume) – additional sources of inequity and alienation.

Moreover, and perhaps even more important, these arrangements all imply the resource users to be the only group that governments need to consider – i.e., the concept usually does not lead to the perception of non-user groups as legitimate 'stakeholders' in the management and resource-allocation process. This assumption contrasts with the concept of 'governance' (see contributions in Kooiman 1992), wherein governments unwilling to accommodate all demands by particular user groups involve groups with different or even opposing interests in the decision-making process, thus forcing the user groups – here, fishers – to justify their privileged access to public resources. In the case of coral-reef fisheries, this might imply, for example, the creation of local management councils in which fishers must negotiate with the operators of SCUBA dive resorts, and perhaps conservationist groups, and where government representatives only set and enforce rules for intergroup negotiations.

The outcome of governance arrangements of this sort, which would always be local in nature, may be not only levels of exploitation that are sustainable and compatible with the interests of different stakeholder groups, but also reduced transaction and enforcement costs for the central government.

Ecology-Based Approaches

Nature is complex, particularly the oceans, and we know very little about the processes that produce and maintain the biomass that we harvest – another set of clichés, part of the smokescreen behind which disciplinary irresponsibility can be hidden. Actually we do know – and have known since the beginning of the twentieth century, when F.I. Baranov developed the principles of quantitative fisheries science (see Baranov 1977), or a least since The Second World War and the giant fishery closure that it entailed (Beverton and Holt 1957) – that excessive fishing reduces stocks and eventually causes them to collapse and that reducing fishing is sufficient, in most cases, for natural stocks to recover (given time).

What has enabled our historic fisheries to last for centuries is that catches were limited relative to stock size; part of the stock was not susceptible to fishing, because it had access to natural refuges. The refuge in the case of northern cod was depth: the historic fishery was an inshore one, and the old, large females whose reproductive output maintained recruitment to the stock were largely inaccessible to the inshore gear dominating the fishery (see Cadigan, this volume). Similarly, most tuna could be caught only if part of their stock strayed inshore, the rest of the stock remaining safe in oceanic waters. In the last decades, technological developments – powerful echo-locating devices, extremely precise satellite positioning, and new gear – have made it possible to locate almost any fish, anywhere, and to exploit what initially were refuges. It is beyond the scope of this chapter to document this change. However, the well-documented recent stock collapses throughout the world would provide much of the evidence, were it to be presented here.

Countering this creeping invasion of natural refuges is possible, and this possibility is probably what lies behind the growing scientific consensus about the efficacy of marine protected areas (MPAs), particularly in their most effective form, as 'no-take' MPAs (Bohnsack 1994; Roberts et al. 1995). Essentially, MPAs work as artificial refuges, by reconciling nature's time scales with those of fishers and markets as required, because even very low fish mortality can drastically reduce the numbers of the large, old, and highly fecund females that contribute most to a stock's recruitment. Thus within a (suitably located and sized) no-take zone, the numbers, and eventually the biomass, of one (or several) previously decimated stocks can recover, even as (regulated) fishing continues outside that zone. Gradually, as stock density and mean ages increase within the zone, it starts exporting eggs and/or later juveniles and adults that contribute to the adjacent fishery, soon offsetting through this export the fishing lost because of the no-take zone itself. A tropical example, documenting the reality of these processes, is found in Russ and Alcala (1994); while Ballantine (1991) provides similar examples from colder waters. The conference proceedings edited by Shackell and Willison (1995) provide a comprehensive overview, including many Canadian examples.

The most important aspect of MPAs, however, may be their feature of simultaneously reducing risk and accommodating imperfect knowledge (Clark 1996) – an intractable problem – using quota-based or other, traditional forms of management (Ludwig, Hillborn, and Walters 1993). Contrary to much of what one may hear about it, lack of knowledge about the dynamics of exploited stocks, as expressed in the above cliché, speaks not against but for the establishment of MPAs. Consider the question whether lack of knowledge about average weights in *Homo sapiens* is used by engineers (who also deal with real people) as a reason for or against building elevators that can accommodate more than their stated maximum number of passengers.

Our Future Is in Places

The text above should have made my sympathies clear: I believe that fisheries management, if it is to lead to anything sustainable, must take into account the places of people in its logic. It must consider, far more than has hitherto been the case, the places of small-scale fisher communities, but also of other stakeholders, and the places of fish, especially places where their populations can recover from fishing. Taking places into account will not, and indeed could not, be done by 'fisheries managers' alone. Rather the public at large, which ultimately owns the resource and whose taxes have so far been misused to subsidize the carnage, must become involved. There are scattered signs of this happening, throughout the world, as a result of the recent massive coverage of global fisheries issues in science and/or nature-oriented magazines (Parfit and Kendrick 1995; Safina, 1995; see Gallaugher and Vodden, this volume), in the general press, and as a result of the trend towards greater accountability of government agencies. The public will not do this en masse, obviously. Rather it will support those people who best express the needs of the time – as happened in the 1970s, when public support swelled for the non-governmental organizations and politicians who advocated an end to whaling. Here again, the conservation movement can be expected to heed the call. It would be good if fisheries scientists put their hearts in the right place as well.

NOTE

1 I thank Dr A. Davis for detailed and helpful comments on the structure and contents of the draft of this chapter.

WORKS CITED

Alverson, D.L., M.H. Freeberg, S.A. Murawski, and J.G. Pope. 1994. 'A Global Assessment of Fisheries Bycatch and Discards.' *FAO Fisheries Technical Paper no. 339.* Rome.

Arneson, R. 1996. 'On the ITQ Fisheries Management System in Iceland.' *Reviews in Fish Biology and Fisheries* 6, no. 1: 63–90.

Ballantine, W.J. 1991. *Marine Reserves for New Zealand.* Leigh Laboratory Bulletin no. 25. University of Auckland.

Baranov, F.I. 1977. *Selected Works on Fishing Gears.* Vol. 3. *Theory of Fishing.* Jerusalem: Israel Program for Scientific Translations.

Beverton, R.J.H., and S.J. Holt. 1957. *On the Dynamics of Exploited Fish Populations.* Fisheries Investigations. Series II. London: Ministry of Agriculture, Fisheries, and Food.

Bohnsack. 1994. 'Marine Reserves: They Enhance Fisheries, Reduce Conflicts, and Protect Resources.' *Naga: The ICLARM Quarterly* 17, no. 3: 4–7.

Charles, A.T. 1995. 'The Atlantic Canadian Ground Fishery: Roots of a Collapse.' *Dalhousie Law Journal* 18, no. 1: 65–83.

Clark, A.T. 1996. 'Refugia.' Paper presented at the National Academy of Sciences International Conference on Ecosystem Management for Sustainable Marine Fisheries, 19–24 Feb. Monterey, Calif.

Clark, C.W. 1990. *Mathematical Bioeconomics: The Optimal Management of Renewable Resources.* 2nd ed. New York: Wiley Interscience.

Davis, A. 1996. 'Barbed Wire and Bandwagons: A Comment on ITQ Fsheries Management.' *Reviews in Fish Biology and Fisheries* 6, no. 1: 97–197.

Finlayson, A.C. 1994. *Fishing for Truth: A Sociological Analysis of Northern Cod Assessments for 1977 to 1990.* St John's: ISER, Memorial University.

Garcia, S., and C. Newton. 1997. 'Current Situation, Trends, and Prospects in World Capture Fisheries.' In E. Pikitch, D.D. Hubert, and M. Sissenwine, eds., *Global Tends in Fisheries Management,* 3–27. American Fisheries Society Symposium 20. Bethesda, Md.: American Fisheries Society.

Hutchings, J.A., and R.A. Myers. 1994. 'What Can Be Learned from the Collapse of a Renewable Resource? Atlantic Cod, *Gadus morhua,* for Newfoundland and Labrador.' *Canadian Journal of Fishieries and Aquatic Sciences* 51: 2126–46.

Kooiman, J., ed. 1992. *Modern Governance–Society Interactions.* London: Sage Publications.

Ludwig, D.R., R. Hilborn, and C. Walters. 1993. 'Uncertainty, Resource Exploitation, and Conservation: Lessons from History.' *Science* 260 (2 April): 17, 36.

McCay, B.J., and J.M. Acheson, eds. 1987. *The Question of the Commons: The Culture and Ecology of Communal Resources.* Tucson: University of Arizona Press.

McGuire, T. 1991. 'Science and the Destruction of a Shrimp Fleet.' *Maritime Anthropological Studies* 4, no. 1: 32–55.

Parfit, M., and R. Kendrick. 1995. 'Diminishing Returns.' *National Geographic* 188, no. 5: 2–37.

Pauly, D. 1988. 'Fisheries Research and the Demersal Fisheries of Southeast Asia.' In J.A. Gulland, ed., *Fish Population Dynamics*, 2nd ed., 329–48. ed. Chichester: Wiley Interscience.

– 1995. 'Anecdotes and the Shifting Baseline Syndrome of Fisheries.' *Trends in Ecology and Evolution* 10, no. 10: 430.

– 1997. 'Small Scale Fisheries in the Tropics: Marginality, Marginalization and Some Implications for Fisheries Management.' In E. Pikitch, D.D. Hubert, and M. Sissenwine, eds., *Global Trends in Fisheries Management*, 40–9. American Fisheries Society Symposium 20, Bethesda, Md.

Pinkerton, Evelyn, ed. 1989. *Co-operative Management of Local Fisheries: New Directions of Improved Management and Community Development*. Vancouver: University of British Columbia Press.

Pinkerton, E., and M. Weinstein, eds. 1995. *Fisheries That Work: Sustainability through Community-Based Management*. Vancouver: David Suzuki Foundation.

Roberts, C., W.J. Ballantine, C.D. Buxton, P. Dayton, L.B. Crowder, W. Milon, M.K. Orbach, D. Pauly, and J. Trexler. 1995. 'Review of the Use of Marine Fishery Reserves in the U.S. Southeastern Atlantic.' NOAA Technical Memorandum NMFS-SEFSC-376. Miami: National Marine Fisheries Service.

Roy, D.J., B.E. Wynne, and R.W. Old, eds. 1991. *Bioscience Society*. Chichester: John Wiley and Sons.

Ruddle, K., and R.E. Johannes, eds. 1985. *The Traditional Knowledge and Management of Coastal Systems in Asia and the Pacific*. Jakarta: UNESCO Regional Office for Science and Technology.

Russ, G., and A. Alcala. 1994. 'Sumilon Island Reserve: Twenty Years of Hopes and Frustrations.' *Naga: The ICLARM Quarterly* 17, no. 3: 8–12.

Safina, C. 1995. 'The World's Imperiled Fish.' *Scientific American* 273, no. 5: 46–53

Shackell, Nancy L., and J.H. Willison. 1995. *Marine Protected Areas and Sustainable Fisheries*. Wolfville, NS: Science and Management of Protected Areas Association.

Walters, Carl. 1994. *Fish on the Line: The Future of Canada's Pacific Fisheries*. Vancouver: David Suzuki Foundation.

19

Conclusion: Lessons Learned

DIANNE NEWELL AND ROSEMARY E. OMMER

The problem in concluding a book such as this is not unlike that of sewing together local ecological knowledge and scientific data in the chapter above by Neis et al: where does the fruitful interface occur that will move us on? Daniel Pauly gets at this issue in his chapter, which recognizes the limits of biological science and the difficulties that social scientists face. The problems that fisheries scientists encounter in being represented in the management of Canada's commercial fisheries is of great concern in both Hutchings's chapter and Gallaugher and Vodden's. The social science and humanities chapters contain an array of methodological perspectives on community experience in managing fisheries, ranging from the intensely local and personal detail to be found in McCay, Newell, Ray, Thoms, and Manore and Van West, through the 'ideal type' of Ommer's 'Rosie's Cove,' to the more quantitative work of Sinclair, Squires, and Downton, and Usher and Tough and the pragmatic, present-day assessments set out by Pinkerton and by Gallaugher and Vodden.

What the chapters here share is a common concern for both the fish and the fishery communities whose existence is challenged by the current crisis in management locally, nationally, and globally. This problem stretches beyond Canada's borders, and beyond the concerns of local community fisheries, so that collectively these chapters address directly and indirectly a real-world crisis of some considerable magnitude. It requires therefore to be thought about by experts in fisheries management, community survival, interdisciplinary scholarship, case law, and new methodologies, and to be informed by local communities and by other interest groups all the way to the level of national government departmental managers and political leaders. Here, in this pioneering, multidisciplinary collection we seek to understand the complexities of lived reality and the multitude of perspectives and endeavours involved. Some of these are discordant.

We call this concluding chapter 'lessons learned.' The first lesson is that there is a remarkable fit among the foundations of small-scale, community-based fisheries, the cultural, economic, and social problems they have faced in the past, and their potential for surviving failure in the present crisis. As suggested in our introduction, and specified in the individual chapters, the 'fit' is to be found in the following areas:

Foundations

• scale of operation and capital investment
• sensitivity to environment
• considerable local ecological knowledge
• a lived culture that tied together people and environment
• flexibility as a way of life, using multiple-resource exploitation (and differing ecological niches) over the seasons and larger annual cycles

Problems

• relative political powerlessness
• difficulties in dealing with increasingly industrial modes of production
• inadequate education, especially in finance and commerce
• dependence structures arising out of possibly well-meaning but destructive state transfer policies

What this all means in policy terms is that, if we seek to foster the continued well-being and liveliood of people in communities such as these, we need:
• a way of dealing positively with a range of part-time occupations based on multiple-species extraction and occupational pluralism
• ways to link up local ecological knowledge and formal science
• ways to generate local economic development (as opposed to simple growth)
• appreciation that 'sustainability' refers to both people and resources, taken together
• awareness that appropriate scales of management must be found that include all participants in the fishery

The second lesson, arising from the first, is that we face a problem of fit between the manner in which we manage our assessments of the health of fisheries – seen as 'the resource' – and the economic viability, or 'health,' of the communities that depend on them for livelihood not only in the sense of 'earning a living' but also in the sense of 'sustaining a way of life.' This is not utopian, rural romanticism, but – as Pinkerton, Gallaugher and Vodden, and Newell show in their different ways – it is pragmatic necessity. 'Visions' of how to proceed – and policy is at best such a vision – require that the people involved share (or at least accept) that vision. Without that, the consequences are precisely those that we have faced in recent years – anomie, cheating, community collapse, despair, division, litigation, out-migration, strikes, and lock-outs. Along with these come, as most of the chapters show and the BC forums highlighted, the problems of inadequately managed resource access – the race for fish, rapid depletion of previously under-used species, increasing adoption of harmful (but short-run 'efficient') technologies, loss of 'last haven' niches for fish, and the taking of excess numbers of juvenile and 'mother' fish, which ultimately lead to stock collapse. None of these evils is confined to offshore, large-scale, industrially organized fisheries. Pauly argues that our future is in places; we must consider the place of small-scale fisher communities and other 'stakeholders' and the place of fish, especially protected locations where stocks can recover from fishing. This is why Hutchings's and Pauly's chapters are so remarkable. They are statements from the natural sciences about the importance of social factors (both inside the Department of Fisheries and Oceans and in the wider community) and social analyses in the management of fisheries.

The third lesson is that it is not enough to consider merely the economics of the firm, large or small, or that of a region even, when managing fish stocks. Indeed, economists' models must be used with great caution, as the best of the models warn, since they are not yet developed to the point that they can incorporate the complexities of the social factors on which this book has focused. Here again, we have an issue that is based partly on scale. The models of fisheries biologists and the economists with whom they work tend to work best at the macro-scale: this is where much of the natural–social science 'interface' has been developing. The local scale – that at which traditional ecological knowledge (TEK) operates – is actually closer to what we need to be able to incorporate into our models. The importance of TEK is made abundantly clear in Neis et al. in particular, but also in Newell, Thoms, and Usher and Tough. We need not only more study of TEK but also work at that scale by economists on such matters as the economics of small-scale, pluralist fishing communities, with their frequent use of informal economies as a strategy for survival. At present, informal economies in fish-based communities are all too often misunderstood as 'people indulging in underground, illegal tax-avoiding behaviour,' or as women doing 'women's work,' or as Aboriginal people behaving 'traditionally.' That is not what is going on, as many of the chapters here show. Instead, people are oiling the wheels of community exchange by in-kind services in circumstances where employment is peripatetic at best, resource bases are fragile and prone to sometimes severe fluctuation, and alternative economic opportunities are scarce.

Finally, we conclude that scale is the starting-point for resolving many of the issues that confront us in fisheries analysis. This means, of course, not only spatial scale but temporal, and it serves us well to think also of the reach of scholarly disciplines in terms of scale. The result of our focus on scale is a series of questions for fisheries management that need to be considered in relation to one another:

- What is the appropriate geographical scale at which to manage a fishery?
- What is the appropriate temporal scale?
- What is the appropriate (scale of) technology and capital?
- At what scale need scholarly models be built to incorporate essential inputs?

Geographically, where watersheds are biologically appropriate – when there is a natural region of both fish and fishers (of various groupings) – consensus, albeit problem-fraught, can emerge around management. That consensus will be, however, limited at best on the west coast, since Pacific salmon may hatch, spawn, and die in a particular watershed, but, in the years between birth and death, they become a 'straddling stock,' crossing international boundaries, moving through many environments, and becoming vulnerable not only to natural and human-made hazards but also to international irresponsibilities in fishing practices, ranging all the way from excessive harvesting to the occurrence of 'ghost' (abandoned) nets, which sink to the ocean floor, where they inadvertently trap fish and marine mammals. In rivers and lakes, however, there is real potential for developing some such watershed co-management technique, as Pinkerton shows for the Skeena River watershed. Where watersheds are inappropriate for all stages of the fish's life cycle – as in the coastal waters of eastern Canada – the ecological concept of assemblages, as laid out above by Villagarcia et al., is probably most appropriate, though it perhaps lacks the kind of landward rootedness that would provide an easy way of dealing with the human side of the equation.

It is precisely in the area of ecological rootedness that we think that historians and anthropologists have something especially concrete to offer – a record of what local cultures have used as their organizing principles, with which therefore there should be some degree of historical and ongoing connectedness. The historical record suggests that there is a way of thinking geographically about fisheries management that could be shared by federal bureaucrats, fisheries biologists, and local communities. It would, however, involve the cooperation of social scientists in a much more comprehensive manner than has occurred heretofore – it would actually involve a whole change in management culture. Its great strength is that it could apply in many fisheries and would work at the level at which the way of life of fishing communities has functioned over the long term. What it cannot yet handle, but, with the application of considerable effort and negotiation, might be developed to cope with, are the separate issues of sports fisheries, First Nations' 'food' fisheries, and offshore industrial fisheries – again, three different scales.

Temporal scale is something that this volume has not confronted directly, though it is hinted at in the chapters on the long-term strategies of local communities, which have always had to deal with vulnerability to resource and economic fluctuations. That these strategies are now failing, and many communities suffering as a result, does not speak to flaws in strategy so much as to a changing world that does not, among other things, allow informal economies or seasonal and multiple-resource exploitation to flourish.

This takes us to a final, crucial question for Canada's marine fisheries. Has recent state policy, with its focus on shorter-term economic growth (not development) effectively destroyed the very structures that kept many fisheries viable from both the human and fish-biology perspective? The temporal scale at which industrial fisheries operate is short – based on capital flows and market fluctuations. Their flexibility lies in their geographical reach – if northern cod fail, multinationals can still fish off Namibia and make a profit; and the world's fisheries are not yet at the point where that strategy will fail as a result of global stock collapse or global irresponsible management. State policy, which might be a saviour here, has not so far proved willing to take the long-term view and protect both stocks and small-scale fisheries. In Canada, in 1992 on the east coast and in 1996 on the west coast, DFO continued to advocate quota limitation and fleet reduction through cutting out of the smaller, less (in the short-term) lucrative fishers (see *Globe and Mail* 20 July 1996). The department concedes that there will be major unemployment as a result but explains that the policy is 'about fish.' It adds that there are too many small-scale fishers catching too few fish – the same complaint as was heard on the east coast until a few years ago. But we now know, and everyone concedes, that the east coast problem was partly one of too much 'effort' – overcapitalization of the offshore fleets is now admitted to have been a major part of the problem. National governments, Pauly reminds us, created centralized agencies to regulate fisheries, with a mandate that favoured industrial (distant-water) fleets over smaller, locally based artisanal fleets.

This lesson, it would appear, has been only partially learned. DFO knows that it must protect fish, but, since it believes that it must therefore not succumb to the politics of unemployment, as its federal and provincial ministerial masters in the past required (this happened in Newfoundland), it seems to have concluded that it is not able to protect local, small-scale commercial fishers. The employment consequences of concurrent DFO management policies are likely to be serious for many communities up and down the BC coast, while the larger corporations will most probably adjust by buying up many of the restricted licences and concentrating capital still further. That the Pacific coast case is becoming too much like that of Newfoundland was obvious to those who attended the recent series of coastal community forums in British Columbia. Under existing policies, true economic development will falter, even as economic growth for a few companies and (perhaps) their employees and (again, perhaps) a few inshore individuals, such as the Fogo Islanders whom McCay interviewed, surges forward, for the time being. What will have been saved, temporarily, may be select fish stocks and a few major stakeholders, though there is no guarantee of this result. What continues to be lost, however, is a great deal of local ecological wisdom, the livelihood of many committed men, women, and families, valuable community and regional ways of life, and an integral part of Canadian culture.

Contributors

Sean T. Cadigan is a social scientist in Ocean Studies at Dalhousie University, where he teaches in the Department of History and the Marine Affairs Program. A native Newfoundlander with a doctorate in history from Memorial University, he is an associate of Memorial's Eco-Research Project and author of *Hope and Deception in Conception Bay: Merchant–Settler Relations in Newfoundland, 1785–1855.*

Lynn Downton, a graduate student in Social Work at Memorial University, was a field research assistant with its Eco-Research Project.

Lawrence F. Felt, professor of sociology, Memorial University, has interests in rural communities and resource dependence, and in the sociology of fisheries science. He is a graduate of Northwestern University, and his recent fields include the human dimensions of marine management, informal economies in rural communities, and management of recreational fisheries. He was one of the authors of *Living on the Edge* and *The North Atlantic Fisheries* and is a member of the Eco-Research Project at Memorial.

Johanne Fischer was a member of the 'Traditional Ecological Knowledge' (TEK) team with the Eco-Research Project, Memorial University A German-trained biologist with an interest in human ecology, she works in fish ecology, especially on tropical freshwater fish, and has had extensive field experience in Nicaragua. She is currently deputy director of the EuroGOOS Office of the Southampton Oceanography Centre in England.

Patricia Gallaugher is a fish cardiovascular physiologist who directs the Science Programs in Continuing Studies, Simon Fraser University, where she is also attached to the Faculty of Science and the Institute of Fisheries Analysis. Gallaugher was principal organizer of the BC coastal community forums held in 1995 and 1996. During her college days she wore 'the boots and scarf,' as a fish filetter for her father's packing company, Royal Fisheries, in Prince Rupert, BC, and later taught for fifteen years in biological sciences at Memorial University.

Richard L. Haedrich is a biological oceanographer and ichthyologist with broad research experience in the systematics and biology of fish. His degrees are from Harvard University, and he spent many years at the Woods Hole Oceanographic Institution before becoming a professor of biology at Memorial University. He is an expert in the biogeography and ecology of deep-sea fishes. With his graduate students, his research of late has turned to the community ecology of both commercial groundfish and the recreational salmonid fisheries. He also fly-fishes and plays trombone in a jazz band.

Jeffrey A. Hutchings is an assistant professor of biology at Dalhousie University and a recent winner of its Faculty of Science Killam Prize. Following graduate studies in salmonid evolutionary ecology at Memorial University of Newfoundland and post-doctoral research on life-history evolution at the University of Edinburgh, Hutchings initiated research into the cause(s) of the collapse of Newfoundland's northern cod as a research fellow at the federal Department of Fisheries and Oceans, Newfoundland region. His essay with Ransome A. Myers, 'What Can Be Learned from the Collapse of a Renewable Resource?: Atlantic Cod, *Gadus morhua*, of Newfoundland and Labrador,' is a seminal contribution to the debate that has influenced fisheries policy.

Bonnie J. McCay is a professor of anthropology and ecology, at Rutgers, the State University of New Jersey, and a graduate of Columbia University. Her Canadian experiences have been central to her life and career – fieldwork on the Musqueum Indian Reserve, Vancouver, and two years and twenty-five summers in Newfoundland. Her research and writing focus on the problems of marine fisheries management from a variety of perspectives, including cultural anthropology and common-property techniques. She is co-editor of an influential collection, *The Question of the Commons*, and a seasonal resident of Joe Batt's Arm, Newfoundland.

Jean L. Manore holds a SSHRC postdoctoral fellowship at Trent University and the University of Calgary. She recently completed her doctoral thesis in history at the University of Ottawa, 'Cross-Currents: The Development of the

Hydro-Electric System in Northeastern Ontario,' which examines how the interactions among system-builders, politicians, loggers, miners, Aboriginal peoples, and the environment shaped power developments. Her great-uncle John Manore is the subject of the chapter included here on the poundnet fisheries of Grand Bend.

Barabara Neis is associate professor and chair in the Department of Sociology, Memorial University. Neis has published numerous articles and reports of her investigations of diverse aspects of the Newfoundland fishery, which she began in 1976. She has co-edited *The Lives and Times of Women in Newfoundland and Labrador: A Collage* and *Invisible: Issues in Women's Occupational Health.* She currently is co-investigator in the traditional ecological knowledge (TEK) component of the Eco-Research Project at Memorial.

Dianne Newell is a professor of history and member of the Fisheries Centre at the University of British Columbia. She has published work on the history of Canada's Pacific coast fisheries, including (as editor) *The Development of the Pacific Salmon-Canning Industry: A Grown Man's Game* and *Tangled Webs of History: Indians and the Law in Canada's Pacific Coast Fisheries. Tangled Webs* won book awards from the Canadian Nautical Research Association and the Canadian Historical Association. She is completing an environmental and labour history of the salmon-cannery camps: *Work at the Rough Edge of the World.*

Rosemary E. Ommer is a professor of history at Memorial University of Newfoundland, where she was also director of the Eco-Research Project, which examined the sustainability of outport communities in the province. Her publications on community and business perspectives on the Atlantic fishery include *From Outpost to Outport* and (as editor) *Merchant Credit and Labour Strategies in Historical Perspective.* Currently she sits on the Board of the Canadian Global Change Program and is co-chair of its Human Dimensions Committee; she is an adjunct professor at the Fisheries Centre, University of British Columbia, and a Visiting Fellow at the Centre for Studies in Religion and Society at the University of Victoria.

Daniel J. Pauly is a professor with the Fisheries Centre and Department of Zoology at the University of British Columbia and principal science adviser at the International Centre for Living Aquatic Resource Management (ICLARM), Manila, Philippines. He has degrees in fisheries biology, zoology, and oceanography from the University of Kiel, Germany, and his areas of research are tropical fisheries management and ecosystem modelling. His recent book, *On the Sex of Fishes and the Gender of Scientists: A Collection of Essays in Fisheries Sciences*, summarizes much of his large body of scholarship.

Evelyn Pinkerton, a maritime anthropologist, is associate professor at the School of Resource and Environmental Management, Simon Fraser University. She has published extensively on innovative institutions for the sustainable management of fisheries and forests, especially on Canada's Pacific coast, and has been involved with many colleagues in developing the theory and practice in common-property resource management. Her recent books include (as editor) *Cooperative Management of Local Fisheries* and (as co-author) *Fisheries That Work*.

Arthur J. Ray is a professor of history at the University of British Columbia specializing in the historical geography of Native peoples of Canada, and he is a recent winner of a UBC Killam Research Prize. His degrees are from the University of Wisconsin, Madison, and among his numerous books are *Indians in the Fur Trade*, *The Fur Trade in the Industrial Age*, and, most recently, *I Have Lived Here since the World Began: An Illustrated History of Canada's Native Peoples*. He also works as a consultant on Native claims cases in Canada.

David C. Schneider is a professor of biology with the Ocean Sciences Centre, Memorial University, and a graduate of the State University of New York, Stoney Brook. A marine biologist who researches the spatial dynamics of large marine organisms, ranging from fish and birds to benthic organisms that inhabit sand and mud, he has served on several independent assessment panels for government and industry.

Peter R. Sinclair is a sociologist and University Research Professor at Memorial University, where he is also a member of the Eco-Research Project. A graduate of Edinburgh University, Sinclair researches formal and informal practices in rural Newfoundland and marine biology and its relationship to fisheries policy. He wrote *From Traps to Draggers* and *State Intervention and the Newfoundland Fisheries* and co-wrote *Village in Crisis; A Question of Survival; Living on the Edge;* and *Aquacultural Development*.

Heather Squires served as a field research assistant with the Eco-Research Project at Memorial University while a master's candidate in environmental Studies, York University, Toronto.

J. Michael Thoms is a doctoral candidate in the Department of History, University of British Columbia, where he is researching the history of Indians and the law in Ontario fisheries. His contribution to this volume is based on community research for his master's thesis in Native studies at Trent University in Ontario – 'Illegal Conservation: Two Case Studies of Conflict between Indigenous and State Natural Resource Management Paradigms,' on the Red Rock First Nation's hunting and fishing near Nipigon, Ontario, and a northern Thailand indigenous community's agriculture (dry rice) and hunting traditions.

Frank J. Tough is a historical geographer and director of the School of Native Studies, University of Alberta. He specializes in archival records revealing government policies towards Aboriginal harvesting of natural resources and has provided expert testimony in court cases pertaining to Aboriginal and treaty rights. With graduate degrees in historical geography from McGill and York (Toronto) universities, he also specializes in post-Confederation Indian and Métis economic history of the Canadian west, including the history of freshwater fisheries. He is author of *'As Their Natural Resources Fail': Native People and the Economic History of Northern Manitoba, 1870–1930*, which won book prizes from the Manitoba Historical Association and the Canadian Historical Association.

Peter J. Usher is an Ottawa-based consultant who specializes in Native claims research. A graduate in geography from the University of British Columbia, Usher has published innovative work on systems of resource management and property arrangements in Aboriginal societies. He has extensive experience in the subarctic and arctic regions of Canada.

Marimar G. Villagarcía is a marine scientist with degrees from the University of Las Palmas, Canary Islands, where she completed her PhD in marine science in the Mathematics Department, and Memorial University. She is currently doing postdoctoral work with the European Community project CANIGO based at the Instituto Canario de Ciencias Marinas (ICCM). She studies the interaction between mathematics and marine ecology, with special focus on spatial and temporal variability as revealed through mathematical analyses of marine communities. Her MSc thesis at Memorial involved a multivariate analysis of fish communities, which forms the basis of her contribution to this volume.

Kelly M. Vodden is a doctoral candidate in geography at Simon Fraser University, where she is also a researcher with both the Institute of Fisheries Analysis and the Community Economic Development Centre. She has been an environmental business and community development consultant since graduating in business administration from the University of Western Ontario, is active in her own coastal fishing community, Steveston, BC, and is a founding member of the Coastal Communities Conservation Society. She helped coordinate the BC coastal forum series, 1995–6, and works with fisheries communities in Alert Bay.

John J. Van West is an anthropologist on the staff of the Ontario Native Affairs Secretariat. At the University of Toronto he wrote a doctoral dissertation on independent fishers in the Port Dover area of Ontario, and more recently he has published on Native freshwater fisheries in Ontario.